Unconventional Hydrocarbon Resources

Unconventional Hydrocarbon Resources

Prediction and Modeling Using Artificial Intelligence Approaches

Edited by Dr. Sid-Ali Ouadfeul
Algerian Petroleum Institute
Sonatrach, Algeria

WILEY

Published by John Wiley & Sons, Inc., Hoboken, New Jersey.
Published simultaneously in Canada.

For general information on our other products and services or for technical support, please contact our Customer Care Department within the United States at (800) 762-2974, outside the United States at (317) 572-3993 or fax (317) 572-4002.

Wiley also publishes its books in a variety of electronic formats. Some content that appears in print may not be available in electronic formats. For more information about Wiley products, visit our web site at www.wiley.com.

Library of Congress Cataloging-in-Publication Data:
9781119389361

Cover Design: Wiley
Cover Image(s): © Ra17/Shutterstock, Artistdesign29/Shutterstock

Set in 9.5/12.5pt STIXTwoText by Straive, Pondicherry, India
Printed and bound by CPI Group (UK) Ltd, Croydon, CR0 4YY

C9781119389361_250723

Contents

Preface

Unconventional Hydrocarbon Resources: Prediction and Modeling Using Artificial Intelligence Approaches highlights novel concepts and techniques in tight reservoir exploration and discusses contribution of artificial intelligence such as neural network, machine learning, and fuzzy logic toward the exploration of these reservoirs. The book contains twenty-one chapters, and each chapter covers examples and applications to real data from different hydrocarbon fields such as the Barnett Shale (United States), Eagle Ford (United States), Baken Oil (United States), Oued Mya (Algeria), Hassi Terfa field (Algeria), Berkine Basin (Algeria), and Williston Basin (United States).

Volume highlights include geophysics of the tight and shale plays, petrophysics of the shale and tight sand gas reservoirs, geomechanic of shale gas reservoirs, wellbore stability in shale gas reservoirs, rock typing of tight sand reservoirs, and fluid flow and permeability of unconventional hydrocarbon reservoirs.

This book will be a valuable resource of information for students, instructors, and researchers working on unconventional hydrocarbon exploration, geophysics, geoengineering, and energy.

Prof Sid-Ali Ouadfeul
Algerian Petroleum Institute, Avenue 1er Novembre,
Boumerdes, 35000, Algeria.

1

Predrill Pore Pressure Estimation in Shale Gas Reservoirs Using Seismic Genetic Inversion with an Example from the Barnett Shale

Sid-Ali Ouadfeul[1], *Mohamed Zinelabidine Doghmane*[2], *and Leila Aliouane*[3]

[1] *Geophysics, Geology and Reservoir Engineering Department, Algerian Petroleum Institute, Sonatrach, Boumerdes, Algeria*
[2] *Department of Geophysics, FSTGAT, University of Science and Technology Houari Boumediene, Algiers, Algeria*
[3] *LABOPHYT, Faculty of Hydrocarbons and Chemistry, University M'hamed Bougara of Boumerdes, Boumerdes, Algeria*

KEYPOINTS

- Artificial neural network used to estimate pore pressure prior to drilling using seismic genetic inversion.
- Stacked 3D seismic amplitudes near three wells used as input for the multilayer perceptron neural network.
- The weights of connections between neurons are optimized during training, and the entire seismic cube is transmitted through the neural network.
- Linear relationship between depth and velocity produced using sonic well-log data of a vertical well, used as a vertical trend in the Eaton model.
- Pore pressure derived from the Eaton model using acoustic impedances, and the suggested method is used to determine the pore pressure in the lower Barnett Shale for hydraulic fracture planning and simulation as well as wellbore stability.

1.1 Introduction

Pore pressure estimation is a crucial phase in oil and gas exploration, and pore pressure maps and profiles are essential for drilling, reservoir stimulation, and wellbore stability. The study by Zhang (2011), who examined the fracture gradient prediction methods and provided the minimum and maximum fracture pressures, is one of the many models that have been put out in the literature to estimate the pore pressure.

Bahmaei and Hossein have predicted the pore pressure using seismic velocity modeling, a case study from the Sefid-Zakhor gas field in Southern Iran is shown, the manufactured sonic logs are modified using the check shot interval velocity of Sefid-Zakhor well no. 1, and the final acoustic impedance model is converted to the velocity model by removing density. Finally, the velocity model is converted to pore pressure The results of the pore pressure model are validated by pore

Unconventional Hydrocarbon Resources: Prediction and Modeling Using Artificial Intelligence Approaches, First Edition. Edited by Sid-Ali Ouadfeul.

pressure data obtained by the modular formation dynamics tester (MDT) well test tool. Generally, the results show the normal trend for pore pressure in the area, except on the left side of the anticline in the 2D seismic section, because of tectonic uplifting.

The most popular empirical techniques for predicting anomalous pore pressure from well logs are then discussed. Additionally, employing depth-dependent normal compaction equations for pore pressure prediction in subsurface formations, Eaton's resistivity and sonic methods are modified. The modified techniques offer a considerably simpler method for managing typical compaction trendlines. The theoretical pore pressure modeling is essential to comprehend the mechanism of the anomalous pressure creation in addition to the empirical methods. It is suggested to use a theoretical pore pressure-porosity model.

In a paper about the predrill pore pressure prediction using seismic data, Sayers et al. (2002) demonstrated that a predrill estimate of pore pressure can be obtained from seismic velocities using a velocity-to-pore-pressure transform. However, the seismic velocities must be derived using techniques having sufficient resolution for well-planning purposes. There are noticeable disparities between the velocity field acquired using reflection tomography and that obtained using a traditional method based on the Dix equation for a deepwater Gulf of Mexico example.

Estimating the pore pressure from well-log data is one of the key jobs in shale gas reservoir characterization (Huffman 2002; Sayers et al. 2002; Sayers 2006). Several methods have been put out in the literature to do this.

A predrill estimate of the formation's pore pressure can be determined using seismic interval velocities by using a velocity-to-pore-pressure transform, according to Sayers et al. (2006)'s study on well-constrained seismic estimation of pore pressure with uncertainty. However, it is important to calculate seismic velocities using a technique that offers adequate spatial resolution for predrill well planning. Seismic interval velocities and well velocities can be combined to create a refined velocity field with an uncertainty that respects the well velocities that are currently available and can be used to accurately forecast pore pressure in a given area of interest. In this article, we propose a technique for pore pressure determination using seismic data generated from Eaton's model. Let us begin by outlining the suggested approach.

1.2 Methods and Application to Barnett Shale

1.2.1 Geological Setting

During a period of marine incursion brought on by the closure of the Lapetus Ocean Basin in the late Mississippian Age, the Barnett Shale was deposited across present-day North Central Texas. The Ouachita Thrust belt started to advance into the modern-day North Texas region toward the end of the Pennsylvanian. The sinking of the South American plate beneath the North American plate is what gave rise to the thrust belt. The foreland basin along the front of the Ouachita Thrust was produced by the thrust's emergence. Early investigations into the basin ascribed Barnett's thermal maturation to its history of burial and the thermal regimes connected to its depth of burial. As additional information became available, explorationists started to question this theory. An alternative theory put out by Bowker (2003), previously of Mitchell Energy/Devon, said that the Ouachita Thrust was responsible for the displacement of hot fluids from east to west, which fueled the maturation process. Figure 1.1 depicts the geologic column of the Mississippian and Pennsylvanian ages. The lower Barnett is the target for our shale gas reserve, and its top is 6650 ft deep (Givens and Zhao 2014).

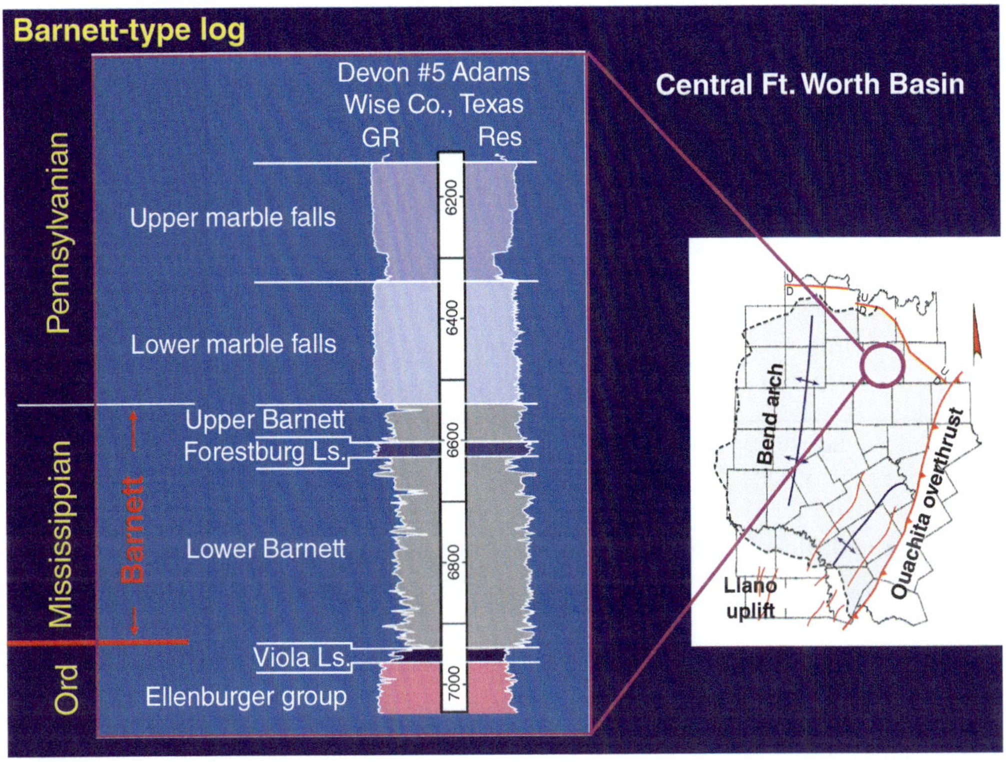

Figure 1.1 Stratigraphic column of Barnett. *Source:* Browning and Martin (1980).

1.2.2 Methods

Eaton's (1975) approach determines the vertical component of the effective stress from the seismic velocity v using the flowing equation and is typically used for pore pressure estimation (Eaton 1975; Bowers 1995):

$$\sigma = \sigma_{\text{Normal}}\left(\frac{V}{V_{\text{Normal}}}\right)^n \tag{1.1}$$

σ_{Normal} and V_{Normal} are the vertical effective stress and seismic velocity expected if the sediment is normally pressured, and n is an exponent that describes the sensitivity of velocity to effective stress, $n = 1.2$ (Kumar et al. 2012). The pore pressure is then given by:

$$p = S_v - \left(S_v - p_{\text{Normal}}\right) \Big/ \left(\frac{V}{V_{\text{Normal}}}\right)^n \tag{1.2}$$

According to Eaton's approach, the difference between the measured velocity and the velocity of sediments under normal pressure (V_{Normal}) increases with depth as follows:

$$V_{\text{Normal}} = V_0 + K * Z \tag{1.3}$$

If Z is assumed to be the P wave's acoustic impedance, then Z is the result of the P wave's velocity and density.

$$Z = \rho * V \tag{1.4}$$

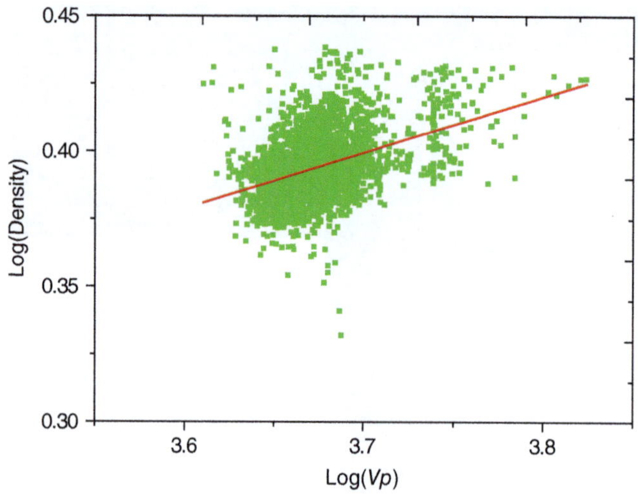

Figure 1.2 Density versus velocity of the P wave in a log-log scale.

According to Gardner's model (Gardner et al. 1974), a power law connects density and velocity:

$$\rho = a * V^b \tag{1.5}$$

Figure 1.2 shows a cross-plot of the density versus the P wave's velocity in log-log scale, where a and b are lithology-related constants.

The following values of a and b were obtained from a linear fit of log of density versus log of velocity of the P wave.

$$a = 1.22, \ b = 0.69 \tag{1.6}$$

The velocity of the P wave V and acoustic impedance are therefore connected by:

$$Z = a * V^{b+1} \tag{1.7}$$

Therefore, the acoustic impedance and the velocity of the P wave are related by:

$$V = \left(\frac{Z}{a}\right)^{\frac{1}{b+1}} \tag{1.8}$$

A normally pressurized formation's velocity is determined by:

$$V = V_0 + K * Z \tag{1.9}$$

where V_0 and K are formation-related constants.

Figure 1.3 displays a cross-plot of the P wave's velocity versus depth. The values of V_0 and K that result from fitting the P wave's velocity against depth linearly are shown here:

$$V_0 = 9656\,\text{ft/s}, \ K = 0.89 \tag{1.10}$$

A typically pressurized formation's porous pressure is determined by:

$$p_{\text{Normal}} = Z * \text{Grad}_{\text{Pp}} \tag{1.11}$$

where Z is the depth and Grad_{Pp} is the pore pressure gradient which is equal to 0.52 psi/ft after Bowker (2007). Figure 1.4 displays the pore pressure versus depth in a lower Barnett Shale

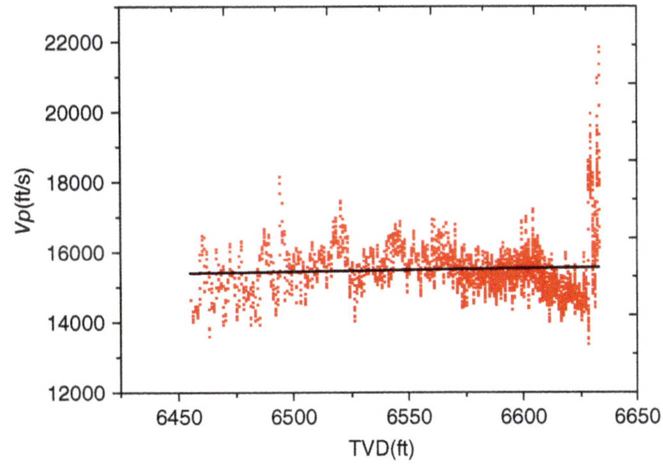

Figure 1.3 Velocity of the *P* wave versus the depth in the lower Barnett Shale gas reservoir.

formation typically pressured formation. In a formation under normal pressure, the vertical tension is given by:

$$S_v = \int_0^z \rho * g dz \tag{1.12}$$

where Z is the depth, ρ is the density, and $g = 9.81\ \text{m/s}^2$ is the gravity field. In a deposit that is typically under pressure, Figure 1.4 also depicts the overburden vertical stress for the lower Barnett

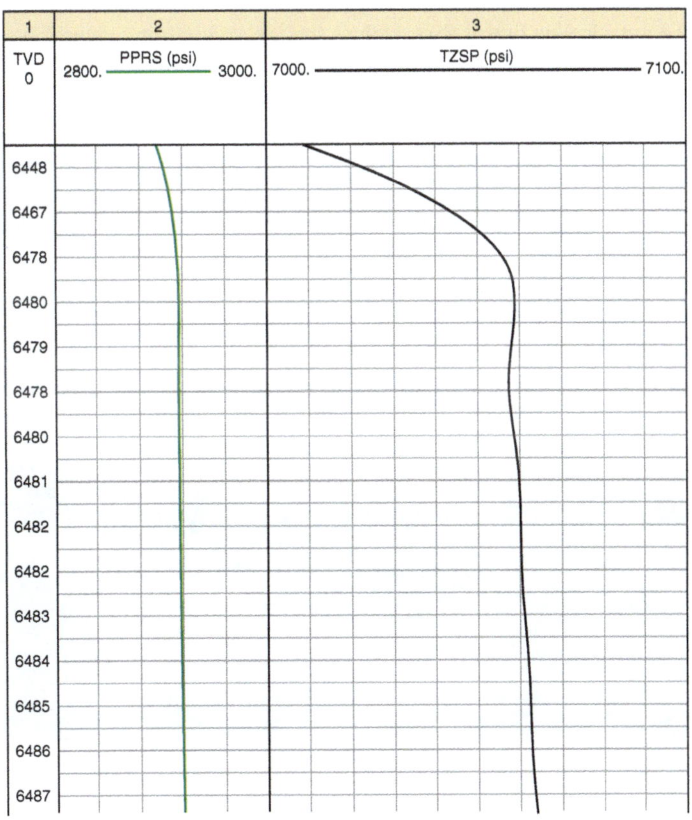

Figure 1.4 (1) True Vertical Depth, (2) Pore pressure versus the depth in normally pressured formation for the lower Barnett Shale formation. (3) Overburden stress versus the depth in the lower Barnett Shale gas reservoir.

Shale gas reservoir. The following equation can be used to derive the pore from the acoustic impedance after accounting for all of these equations and parameters:

$$p = S_{\mathrm{v}} - \left(S_{\mathrm{v}} - p_{\mathrm{Normal}}\right) * \left(\frac{\left(\frac{Z}{a}\right)^{\frac{1}{b+1}}}{\left(V_0 + K * z\right)}\right)^n \tag{1.13}$$

$$p = \int_0^z \rho * g\mathrm{d}z - \left(\int_0^z \rho * g\mathrm{d}z - z * \mathrm{Grad}_{\mathrm{Pp}}\right) * \left(\frac{\left(\frac{Z}{a}\right)^{\frac{1}{b+1}}}{\left(V_0 + K * z\right)}\right)^n \tag{1.14}$$

where n is a constant-related ton = 1.2.

1.3 Data Processing

The raw well-log data of three horizontal wells – 1H, 2H, and 3H, as well as the 3D seismic cube recorded in the Fort Worth Basin – are combined in order to invert the total organic carbon from 03D seismic data. Figure 1.5 displays a 3D seismic data cube that was collected in the area following a stack. The figure also displays the top and bottom of the lower Barnett surfaces as well as the track of the 03 horizontal wells. Density is depicted in Figure 1.6 (RHOB in g/cc) – the P wave's

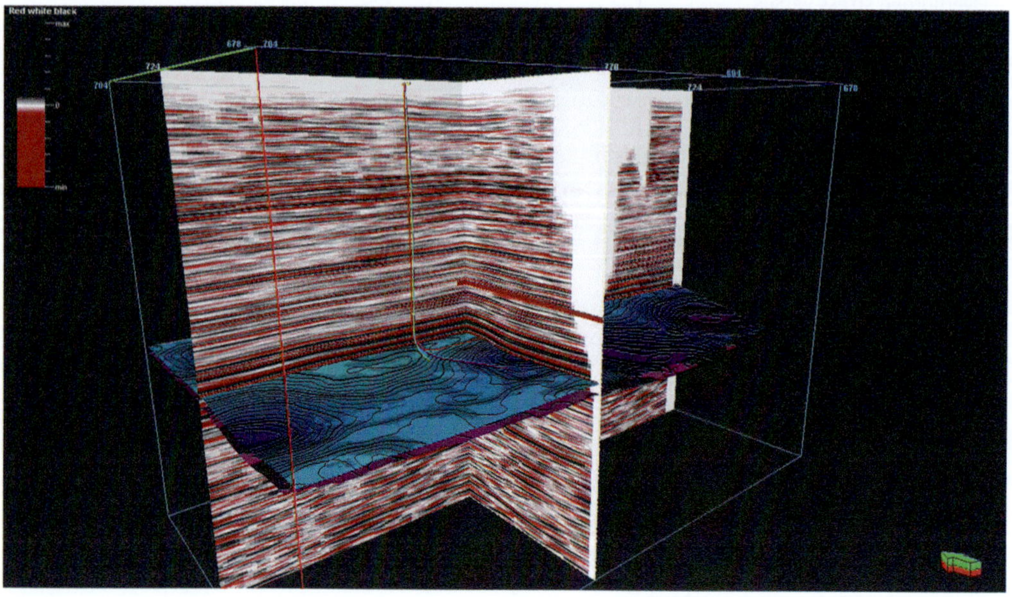

Figure 1.5 3D seismic data cube after stack (Inline 770, Xline 724) with top and lower Barnett surfaces and trajectory of the 03 horizontal wells.

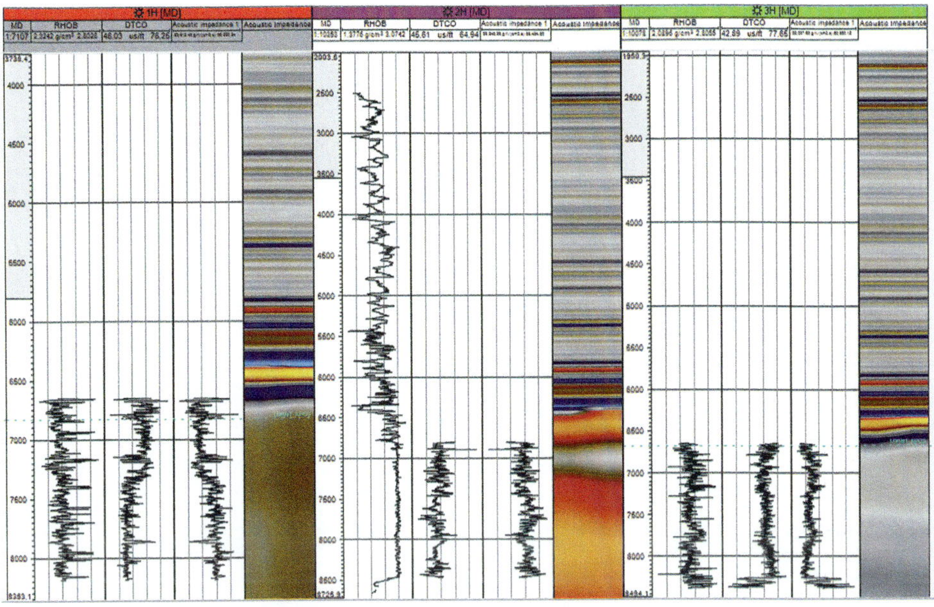

Figure 1.6 Density, slowness of the *P* wave, derived acoustic impedance, and seismic data recorded near wells 1H, 2H, and 3H.

slowness (DTCO in Ms/ft); the *P* wave's derived acoustic impedance for wells 1H, 2H, and 3H; and the seismic information gleaned from the seismic cube that is situated close to the three wells. The first step is to implant a neural multilayer perceptron (MLP) machine, which consists of one hidden layer and three neurons, with the extracted seismic data from Figure 1.5 as an input and the derived acoustic impedances from the raw well-logs data as an output. Only the data from wells 1H and 2H are used at this point in the backpropagation training method as this neural machine is trained in a supervised mode. Figure 1.6 illustrates an example of an inverted seismic cube (inline = 782, crossline = 723, and depth slice at 5988 ft). Weights of connection between neurons are calculated at the conclusion of the training stage, after which the entire 03D seismic cube is propagated through the MLP machine, and an output is calculated using the weights of connection.

Figure 1.7 depicts the estimated pore pressure calculated using the proposed method for the lower Barnett Shale gas reservoir crossed by the horizontal well H01 in an inline tangent to the path of the horizontal well H01. The inverted cube of the acoustic impedance is then used in equation 03 to calculate a cube of the pore pressure for the lower Barnett Shale reservoir.

1.4 Results Interpretation and Conclusions

Figure 1.6 shows that the pore pressure varies between 2880 and 2920 psi, which is consistent with the pore pressure estimate made using the hydrostatic stress depicted in Figure 1.4. This final proof shows that the suggested model can produce accurate outcomes. However, we can see that in other instances, such as the hydrostatic model, the projected pore pressure does not vary exponentially. Additionally, we can explain some depth intervals' large lateral variations in pore pressure (see Figure 1.8) by the distribution of spontaneous cracks and pore pressure. As a result, the suggested model may be effectively employed for shale gas reservoirs, shale play geomechanics, wellbore stability, and modeling and planning of hydraulic fracturing.

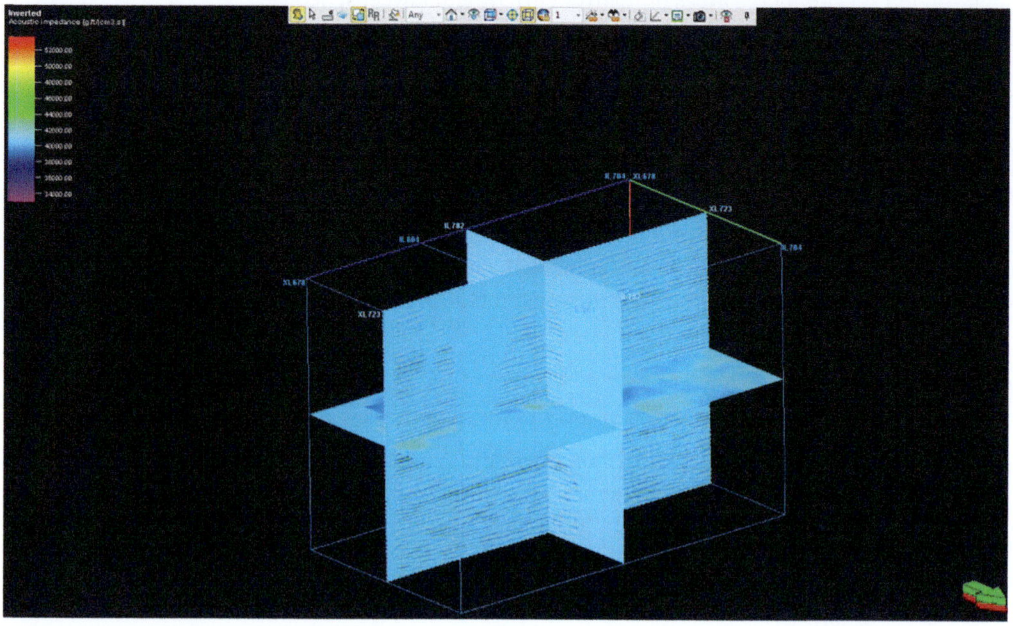

Figure 1.7 Inverted acoustic impedance of the *P* wave.

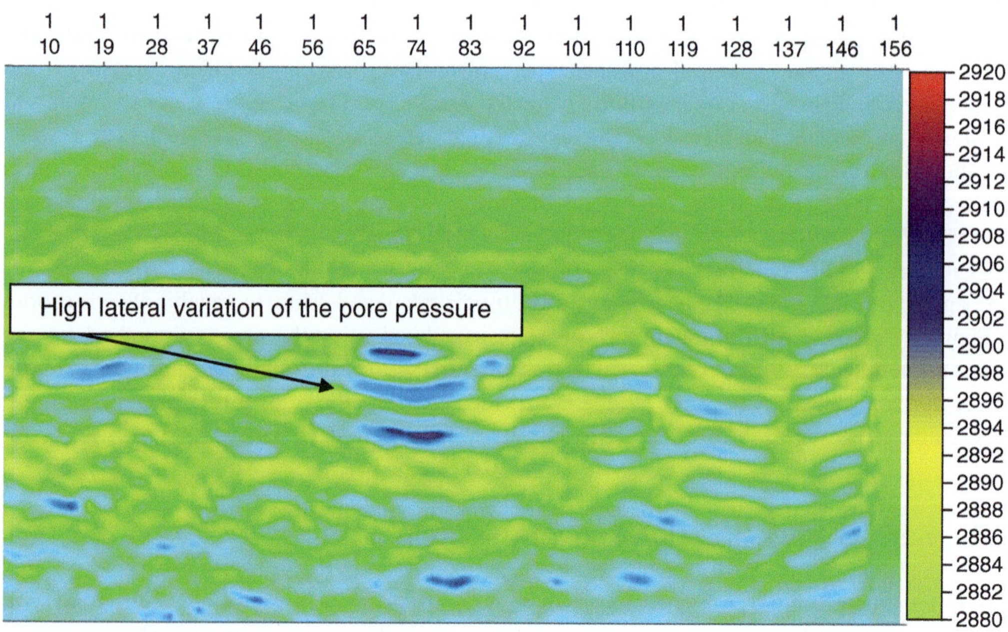

Figure 1.8 Estimated pore pressure using the proposed method (inline) for the lower Barnett Shale gas reservoir crossed by the horizontal well H01.

References

Bowers, G.I. (1995). Pore pressure estimation from velocity data: accounting for pore pressure mechanisms besides undercompaction. *SPE Drilling and Completion* 10: 89–95.

Bowker, K.A. (2003). Recent development of the Barnett Shale Play, Forth Worth Basin. *West Texas Geol Soc Bull* 42: 4–11.

Bowker, K.A. (2007). Barnett Shale gas production, Fort Worth Basin: issues and discussion. *AAPG Bulletin* 91 (4): 523–533.

Browning, D.W. and Martin, C.A. (1982). Geology of North Caddo Area, Stephens County, Texas. In: *Petroleum Geology of the Fort Worth Basin and Bend Arch Area* (ed. C.A. Martin), 315–330. Dallas Geol Soc.03; 2: 6.

Eaton, B.A. (1975). The equation for geopressure prediction form wells logs. SPE, 5544, Dallas, TX (September 1975).

Gardner, G.H.F., Gardner, L.W., and Gregory, A.R. (1974). Formation velocity and density – the diagnostic basics for stratigraphic traps. *Geophysics* 39: 770–780. https://doi.org/10.1190/1.1440465.

Givens, N. and Zhao, H. (2014). The Barnett Shale: not so simple after all, republic energy. *AAPG Annual Meeting Program* 13: A52.

Huffman, A.R. (2002). The future of pressure prediction using geophysical methods. *Memoirs-American Association of Petroleum Geologists* 217–234.

Kumar, B., Niwas, S., Bikram, B., and Mangaraj, K. (2012). Pore pressure prediction from well logs and seismic data. *9 Biennial International Conference and Exposition on Petroleum Geophysics*, Hyderabad.

Sayers, C. (2006). An introduction to velocity-based pore-prediction estimation. *The Leading Edge* 25 (12): 1496–1500.

Sayers, C.M., Johnson, G.M., and Denyer, G. (2002). Predrill pore pressure prediction using seismic data. *Geophysics* 67: 1286–1292.

Sayers, C.M., den Boer, L.D., Nagy, Z.R., and Hooyman, P.J. (2006). Well- constrained seismic estimation of pore pressure with uncertainty. *The Leading Edge* 25 (12): 1524–1526.

Sayers, C.M., Johnson, G.M. Denyer, G. (2002). Predrill pore-pressure prediction using seismic data. *Geophysics* 67 (4): 1286–1292. doi: https://doi.org/10.1190/1.1500391.

Zhang, J. (2011). Pore pressure prediction from well logs: methods, modifications, and new approaches. *Earth-Science Reviews* 108 (1–2): 50–63. ISSN: 0012-8252, https://doi.org/10.1016/j.earscirev.2011.06.001.

Zhang, J., Yin, SX. Fracture gradient prediction: an overview and an improved method. *Pet. Sci.* 14, 720–730 (2017). https://doi.org/10.1007/s12182-017-0182-1.

2

An Analysis of the Barnett Shale's Seismic Anisotropy's Role in the Exploration of Shale Gas Reservoirs (United States)

Sid-Ali Ouadfeul[1], Leila Aliouane[2], Mohamed Zinelabidine Doghmane[3], and Amar Boudella[3]

[1] *Algerian Petroleum Institute, IAP, Boumerdes, Algeria*
[2] *LABOPHYT, Faculty of Hydrocarbons and Chemistry (FHC), University Mhamed Bougara of Boumerdes (UMBB), Boumerdes, Algeria*
[3] *Geophyics Department, FSTGAT, USTHB, Algiers, Algeria*

KEYPOINTS

- Fast and slow shear wave slowness data of the Sonic Scanner tool are used to characterize the Lower Barnett Shale gas reservoir.
- These data are used to determine the fractures azimuth.
- Obtained results demonstrate that the Sonic Scanner tool is not recommended for horizontal wells.

2.1 Introduction

One of the most significant hydrocarbon accumulations in the biosphere are shale gas reservoirs. Exploring these massive reservoirs requires modern technology because they have limited porosity, low permeability, and complex fractures network (Tsvankin et al. 2010; Aliouane and Ouadfeul 2014).

A paper on the characterization and simulation of naturally fractured tight gas sandstone reservoirs of low permeability sandstones of the Frontier Formation in the Green River Basin was published by Harstad et al. in 1995. Characterizing natural fractures in tight shale and sand gas reservoirs is an important and difficult task. They have demonstrated that a thorough comprehension of the traits and fluid-flow behavior of the local natural fracture system that regulates the productivity of the reservoirs is necessary for their successful utilization. As a result, five outcrop studies of the Frontier sandstone in the basin were used to determine the fracture properties. By integrating geomechanics and diagenesis, Olson et al. (2009) have analyzed natural fractures in tight gas sandstones. Ding et al. (2021) have published a paper where a physical model containing two groups of shale blocks as the target formations is constructed in the laboratory. Group S contains six shale blocks with different clay contents, and group N contains six shale blocks with different porosity. The acquired 2D seismic data is used to analyze the seismic responses of two

Unconventional Hydrocarbon Resources: Prediction and Modeling Using Artificial Intelligence Approaches, First Edition. Edited by Sid-Ali Ouadfeul.

corresponding seismic lines. An anisotropic three-term inversion method is applied to one of the seismic inlines to estimate the elastic properties of the target shale blocks. The inversed attributes can be used to reveal the effects of shale clay contents. This study shows the substantial significance of using a physical model to observe the seismic responses of shale properties. The inversion results indicate that the anisotropic three-term inversion method could provide accurate results of elastic properties as well as the *P*-wave anisotropy parameter for shale formations.

The definition of seismic anisotropy is the relationship between acoustic characteristics and the direction of measurement (Tsvankin et al. 2010). Wild (2011) published a paper demonstrating some practical applications of seismic anisotropy. He demonstrated that seismic anisotropy has a significant impact on the well-to-seismic tie used for seismic processing and interpretation. More specifically, the velocity variations caused by anisotropy might be corrected to yield a better well tie or integrated into the processing seismic data. Additionally, he has shown how the amplitude versus azimuth (AVAZ) technique may be used to estimate fracture azimuth from seismic data utilizing seismic anisotropy.

The data from the sonic scanner tool have been used in this chapter to demonstrate the contribution of seismic anisotropy to the characterization of natural fractures. The paper begins by defining seismic anisotropy and the sonic scanner tool. Next, we discuss the geological setting of the Barnett Shale formation and the information the sonic scanner tool collected from a horizontal well drilled in the lower Barnett Shale gas formation. The paper concludes with an interpretation of the results and some generalizations.

2.2 Seismic Anisotropy

Anisotropy is a recognized parameter in geophysics that refers to the variation of wave velocities with the direction of propagation. It is defined as the attribute of being directionally dependent (Thomsen 1986, 2001). An example of a waveform moving through an anisotropic medium is shown in Figure 2.1. As the *S*-wave enters the medium, it splits into horizontal and vertical components; as a result, the horizontal and vertical component waveforms exit the material at separate times. In particular, it is possible to see Shear-wave splitting, also known as *S*-wave splitting, in the anisotropic reservoirs. A shear wave splits into two orthogonal waves when it passes through an anisotropic zone, with one wave moving more quickly than the other. The faster-moving *S*-wave will reach the geophones first.

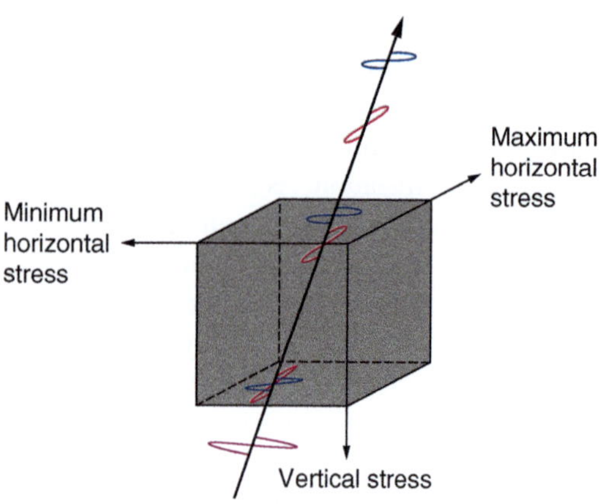

Figure 2.1 Shear wave splitting.

Seismic anisotropy has gained popularity in the oil and gas sector recently. Indeed, if the velocity models employed for the reservoir do not properly account for anisotropy, the anisotropy of sedimentary basins including oil and gas reservoirs might result in increased microseismic event localization inaccuracies (Tsvankin 1997, 2001). Practically speaking, logging techniques like sonic logs are largely used in the field to gather information on formation velocity. Unfortunately, these instruments only measure the velocities along the borehole axis (i.e. vertically), and, hence, frequently ignore horizontal velocities.

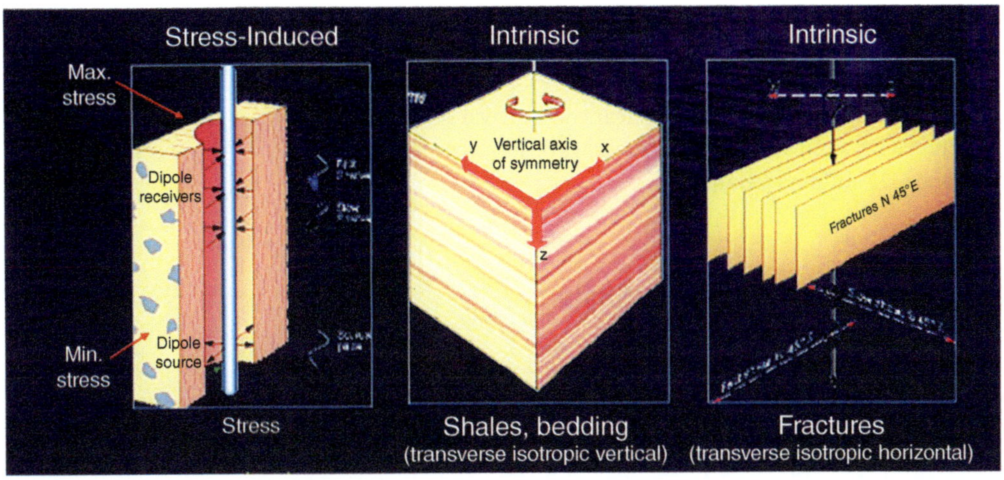

Figure 2.2 Three possible causes of sonic anisotropy (stress-induced anisotropy, VTI, and HTI mediums).

As shown in Figure 2.2, there are three geological models that are very popular in the field of seismic anisotropy research. The first one is known as stress-induced anisotropy and it measures the difference between minimum and maximum stress. The second model is known as the VTI medium, and the third one is known as the HTI medium.

The sonic scanner tool has 13 axial stations that are spaced apart by 6 in, giving the receiver array a total aperture of 6 ft (Figure 2.3). Each of the 13 stations has eight azimuthal receivers spaced every 45° around it, giving the entire receiver array 104 sensors in all. The high-fidelity receivers' ability to maintain a steady response across the tool's entire operating range of pressure and temperature has been confirmed. Additionally, receivers are routinely calibrated for increased signal-to-noise ratio and propagation mode separation accuracy. The two dipole transmitters are referred to as "0°" and "90°"; they produce a dipole mode that is 90° from the tool reference (looking normally downhole in a clockwise direction), which is used for other measurements, and aligned to it, respectively. A "chirp" is used by the frequency drive to create a high-energy, wide-band signal. Figure 2.3 shows the dipole measurement in red and the monopole observations in blue (Arroyo Franco et al. 2006). According to Arroyo Franco et al. (2006), the sonic scanner tool can measure the following sound parameters:

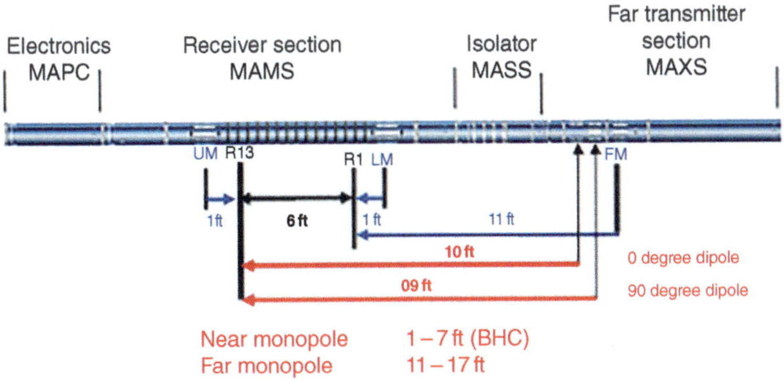

Figure 2.3 Scheme of multi-receiver sonic scanner tool.

- The difference between the cross-line energy's minimum and maximum.
- Azimuth of fast shear waves.
- Slow and fast shear wave slowness.
- Difference between slow and fast shear wave arrival times.
- Difference between slow and fast shear wave slowness.
- C44 and C55 stiffness parameters.

2.3 Application to Barnett Shale

2.3.1 Geological Setting

Due to a major marine transgression brought on by the closure of the Lapetus Ocean Basin, the Barnett Shale was deposited across modern-day North Central Texas during the late Mississippian Age (Figure 2.4). The Ouachita Thrust belt started to advance into what is now North Texas around the end of the Pennsylvanian. The sinking of the South American plate beneath the North American plate is what gave rise to the thrust belt. The foreland basin along the front of the Ouachita Thrust was produced by the thrust's emergence. Early investigations into the basin ascribed Barnett's thermal maturation to its history of burial and the thermal regimes connected to its depth of burial. As additional information became available, explorers started to question this theory. A different scenario was put up by Kent Bowker, formerly of Mitchell Energy/Devon, who claimed that the Ouachita Thrust-related displacement of hot fluids from east to west was what caused the maturation process. A generalized stratigraphic column for the Fort Worth Basin, including the Barnett Shale, is shown in Figure 2.5 (Montgomery et al. 2005).

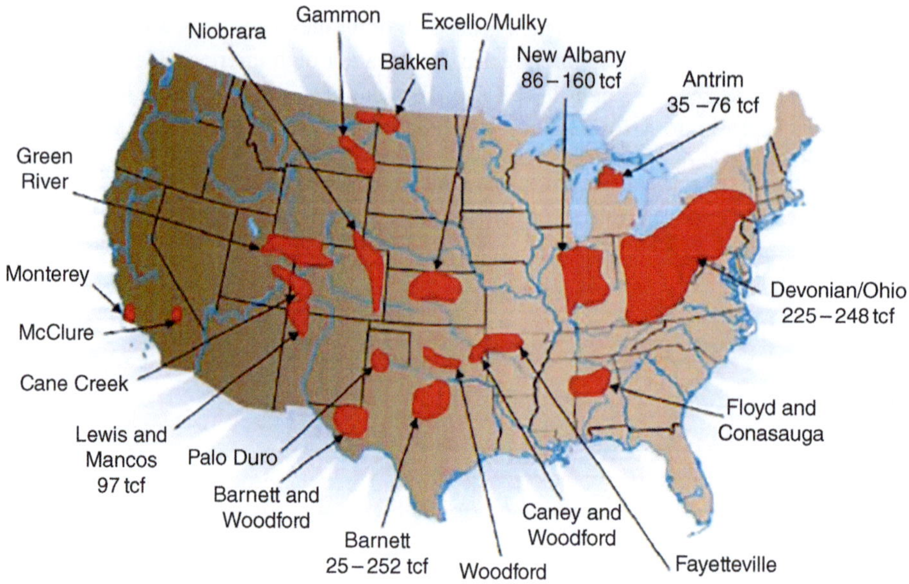

Figure 2.4 Shale gas basins of the United States.

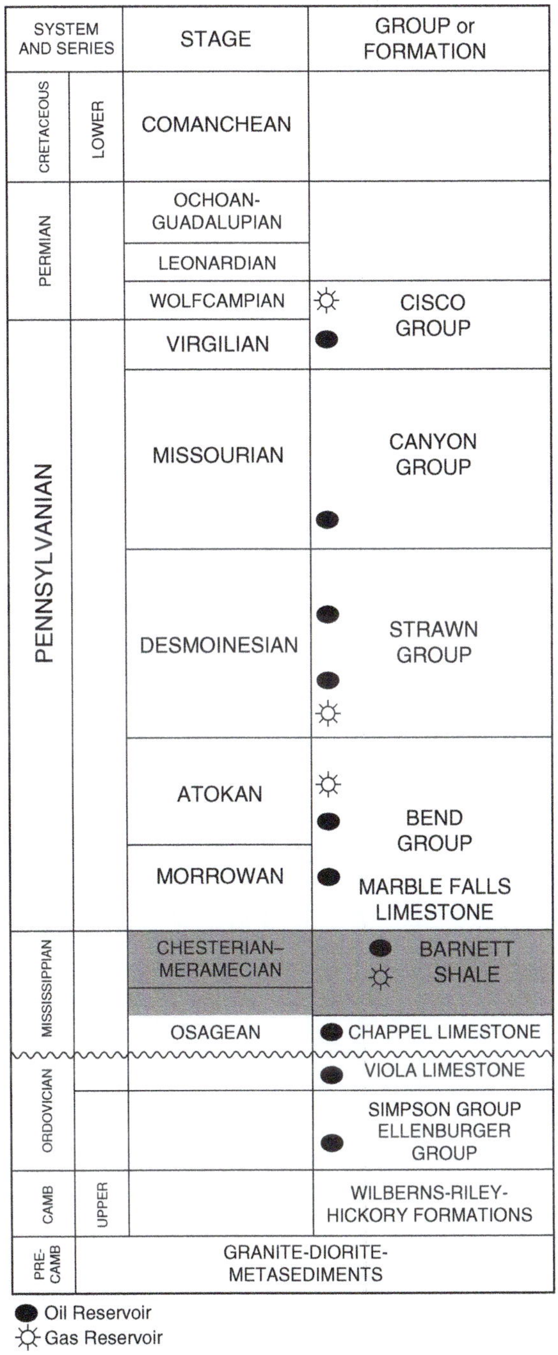

Figure 2.5 A broad stratigraphic column for the Barnett Shale in the Fort Worth Basin Shale. *Source:* Montgomery et al. (2005).

2.3.2 Data Analysis

The data from a horizontal well drilled in the lower Barnett formation is shown in Figure 2.6; track 01 represents the measured depth in feet, and track 02 represents the maximum and lowest energy of the slow shear wave; it has been noticed that there is high anisotropy of energy in the depth intervals [6800 ft, 7200 ft] and [7400 ft, 7800 ft], track 03 presents the slowness of the fast (red color)

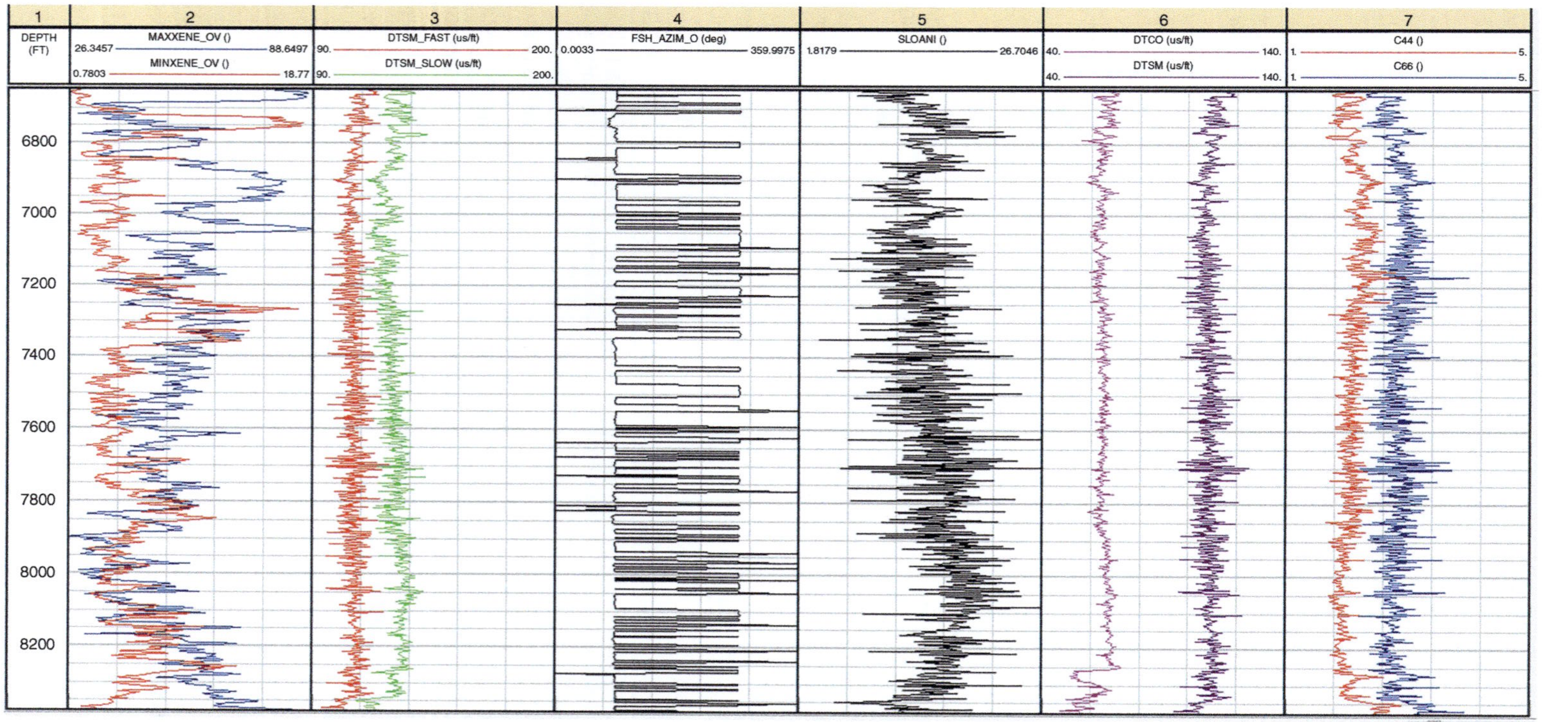

Figure 2.6 Data of the sonic scanner tool, (01) depth (ft), (02) maximum and minimum energy, (03) fast and slow shear wave slowness, (04) fast shear wave azimuth, (05) slowness anisotropy, (06) slowness of the *P* and *S* waves, and (07) C44 and C44 stiffness parameters.

and slow (green color) *S*-wave, track 04 is the azimuth of the fast shear wave, track 05 is the difference between the two slowness, track 07 is the slowness of the *P* and *S* waves, and track 08 is the log of the C44 and C66 stiffness parameters. In the depth intervals [6800 ft–7200 ft], [7300 ft–7800 ft], and [8200 ft–8300 ft], we see strong energy anisotropy, although the entire observed depth [6750 ft–8300 ft] is characterized by high slowness anisotropy. The rapid shear azimuth corresponds to the real natural fracture azimuth if we assume that the lower Barnett Shale deposit is an HTI medium (Ouadfeul and Aliouane 2015).

According to Tsvankin's paper (Tsvankin 2001), track 7 includes the C44 and C66 stiffness parameters, which are directly related to the fast and slow shear wave velocities (see Figure 2.7). According to the following correlations, C44 and C66 are given:

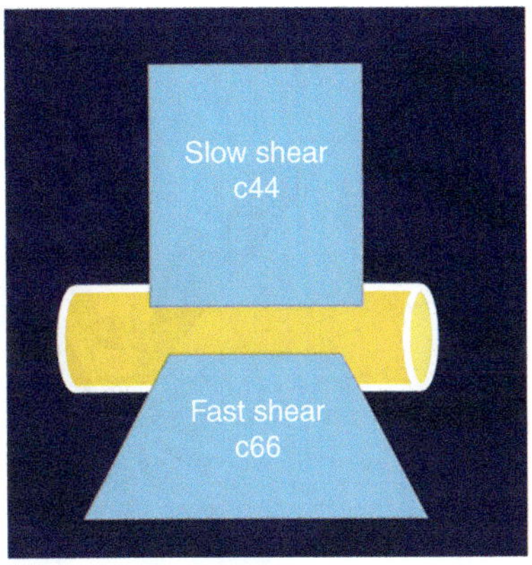

Figure 2.7 Sonic scanner in the studied horizontal well.

$$C44 = \rho * V_{\text{slow shear}}^2 \tag{2.1}$$

$$C66 = \rho * V_{\text{fast shear}}^2 \tag{2.2}$$

where ρ is the density and $V_{\text{slow shear}}$ and $V_{\text{fast shear}}$ are the velocity of the fast and slow shear waves, respectively. The natural cracks of the lower Barnett Shale deposit have four azimuths, which are 0, 80, 90, and 270°, according to fast shear wave azimuth data (see Figure 2.8). When Figures 2.8 and 2.9 are compared, it is evident that the sonic scanner tool is unable to detect other azimuths, such as 30, 40, and 140°. This is supported by the rose diagram of the natural cracks based on Gale et al. (2010)'s research of the lower Barnett Shale formation.

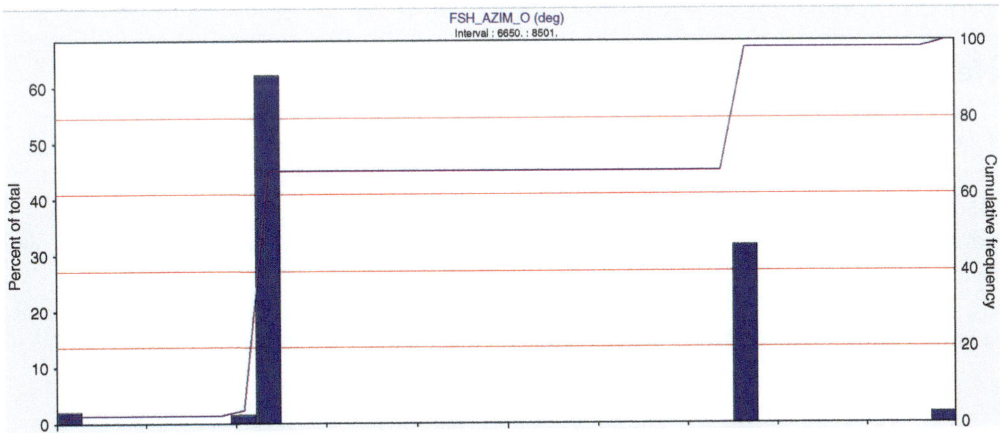

Figure 2.8 Bar chart of the fast shear wave azimuth recorded by the sonic scanner tool.

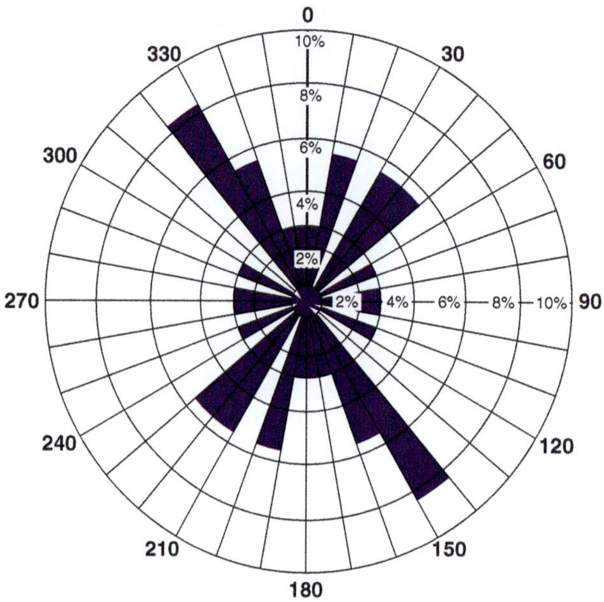

Figure 2.9 Rose diagram of natural fractures azimuth and dip after Gale et al. (2010).

2.4 Conclusions

The role played by the sonic scanner tool in a horizontal well that was drilled in the lower Barnett Shale formation is highlighted in this chapter. The well was logged for the characterization of natural fractures based on the splitting of the shear wave, which produces anisotropy of this wave. Additionally, it can be shown from a comparison of the azimuths of natural fractures provided by the sound scanner tool and those provided by image logs that not all azimuths could be provided by the sonic scanner tool. The acquired results are crucial for characterizing unconventional hydrocarbon resources because they show that extra caution should be taken when using the data from the sonic scanner instrument to explore such vast hydrocarbon reserves in horizontal wells.

References

Aliouane, L. and Ouadfeul, S. (2014). Sweet spots discrimination in shale gas reservoirs using seismic and well-logs data. A case study from the Worth Basin in the Barnett Shale. *Energy Procedia* 59: 22–27. https://doi.org/10.1016/j. egypro.2014.10.344.

Arroyo Franco, J.L., Mercado Ortiz, M.A., De, G.S. et al. (2006). Sonic investigations in and around the borehole. *Oilfield Review* 18 (1): 14–33.

Ding, P.-B., Gong, F., Zhang, F., and Li, X.-X. (2021). A physical model study of shale seismic responses and anisotropic inversion. *Petroleum Science* 18 (4): 1059–1068. https://doi.org/10.1016/j.petsci.2021.01.001.

Gale, F.J., Reed, R.M., Becker, S.P., and Ali, W. (2010). Natural fractures in the Barnett Shale in the Delaware Basin, Pecos Co. West Texas: comparison with the Barnett Shale in the Fort worth Basin. AAPG Convention, Denver, Colorado (7–10 June 2009).

Harstad, H., Lawrence, W.T., and Lorenz, J.C. (1995). Characterization and simulation of naturally fractured tight gas sandstone reservoirs. *SPE Annual Technical Conference and Exhibition*, Dallas, Texas (22–25 October).

Montgomery, S.L., Jarvie, D.M., Bowker, K.A., and Pollastro, R.M. Geochemistry and Geology by Daniel M. Jarvie, Kent A. Bowker, and Richard M. Pollastro(2005). Mississippian Barnett Shale, Fort Worth basin, north-central Texas: gas-shale play withmulti–trillion cubic foot potential. *AAPG Bulletin* 89 (2): 155–175.

Olson, J.E., Laubach, S.E., and Lander, R.H. (2009). Natural fracture characterization in tight gas sandstones: integrating mechanics and diagenesis. *AAPG Bulletin* 93 (11): 1535–1549.

Ouadfeul, S., Aliouane, L., 2015, Contribution of the seismic anisotropy in the understanding of tight sand reservoirs with anexample from the Algerian Sahara, *SEG Technical Program Expanded Abstracts* 2015: pp. 331–335, doi: https://doi.org/10.1190/segam2015-5809635.1.

Thomsen, L. (1986). Weak elastic anisotropy. *Geophysics* 51: 1954–1966. https://doi.org/10.1190/1.1442051.

Thomsen, L. (2001). Seismic anisotropy. *Geophysics* 66: 40–41. https://doi.org/10.1190/1.1444917.

Tsvankin, I. (1997). Anisotropic parameters and *P*-wave velocity for orthorhombic media. *Geophysics* 62: 1292–1309. https://doi.org/10.1190/1.1444231.

Tsvankin, I. (2001). *Seismic Signatures and Analysis of Reflection Data in Anisotropic Media.* Pergamon Press.

Tsvankin, I., Gaiser, J., Grechka, V. et al. (2010). Seismic anisotropy in exploration and reservoir characterization: an Overview. *Geophysics* 75 (5): 75A15–75A29. https://doi.org/10.1190/1.3481775.

Wild, F. (2011). Practical applications of seismic anisotropy. *First Break* 29: 117–124.

3

Wellbore Stability in Shale Gas Reservoirs with a Case Study from the Barnett Shale

Sid-Ali Ouadfeul[1], *Mohamed Zinelabidine Doghmane*[2], *and Leila Aliouane*[3]

[1] *Algerian Petroleum Institute, Sonatrach, Boumerdes, Algeria*
[2] *Department of Geophysics, FSTGAT, University of Science and Technology Houari Boumediene, Algiers, Algeria*
[3] *LABOPHYT, Faculty of Hydrocarbons and Chemistry, University M'hamed Bougara of Boumerdes, Boumerdes, Algeria*

KEYPOINTS

- Geomechanical data such as pore pressure, overburden, horizontal maximum and minimum stress, Young modulus, and Poisson's ratio are used to keep the well-bore stable during the drilling.
- The goal is to provide a mud weight window that will be used during the drilling operation.
- A case study of the Lower Barnett shale is shown.

3.1 Introduction

One of the most difficult aspects of drilling a well is keeping the wellbore steady. According to studies, unplanned events related to wellbore instability account for more than 10% of well expenditures and are estimated to cost the sector more than $1 billion annually. Every stage of the drilling fluids industry, from research and development activities through field application by the mud engineer, places a high focus on preventing shale instability. As more recent technologies are invented and applied, older ones are polished. When a hole is drilled and the formation is exposed to drilling mud, there is a radical shift in the mechanical stress as well as the chemical and physical conditions. Shale sloughing and caving are the most common signs of hole instability and lead to hole expansion, bridges, and fill. The most frequent effects are stopped pipes, sidetracks, challenges with logging and interpretation, challenges with sidewall core recovery, challenges with running casing, subpar cement work, and lost circulation. All of these factors lead to rising expenses, the potential for well or hole loss, or decreased output. Liang et al. (2014) developed a wellbore stability model for shale gas reservoir taking into account the coupling of several weakness planes and porous flow in one of the many papers that have been published on the subject. An article on the borehole stability study of horizontal drilling in shale gas reservoirs was published in 2013 by Yan et al. Darvishpour et al. (2019) have published a paper about wellbore stability analysis to determine the safe mud weight window (SMWW) for sandstone layers, they have shown that the safe drilling-fluid density range for maintaining wellbore stability is determined and simulated using FLAC3D software and a finite volume model established with drilled strata geomechanical features. The initiation of the plastic condition is used to

Unconventional Hydrocarbon Resources: Prediction and Modeling Using Artificial Intelligence Approaches, First Edition. Edited by Sid-Ali Ouadfeul.

determine the SMWW in specific sandstone layers. The effects of rock strength parameters, major stresses around the wellbore, and pore pressure on the SMWW are investigated for this wellbore. Sensitivity analysis reveals that a reduction in cohesion and internal friction angle values leads to a significant narrowing of the SMWW. On the other hand, the reduction of pore pressure and the ratio between maximum and minimum horizontal stresses causes the SMWW to widen significantly. The ability to readily quantify changes in SMWW indicates that the developed model is suitable as a well-planning and monitoring tool. Almalikee and Al-Najm have published a paper showing the wellbore stability analysis and application to optimize high-angle wells design in the Rumaila oil field, Iraq. The results gave a prediction of the appropriate mud density that can control the wellbore shear failure for any well type in order to eliminate or minimize wellbore instability and indicated that the preferred direction to drill deviated and horizontal wells should be parallel to maximum horizontal stress (NE–SW). The study also introduced the recommended drilling mud weight for vertical and deviated wells in the Rumaila oil field for the sake of gaining wellbore stability which requires raising the drilling mud weight to more than 1.27 g/cm^3 starting from the bottom of the Sadi Formation to the well total depth. The results also showed that wells with low inclination (less than 30°) are less stable than highly inclination wells (more than 30°).

In this essay, we present a case study of a wellbore stability analysis, with the aim of recommending a mud weight window for the drilling engineer to utilize. The pore pressure is calculated using Eaton's model. The paper begins by discussing wellbore stability issues, followed by a description of Eaton's model and a discussion of fundamental wellbore stability concepts like breakouts and drilling-induced fractures (DIFs). An analysis is done on data from a horizontal well that was drilled in the Lower Barnett Shale.

3.2 Wellbore Stability

A cylinder of rock is swapped out with a cylinder of fluid with a varied density as we drill the wellbore. As a result of the mud's inability to withstand shear stress, which the rock can, in situ tensions are redistributed close to the hole, creating a stress concentration. The in situ stresses are balanced, and the rock's strength is sufficient to withstand the in situ shear stresses. The rock may fail or distort because it is not sturdy enough to withstand the stress concentration (Almalikee and Al-Najm 2019). Causes of wellbore instability include:

3.2.1 Mechanical Stress

- Tension failure – fracturing and lost circulation
- Compression failure – spalling and collapse or plastic flow
- Abrasion and impact

3.2.2 Chemical Interactions with the Drilling Fluid

- Shale hydration, swelling, and dispersion
- Dissolution of soluble formations

3.2.3 Physical Interactions with the Drilling Fluid

- Erosion
- Wetting along pre-existing fractures (brittle shale)
- Fluid invasion – pressure transmission

To accurately analyze a condition and put a corrective plan in place, the drilling fluids engineer must have a thorough understanding of shale and wellbore instability. The assessment and treatment of wellbore instability require a systematic strategy that integrates a number of disciplines. In other words, a mud engineer's knowledge extends beyond drilling muds. It is essential to have a solid working knowledge of every aspect of the operation as well as a fundamental understanding of mechanics, geophysics, and the chemistry of water and clay. In order to address wellbore instability, a variety of potential causes must be considered. The most likely failure mode can be identified and a suitable reaction can be used to stop or tolerate the instability by assessing these associated conditions. These comprise mechanical issues like:

- Issues with cleaning holes
- Wellbore erosion
- Physical impact damage
- Mud weights and pore pressures
- Surge and swab pressures
- Wellbore stresses

Additionally, chemical conditions should be assessed, such as:

- Reactivity of the failing formation
- Chemical compatibility of the mud system
- Possible wellbore dissolution

3.3 Pore Pressure Estimation Using the Eaton's Model

Estimating the pore pressure from well-log data is one of the key jobs in shale gas reservoir characterization (Huffman 2002; Sayers et al. 2002, 2006). Several methods have been put out in the literature to do this. We can use the vertical model, which employs the P wave's slowness and a slowness curve with a constant gradient, as an example. Eaton's (1975) approach calculates the vertical component of the effective stress from the seismic velocity v using the flowing equation, and is typically employed for the computation of the pore pressure (Eaton 1975; Bowers 1995):

$$\sigma = \sigma_{\text{Normal}} \left(\frac{V}{V_{\text{Normal}}} \right)^n \tag{3.1}$$

σ_{Normal} and V_{Normal} are the vertical effective stress and seismic velocity expected if the sediment is ordinarily under pressure, and the exponent n indicates how sensitive the speed is to the effective stress. Then, the following provides the pore pressure:

$$p = S_v - \frac{S_v - p_{\text{Normal}})}{\left(\frac{V}{V_{\text{Normal}}} \right)^n} \tag{3.2}$$

According to Eaton's approach, the difference between the measured velocity and the velocity of sediments under normal pressure (V_{Normal}) increases with depth as follows:

$$V_{\text{Normal}}(z) = V_0 + k * z \tag{3.3}$$

As was previously demonstrated, an accurate and consistent measurement of the *P* wave's slowness is necessary for Eaton's model. In some circumstances and as a result of certain acquisition issues, such as poor estimates of the compression wave's slowness, this will cause issues with pore pressure measurement. We recommend using fuzzy logic to predict the pore pressure from well-log data for this reason.

3.4 Shale Play Geomechanics and Wellbore Stability

A cylinder of rock is swapped out for a cylinder of fluid with a variable density when we drill the wellbore. The rock can withstand shear and other types of stress but the mud cannot withstand shear stress. As a result, there was stress concentration close to the hole where the in situ loads were distributed differentially. Because the in situ stresses are balanced, the rock's strength is sufficient to withstand the in situ shear stresses. The rock may not be robust enough to withstand the stress concentration, failing or deforming as a result. For a vertical well, there are three main stresses at the wellbore: the axial stress (σ_a), the tangential stress (σ_θ), and the radial stress (σ_r).

Figure 3.1 illustrates the effective stresses at the borehole wall parallel to the major stress if we assume that σ_A is the greatest horizontal stress and σ_B is the minimum horizontal stress. These stress concentrations may produce distinctive deformation patterns if they exceed the strength of the rock surrounding the hole. Figure 3.2 depicts the conventional wellbore deformation in which the crack opens in the direction of least stress and propagates in the direction of maximum stress. When the rock's low tensile strength (typically 10% of unconfined compressive strength) is exceeded by the hoop stress, which is caused by significant wellbore pressure (e.g. leak-off test, excessive mud weight, and hydraulic fracturing).

- Significant variation between normal stresses and wellbore axis

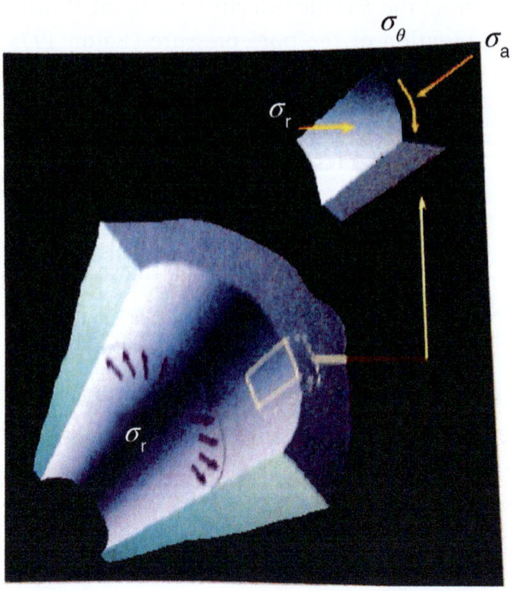

Figure 3.1 Main tensions at the wellbore.

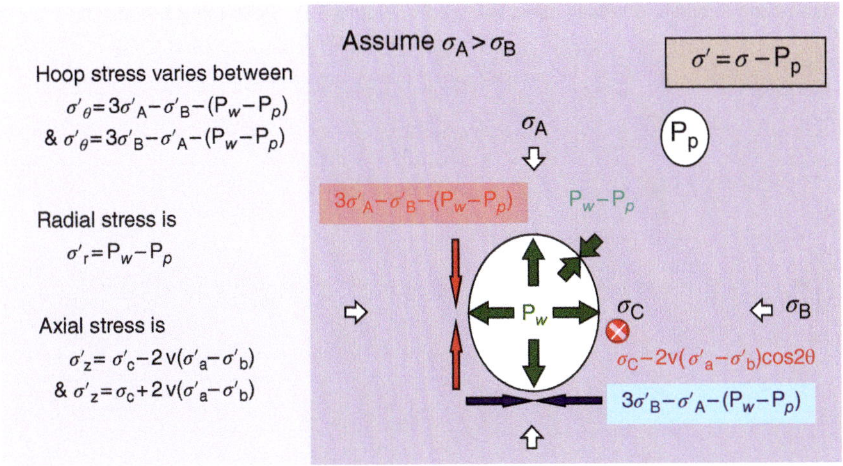

Figure 3.2 Effective stresses for a wellbore parallel to the primary stress at the borehole wall.

DIFs can result in wellbore breathing but do not always spread away from the wellbore, making them less significant failures. They are an excellent source of knowledge. The DIFs in rock failure mode 01 are depicted in Figure 3.3. The radial stress should not be extremely low compared to the maximum hoop stress to prevent breakouts (see Figure 3.4), which means:

Figure 3.3 Classical wellbore deformations.

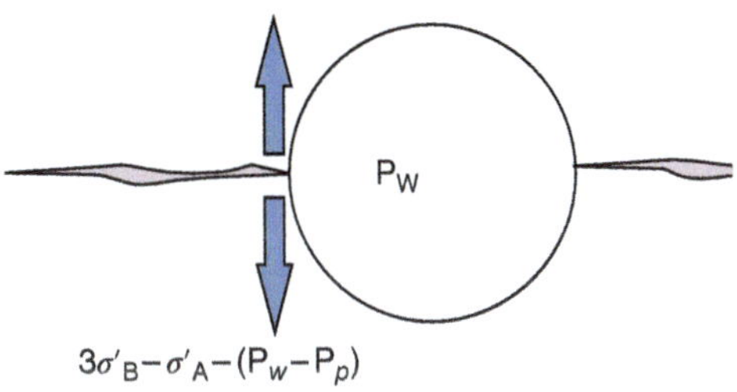

Figure 3.4 Rock failure mode 01: drilling-induced fractures.

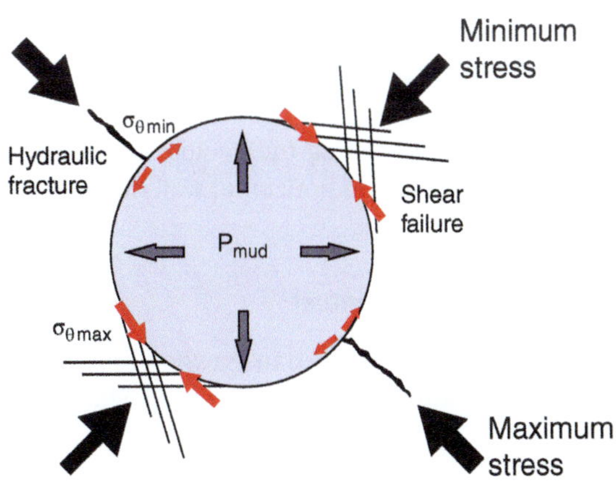

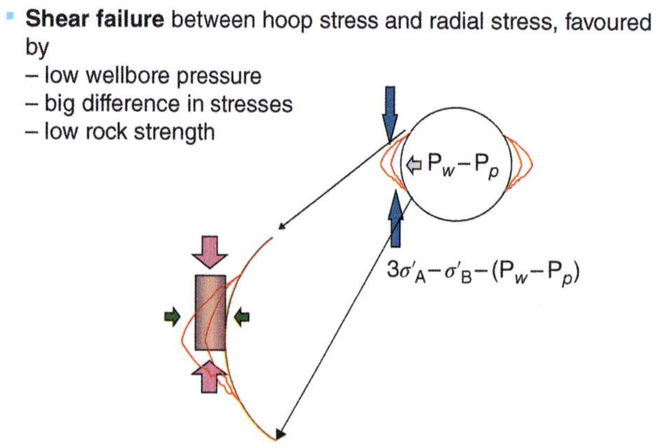

Figure 3.5 Rock failure mode 02: breakouts.

$$P_w - P_p \approx 3\sigma'_A - \sigma'_B - (P_w - P_p) \tag{3.4}$$

The mud weight window that checks the following conditions must be predicted by the wellbore stability:

- Not high enough to cause differential sticking, lost circulation, or tensile fracture.
- Not so low as to cause a shear failure of some sort.

The determination of a rock failure criterion is a crucial step in the investigation of wellbore stability. The Mohr-Coulomb criterion is the wellbore stability analysis criterion that is most frequently applied (Chen et al. 2008, 2014; Cheng et al. 2006). According to Al-Ajmi and Zimmerman (2006), the Mohr-Coulomb criterion only considers the greatest principal stress and the minimum principal stress, which presupposes that the intermediate stress has no bearing on rock strength. The criteria for non-deviated well rock failure are shown in Figure 3.5, and they include:

- The wellbore pressure required to prevent breakout.
- The maximum wellbore pressure for preventing fractures caused by drilling.

The wellbore state vs. mud weight is shown in Figure 3.6 and ranges from collapse to lost circulation.

3.5 Application to Barnett Shale

Let us start by discussing the geological context of the Barnett Shale, to which the proposed approach is applied to a horizontal well that was dug there.

3.5.1 Geological Context

During a period of marine transgression brought on by the closure of the Lapetus Ocean Basin in the late Mississippian Age, the Barnett Shale was deposited across present-day North Central Texas. The Ouachita Thrust belt started to advance into the modern-day North Texas region

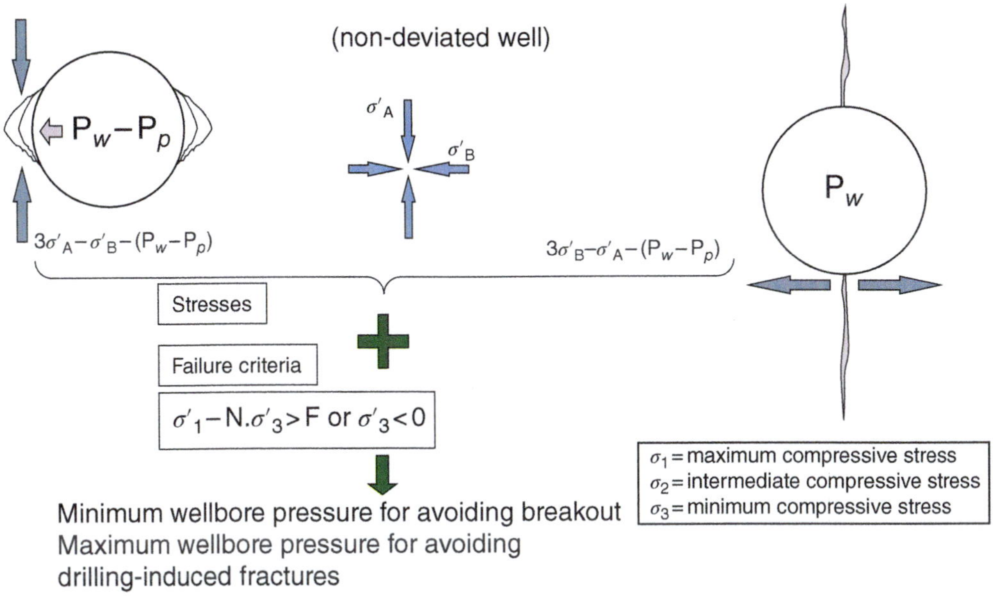

Figure 3.6 Mud weight for breakouts and drilling-induced fractures.

towards the end of the Pennsylvanian. The sinking of the South American plate beneath the North American plate is what gave rise to the thrust belt. The foreland basin along the front of the Ouachita Thrust was produced by the thrust's emergence. Early investigations into the basin ascribed Barnett's thermal maturation to its history of burial and the thermal regimes connected to its depth of burial. Figure 3.7 depicts the geologic column of the Mississippian and Pennsylvanian periods. The lower Barnett Shale gas reservoir is our goal, and its top is at 6650 m (Givens and Zhao 2014).

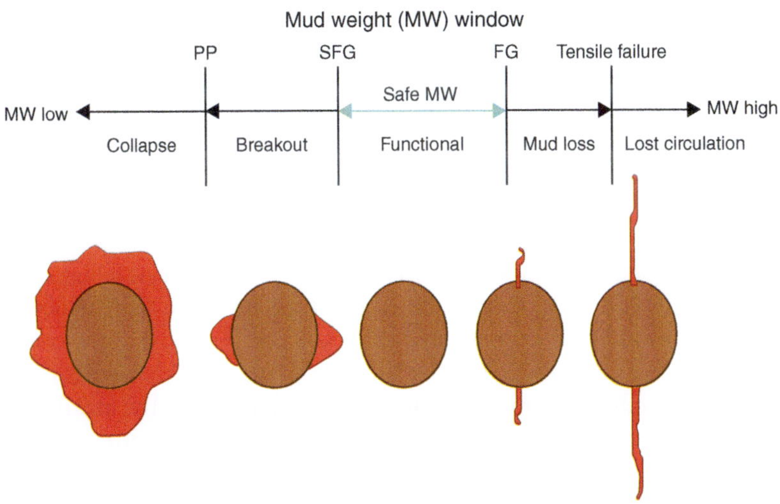

Figure 3.7 Wellbore state versus the mud weight window.

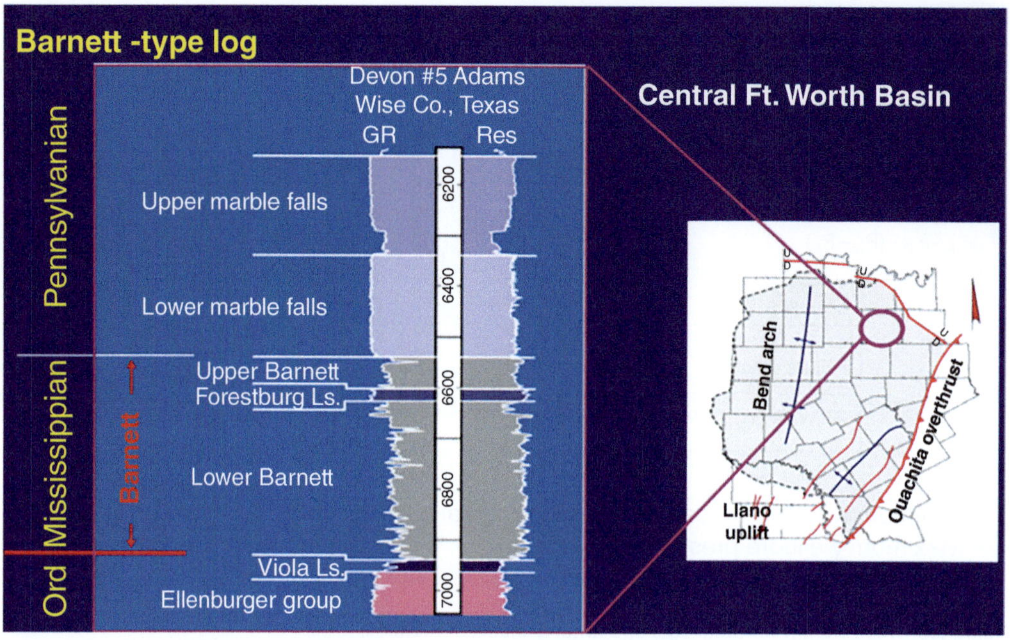

Figure 3.8 Stratigraphic column of Mississippian and the Pennsylvanian ages. *Source:* Browning (1982).

3.5.2 Data Processing

The data from the recorded and deduced well-logs, comprising the *P* and *S* wave sluggishness, neutron porosity, and confining stress, are shown in Figure 3.8. The highest and minimum horizontal stresses are the friction angle. Archimedes' model is used to determine the overburden stress. The Eaton's model is used to calculate the pore pressure. Results obtained are a forecast of the mud weight that can be employed while drilling to prevent any wellbore mechanical instability (Figure 3.9). If any breakouts or mud loss occurs when drilling, the mud weight can be altered.

3.6 Conclusion

Ignoring well-bore stability during a shale play's exploration and drilling can result in negative outcomes like mud loss, drilling-induced cracks with high breakout rates, and collapse. Geomechanics is mandatory in shale gas unconventional resources. Establishing a mud-weight envelope model is necessary to prevent this undesirable outcome. This necessitates thorough familiarity with a wide range of geomechanical parameters, including the maximum and minimum horizontal stress, overburden stress, enclosing cohesive strength, friction angle, etc.

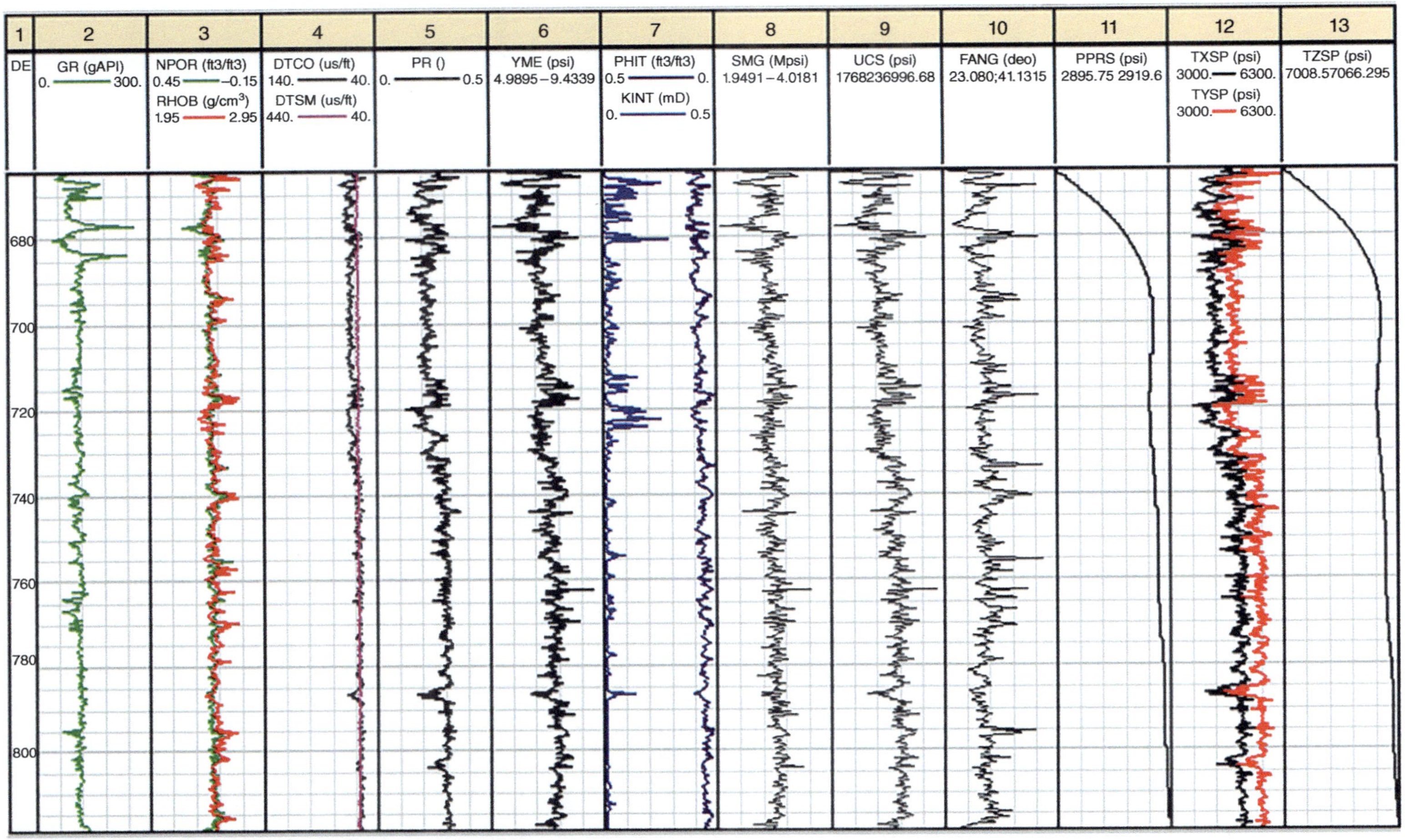

Figure 3.9 Raw well-logs data and the derived pressure using Eaton's model.

References

Al-Ajmi, A.M. and Zimmerman, R.W. (2006). Stability analysis of vertical boreholes using the Mogi–Coulomb failure criterion. *International Journal of Rock Mechanics and Mining Sciences* 43 (8): 1200–1211.

Almalikee, H.S. and Al-Najm, F.M. (2019). Wellbore stability analysis and application to optimize high-angle wells design in Rumaila oil field, Iraq. *Modeling Earth Systems and Environment* 5: 1059–1069. https://doi.org/10.1007/s40808-019-00591-1.

Bowers, G.I. (1995). Pore pressure estimation from velocity data: accounting for pore pressure mechanisms besides undercompaction. *SPE Drilling and Completion* 10: 89–95.

Browning, D.W. (1982). Geology of North Caddo Area Stephens County, Texas. Petroleum Geology of the Fort Worth Basin and Bend Arch Area, 315–329. Dallas Geological Society.

Chen, M., Jin, Y., and., Zhang, G.Q. 2008. *Petroleum Related Rock Mechanics*, Beijing: Science Press.

Chen, P., Ma, T.S., and Xia, H.Q. (2014). A collapse pressure prediction model of horizontal shale gas wells with multiple weak planes. *Natural Gas Industry* 34 (12): 87–93.

Cheng, Y.F., Wang, Y.J., Zhao, T.C., and Shen, H.C. (2006). Strength analysis of formation rock for porous media coupled with mechanical-chemical effects. *Chinese Journal of Rock Mechanics and Engineering* 25 (9): 1912–1916.

Darvishpour, A., Seifabad, M.C., Wood, D.A., and Ghorbani, G. (2019). Wellbore stability analysis to determine the safe mud weight window for sandstone layers. *Petroleum Exploration and Development* 46 (5): 1031–1038. https://doi.org/10.1016/S1876-3804(19) 60260-0.

Eaton, B.A. (1975). The equation for geopressure prediction from well logs. Paper presented at the Fall Meeting of the Society of Petroleum Engineers of AIME, Dallas, Texas (September 1975). https://doi.org/10.2118/5544-MS.

Givens, N. and Zhao, H. (2014). The Barnett Shale: not so simple after all, *Republic Energy 2014. AAPG Annual Meeting*, Dallas, Texas (18–21 April 2004).

Huffman, A.R. (2002). The future of pressure prediction using geophysical methods. *The Leading Edge* 21 (2): 199–205. doi: https://doi.org/10.1190/1.1452613.

Liang, C., Chen, M., Jin, Y., and Yunhu, L. (2014). Wellbore stability model for shale gas reservoir considering the coupling of multi-weakness planes and porous flow. *Journal of Natural Gas Science and Engineering* 21: 364–378.

Sayers, C.M., Johnson, G.M., and Denyer, G. (2002). Predrill pore pressure prediction using seismic data. *Geophysics* 67: 1286–1292.

Sayers, C.M., Den Boer, L.D., Nagy, Z.R., and Hooyman, P.J. (2006). Well- constrained seismic estimation of pore pressure with uncertainty. *The Leading Edge* 25: 1524–1526.

4

A Comparison of the Levenberg-Marquardt and Conjugate Gradient Learning Methods for Total Organic Carbon Prediction in the Barnett Shale Gas Reservoir

Sid-Ali Ouadfeul[1], Mohamed Zinelabidine Doghmane[2], and Leila Aliouane[3]

[1] *Algerian Petroleum Institute, Sonatrach, Boumerdes, Algeria*
[2] *Department of Geophysics, FSTGAT, University of Science and Technology Houari Boumediene, Algiers, Algeria*
[3] *LABOPHYT, Faculty of Hydrocarbons and Chemistry, University M'hamed Bougara of Boumerdes, Boumerdes, Algeria*

KEYPOINTS

- Data of horizontal wells drilled in the Lower Barnett shale gas reservoir are used to predict the total organic carbon.
- The multilayer perceptron neural network is training in a supervised mode.
- The Shmoker's TOC is used as a desired output.
- The conjugate gradient and Levenberg-Marquardt are used as training algorithms.
- Obtained results demonstrate that the Levenberg-Marquardt training algorithm is recommended to predict the TOC from well-logs data.

4.1 Introduction

Artificial neural networks (ANNs) have grown to be a very popular and effective tool in reservoir characterization. The multilayer perceptron (MLP) and self-organizing map (SOM) have been proposed by Ouadfeul and Aliouane (2012) to improve lithofacies classification from well-logs data. Wang et al. (2013) suggested a hybrid framework for reservoir characterization utilizing fuzzy ranking and an ANN. In this case study of reservoir porosity characterization that is being presented, three different oil wells from southwestern Alberta, Canada, were employed. The findings obtained show that the suggested procedure can produce accurate results. ANNs have been used by Aminian and Ameri (2005) to characterize reservoirs with little data. In articles published in 2014 and 2015, Ouadfeul and Aliouane presented a preliminary neural network evaluation of shale gas resources. A paper by Yeon et al. (2009) demonstrates how a neural network discharge model improves total organic carbon (TOC) predictions.

Wibowo et al. (2022) have published a paper showing TOC prediction using machine learning methods based on well logs data. The purpose is to predict the TOC value using a well log by applying the machine learning method with the MLP ANN technique. Eighteen data samples from the Talang Akar Formation were used for training and testing the MLP-ANN model. The well log data used to predict TOC are density log (RHOB), transit time (DT), deep resistivity (ILD), gamma rays (GR), and neutron porosity (NPHI), and produce a high correlation (R2 0.87 and the mean absolute percentage

Unconventional Hydrocarbon Resources: Prediction and Modeling Using Artificial Intelligence Approaches, First Edition. Edited by Sid-Ali Ouadfeul.

error (AAPE) 10%) against the resulting MLP-ANN model. The TOC prediction technique carried out will help a geophysicist (geophysicist and reservoir geology) evaluate the source rock in an oil and gas field without the need to have a large number of source rock sample data.

In this study, we demonstrate the use of an MLP neural network to forecast the TOC content of the lower Barnett Shale gas reservoir. Raw well-log data recorded in a horizontal well serves as the ANN machine's input, while Schmoker's method-calculated TOC serves as its output (Schmoker 1980). The results of the Levenberg-Marquardt and conjugate gradient training algorithms are contrasted.

The Levenberg-Marquardt training algorithm and the geological setting of the Barnett Shale are described at the outset of the research. Following the processing of data from two horizontal wells using an ANN, the paper concludes with an interpretation of the findings.

4.2 Levenberg-Marquardt Learning Algorithm

An ANN is often an adaptive system that modifies its structure in response to information coming from the outside or inside of the network during the learning phase. The goal of the learning process is to identify a set of connections W that provides a mapping that closely matches the training set (Souza 2010). Additionally, neural networks can be thought of as fundamentally nonlinear functions having the following structure (Souza 2010):

$$F(W,X)=Y \tag{4.1}$$

where X is the input vector given to the network, W are the network's weights, and Y is the matching output vector that the network has attempted to estimate or forecast. A usual ordering of the weight vector W is layer first, followed by neurons, and then the weights of each neuron plus its bias. The Levenberg-Marquardt technique is a straightforward but reliable approach to function approximation. It essentially entails resolving the following equation (Levenberg 1994; Marquardt 1963):

$$\left(J^T J+\gamma I\right)\delta=J^T E \tag{4.2}$$

where E is the error vector comprising the output errors for each input vector used for training the network, J is the Jacobian matrix for the system, γ is the Levenberg's damping factor, δ is the weight update vector that we wish to find. The value δ indicates how much we should adjust our network weights in order to have a (potentially) better result. The J^TJ matrix is another name for the roughly Hessian matrix.

At each iteration, the damping factor γ is changed to direct the optimization process. When E is rapidly reduced, using a smaller number will move the algorithm closer to the Gauss-Newton algorithm; but when iteration fails to sufficiently reduce the residual, increasing will move the process one step closer to the gradient descent direction.

The Levenberg-Marquardt simply entails computing equation (4.2) with various values of γ until the sum of squared error decreases, as was previously stated. So, each learning iteration (epoch) will consist of the following basic steps (Özgür 2004; Souza 2010):

1) Determine the Jacobian (by using finite differences or the chain rule).
2) Calculate the error gradient in 2.

$$g=J^T E \tag{4.3}$$

3) Using the Jacobian Cross Product, Estimate the Hessian

$$H = J^T J \tag{4.4}$$

4) Solve $(H + \gamma I)\delta = g$ to find δ
5) Utilizing the value of δ, update the network weights w.
6) Update the weights and recalculate the sum of squared errors.
7) Discard the new weights, raise γ, and proceed to step 4 if the total of squared errors has not decreased.
8) Else decrease γ and stop.

4.3 Application to Barnett Shale

4.3.1 Geological Setting

In the Fort Worth Basin, there is a peculiar reservoir called the Barnett Shale. The Ouachita Thrust Belt, a structural arch (bend arch), and the Llano Precambrian uplift surrounded the foreland basin, which was formed along the convergent margin of the tectonic plates (see Figure 4.1). The basin is asymmetrical, with the thrust front's adjacent axis being the structurally deepest. An island-arc system that surrounded the Fort Worth Basin contributed some coarse-grained sediment to the Barnett Shale formation. There are some naturally occurring fractures in the Barnett that are more prevalent in the limestone interbeds and have a tendency to be mineralized (Aguilar and Verma 2014). The tectonic movements connected to the Ouachita orogeny created these sets (Bruner and Smosna 2011). Karsts in the underlying Ellenberger cause localized subsidence in the Barnett Shale, high-angle faults, and karst chimneys. These karst failures have less of an impact on the Barnett in our study region (Loucks and Ruppel 2007). The lower Barnett is deposited over this unconformity because the Viola limestone has an erosional surface toward the NE. Additionally, the Forestburg limestone, which thins and vanishes toward the SE, divides the Barnett Shale into two units: the lower and the upper half (see Figure 4.2). According to Roderick Perez's (2003) analysis of Figure 4.3's mineralogy and TOC, we can see that the Lower Barnett is distinguished by high to good TOC. The lower Barnett Shale is characterized by a high quartz content, which is advantageous for hydraulic fracture, according to the mineralogy tracks. Although the amount of calcite is quite little, the clay concentration is typical. Because of its high brittleness index and other favorable characteristics, the lower Barnett Shale is an excellent Shale play (Jarvie et al. 2007).

4.3.2 Data Processing

In order to process data from two horizontal wells drilled in the lower Barnett Shale formation, neural networks were used. Figure 4.4 displays the raw well-logs data that were recorded in the well WELL01, which are as follows: The processed depth is [6500–8800 ft] because of the natural gamma ray, neutron porosity, and the slowness of P and S waves. Figure 4.5 displays the corresponding Schmoker's TOC and the raw bulk density log (Schmoker 1979, 1980). Using Schmoker's TOC log as the output and the four raw well-logs data as the input, an MLP neural network machine is trained under supervision using the Levenberg-Marquardt training algorithm. The ANN machine is trained for each experience, the weights of the connections are optimized, and the calculated output after propagation of the input in the machine is compared with the desired output. The neural machine has three layers, and the number of neurons in the hidden layer is fixed after many

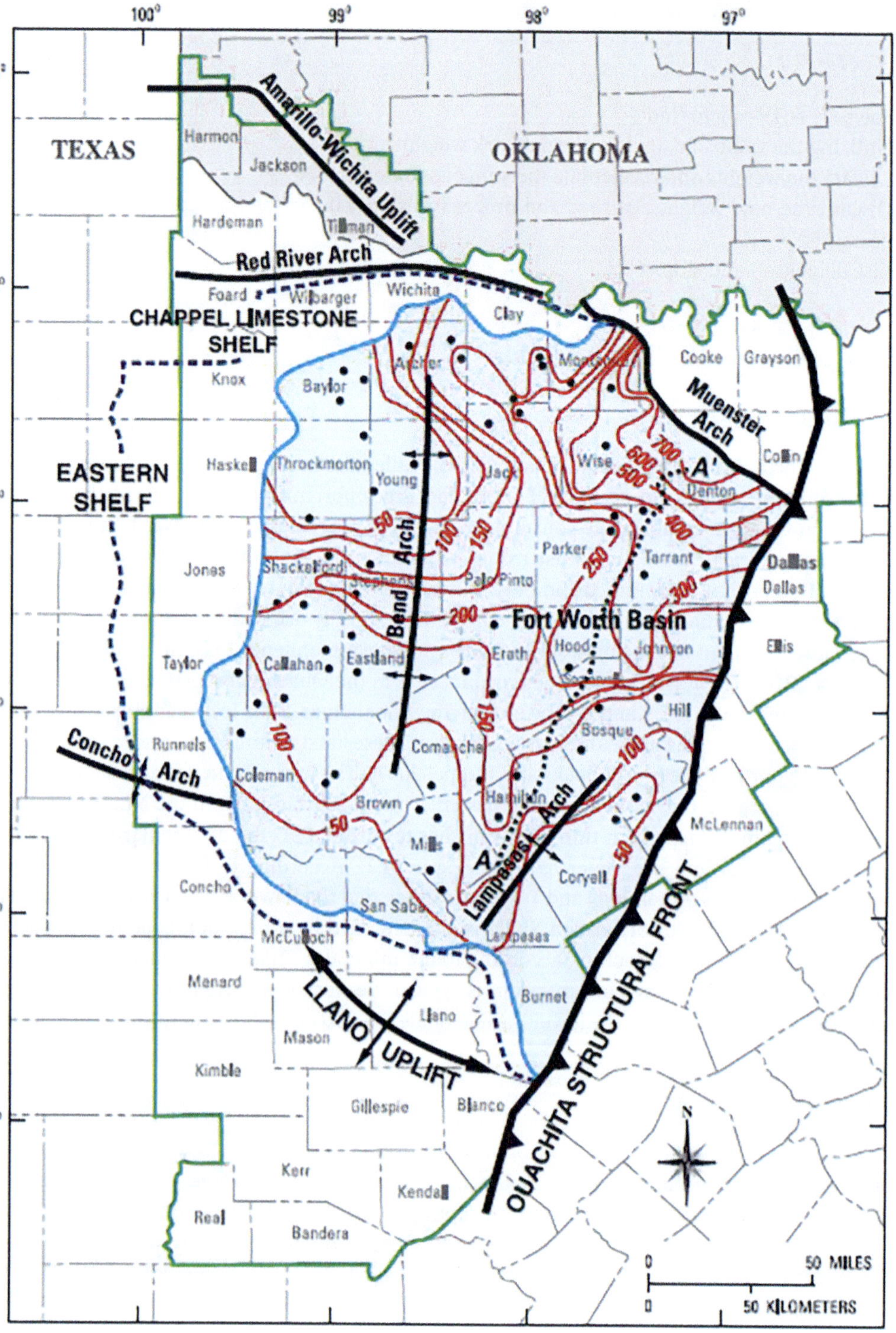

Figure 4.1 A Barnett Shale Isopach map illustrating thickness. *Source:* Bruner and Smosna (2011).

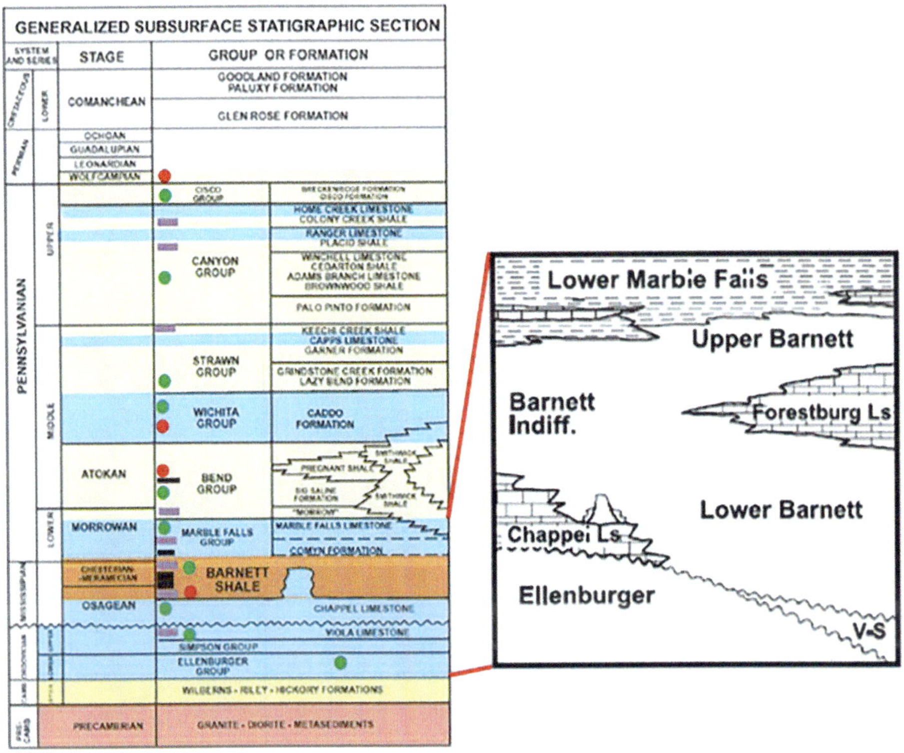

Figure 4.2 Barnett Shale stratigraphy. *Source:* Loucksand Ruppel (2007).

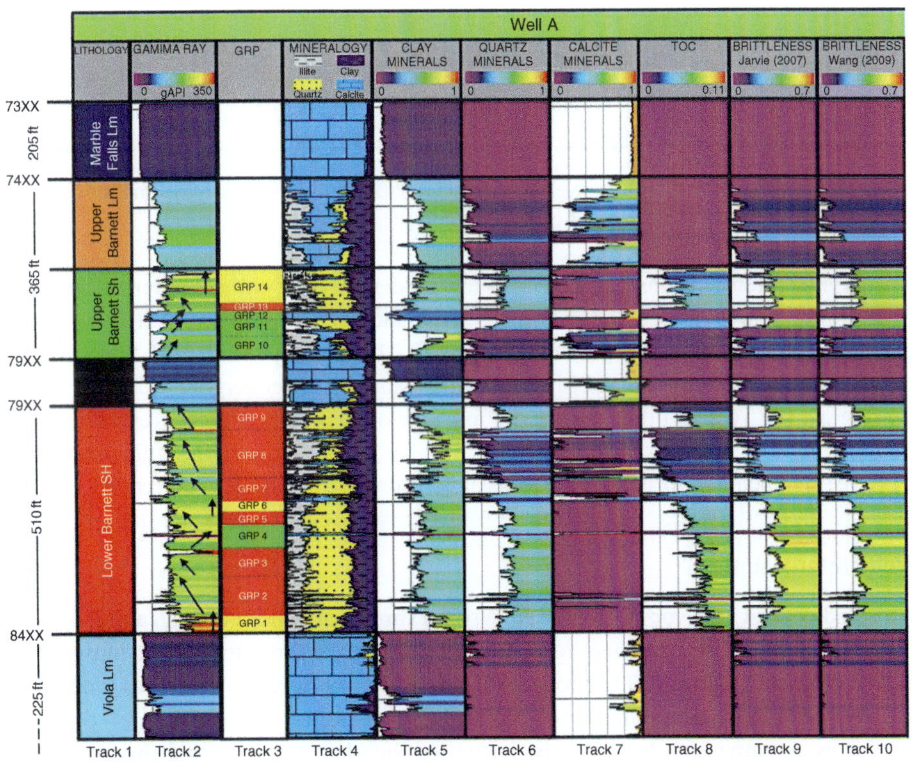

Figure 4.3 The lower Barnett's mineralogy and TOC content. *Source:* Perez (2003).

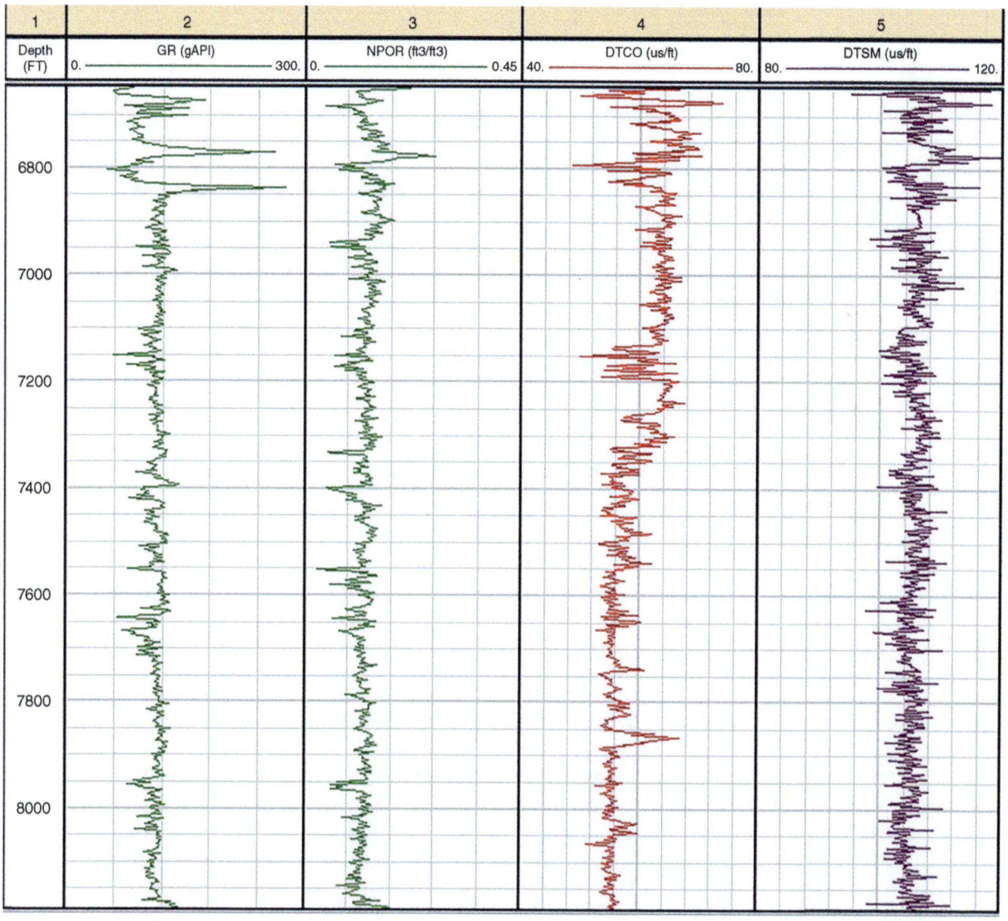

Figure 4.4 Data from a horizontal well's raw well-logs (WELL01): (1) Measured depth, (2) Gamma ray, (3) Neutron porosity, (4) Slowness of the *P* wave, (4) Slowness of the *S* wave.

numerical experiences. Ten neurons are shown in the hidden layer of the MLP neural network machine in Figure 4.6. The estimated TOC using this MLP machine is contrasted with the required TOC in Figure 4.7. Data from another horizontal well, WELL02, is propagated through this machine to test the effectiveness of the constructed MLP machine using the Levenberg-Marquardt training algorithm to forecast the TOC in other wells close to the pilot well WELL01. Figures 4.8 and 4.9 compare the computed TOC for the same well using Schmoker's approach with the anticipated TOC predicted by the MLP machine for the raw well-logs data recorded in WELL02, respectively.

4.3.3 Results Interpretation

The calculated and desired TOC for the pilot well (WLL01) are shown in Figure 4.7, and it is clear that the MLP machine with the Levenberg-Marquardt training algorithm produces an output that is very similar to the desired output. The root mean square error between the two logs is equal to 1.06, which demonstrates the neural machine's successful learning.

Figure 4.9 shows a comparison between the output of the MLP machine and the calculated TOC using Schmoker's model. The root mean square between the two logs is 2.73, demonstrating the ability of the Levenberg-Marquardt algorithm to predict the TOC in other wells close to the pilot well. The weights of the connection are used to propagate the raw well-logs data of the

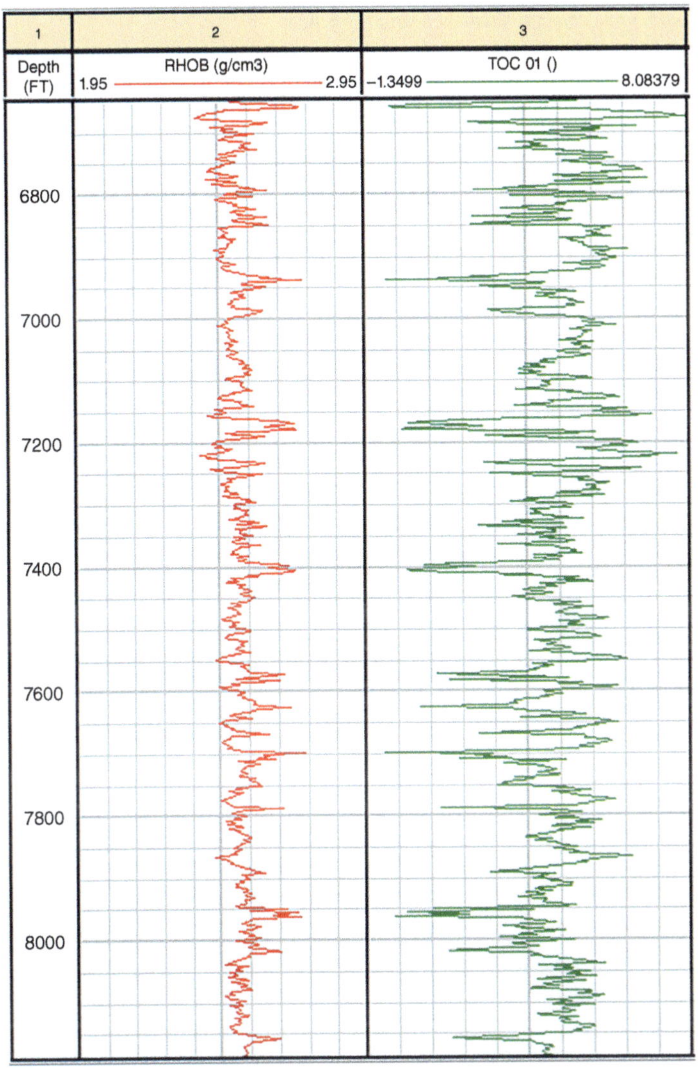

Figure 4.5 Measured bulk density log for the horizontal well WELL01 and the corresponding Schmoker's TOC.

Figure 4.6 10 Neurons in the buried layer of a multilayer perceptron (MLP) neural network computer.

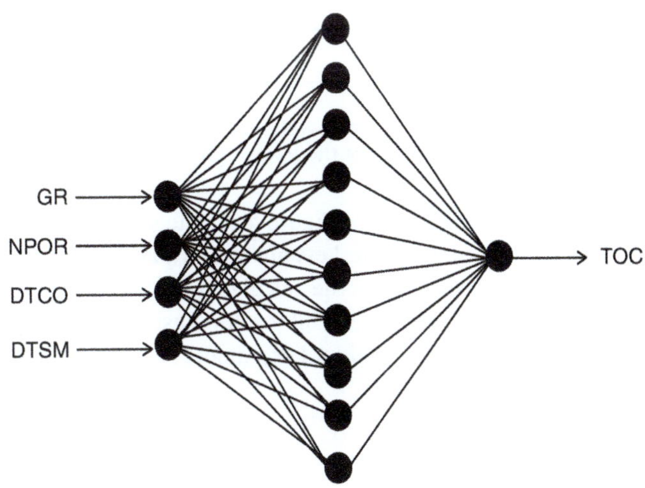

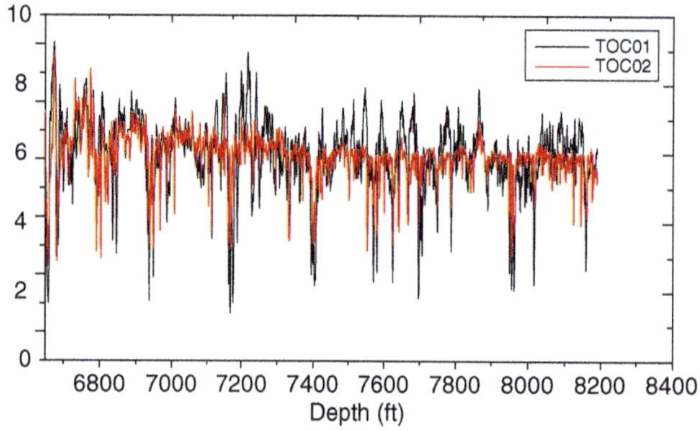

Figure 4.7 Comparing the calculated total organic carbon (TOC01) for the training well WELL01 with the intended TOC using the MLP machine with Levenberg-Marquardt (TOC02).

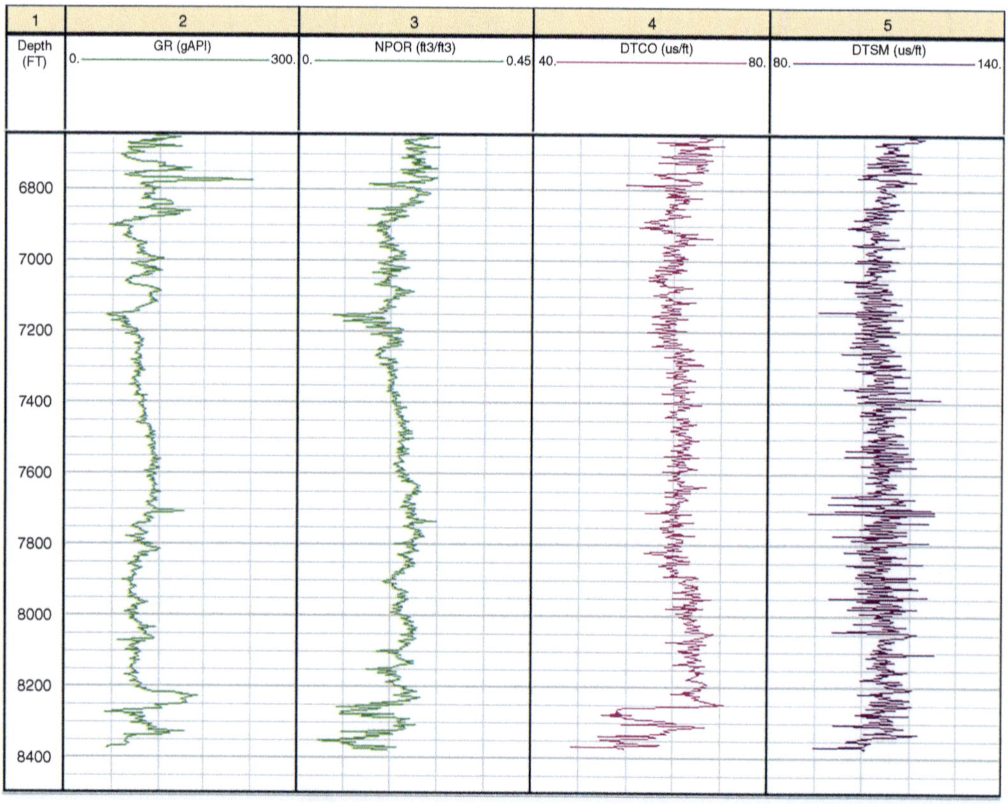

Figure 4.8 Data from a horizontal well's raw well-logs (WELL02): (1) Measured depth, (2) Gamma ray, (3) Neutron porosity, (4) Slowness of the *P* wave, (5) Slowness of the *S* wave.

horizontal well WELL02 (see Figure 4.8). The same process is carried out again with the conjugate gradient to contrast the Levenberg-Marquardt algorithm with another MLP training algorithm. Figure 4.10 illustrates the calculated TOC following propagation of the raw well-logs data of well WELL02 through the MLP machine using the conjugate gradient training algorithm and the estimated TOC using Schmoker's model. The root mean square between the two outputs is

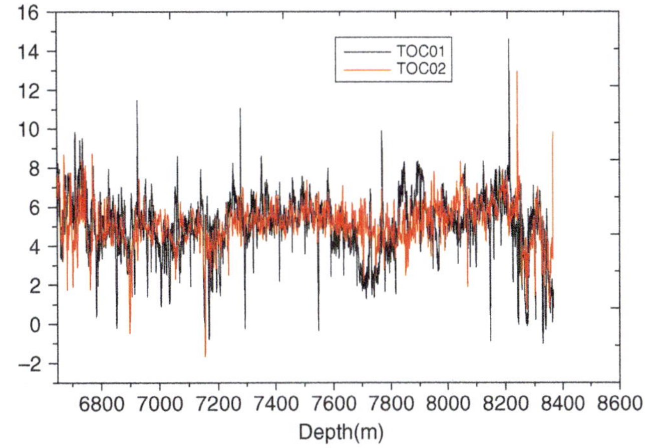

Figure 4.9 Schmocker's TOC (black color) and predicted TOC (red color) for the horizontal well WELL02 were compared using the MLP machine and Levenberg-Marquardt training algorithm.

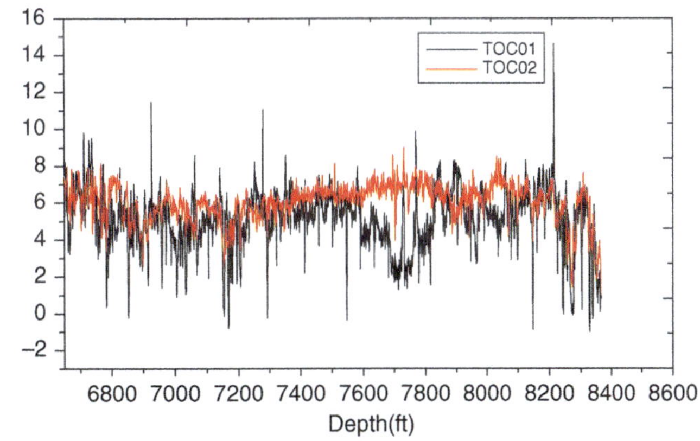

Figure 4.10 Schmocker's TOC (black color) and predicted TOC (red color) for the horizontal well WELL02 were compared using the MLP machine and conjugate gradient training algorithm.

3.64, highlighting the conjugate gradient training algorithm's inability to predict the TOC in adjacent wells. We can also see the discrepancy between the desired and calculated TOC in the depth (see Figure 4.10).

4.4 Conclusions

This research uses an MLP machine with a single hidden layer to provide a preliminary forecast of the TOC using data from raw well logs. The Levenberg-Marquardt and conjugate gradient learning algorithms are compared in this work, and the findings highlight the Levenberg-Marquardt's strength.

The suggested approach has a significant potential for resolving Schmoker's model ambiguity in the situation of a lake with bulk density log measurement. For shale gas reservoir characterization tasks like kerogen volume estimation, lithofacies classification, basin modeling, and sweet spots discriminating, the machine as it is now implemented is highly helpful. To test the effectiveness of the ANN, we advise applying it to data from various source rocks.

References

Aguilar, M. and Verma, S. (2014). TOC and fracture characterization of the Barnett Shale with predicted gamma ray and density porosity volumes. *SEG Technical Program Expanded Abstracts* 2014: 2662–2666.

Aminian, K. and Ameri, S. (2005). Application of artificial neural networks for reservoir characterization with limited data. *Journal of Petroleum Science and Engineering* 49 (3–4): 212–222.

Bruner, K.-R. and Smosna, R. (2011). A comparative study of the Mississippian Barnett Shale, Fort Worth Basin, and devonian marcellus shale, appalachian basin, DOE/NETL-2011/1478.

Jarvie, D.M., Hill, R.J., Ruble, T.E., and Pollastro, R.M. (2007). Unconventional shale - gas systems: the Mississippian Barnett Shale of north - central Texas as one model for thermogenic shale - gas assessment. *AAPG Bulletin* 91: 475–499. https://doi.org/10.1306/12190606068.

Levenberg, K. (1994). A method for the solution of certain problems in least squares. *Quarterly of Applied Mathematics* 2: 164–168.

Loucks, R.-G., and. Ruppel, S-C., 2007, Mississippian Barnett Shale: lithofacies and depositional setting of a deep-water shale-gas succession in the Fort Worth Basin, Texas, *AAPG Bulletin*, v. 91 no. 4 p. 579–601.

Marquardt, D. (1963). An algorithm for least-squares estimation of nonlinear parameters. *SIAM Journal on Applied Mathematics* 11: 431–441.

Ouadfeul, S. and Aliouane, L. (2012). Lithofacies classification using the multilayer perceptron and the self-organizing neural networks. *Neural Information Processing*, Volume 7667 of the series Lecture Notes in Computer Science, Doha, Qatar (12–15 November 2012), pp 737–744.

Ouadfeul, S. and Aliouane, L. (2014). Shale gas reservoirs characterization using neural network. *Energy Procedia* 59: 16–21.

Özgür, K. (2004). Multi-layer perceptrons with Levenberg-Marquardt training algorithm for suspended sediment concentration prediction and estimation. *Hydrological Sciences–Journal–des Sciences Hydrologiques* 49 (6): 1024–1040.

Perez, R. (2003). Brittleness estimation from seismic measurements in unconventional reservoirs: application to the Barnett Shale. Ph.D. Dissertation. The University of Oklahoma.

Schmoker, J. (1979). Determination of organic content of Appalachian devonian shales from formation-density logs. *American Association of Petroleum Geologists Bulletin* 63 (1979): 1504–1537.

Schmoker, J. (1980). Organic content of devonian shale in western Appalachian bas. *American Association of Petroleum Geologists Bulletin* 64 (1980): 2156–2165.

Souza, C.D. (2010). Neural network learning by the Levenberg-Marquardt algorithm with bayesian regularization (part 1). http://www.codeproject.com/Articles/55691/Neural-Network-Learning-by-the-Levenberg-Marquardt (accessed May 2016).

Wang, B., Wang, X., and Chen, Z. (2013). A hybrid framework for reservoir characterization using fuzzy ranking and an artificial neural network. *Computers & Geosciences* 57: 1–10.

Wibowo, R.C., Dewanto, O., and Sarkowi, M. (2022). Total organic carbon (TOC) prediction using machine learning methods based on well logs data. *AIP Conference Proceedings* 2563: 030005. https://doi.org/10.1063/5.0103209.

Yeon, IS., Jun, KW., Lee, HJ., The improvement of total organic carbon forecasting using neural networks discharge model, *Environmental Technology* 2009;30(1):45–51. doi: https://doi.org/10.1080/09593330802468780.

5

Identifying Sweet Spots in Shale Reservoirs

Susan Smith Nash

American Association of Petroleum Geologists, Tulsa, OK, USA

KEYPOINTS

- Sweet spot identification in shale reservoirs can involve both large-scale and field-level studies
- Sweet spot workflows include integration of multiple data sets and the use of advanced analytics
- Workflows for sweet spot identification can dramatically improve results

5.1 Introduction

A "sweet spot" is an informal term that indicates a part of a reservoir or a field that is preferentially enriched with hydrocarbons and can be produced effectively, often due to reservoir quality enhancements such as fractures, enhanced permeability, total organic content (TOC), and relatively high oil saturation. Sweet spots tend to be discrete "pods" within a heterogeneous unit, and thus being able to target them for pinpoint drilling and hydraulic fracturing can dramatically increase the economic viability of a play, particularly in the case of fine-grain reservoirs such as shale plays.

Sweet spots change over time due to production and also injection operations because of induced fractures, fluid flow, and adsorption.

Sweet spot hunting has been made more effective with the advent of distributed cloud computing and the use of Big Data to develop models that integrate a wide array of historical as well as continuous data (such as in the case of production or monitoring) (Dubey et al. 2017).

5.2 Materials and Methods

This paper examines published articles, results, and case studies in three basic categories:

1) Theoretical/process-focused: The United States Geological Survey (USGS) has published an extensive series of studies in its World Petroleum Assessments. In it, it uses a theoretical

Unconventional Hydrocarbon Resources: Prediction and Modeling Using Artificial Intelligence Approaches, First Edition. Edited by Sid-Ali Ouadfeul.

framework, the total petroleum system (TPS). In addition, there have been a number of new integrative approaches to identifying sweet spots which rely on ongoing history matching and input from passive seismic in order to identify emerging fracture networks.

2) Case studies: Examples of the successful implementation of integrative reservoir models, both in the regional and field-level scales were used.
3) Laboratory studies and analytics: petrophysics, reservoir models, Big Data, and Deep Learning "smart" models: New abilities to measure fractures and pores, identify trace elements and isotopes in reservoir fluids, and integrate large and constantly changing data sets in order to develop a "smart" reservoir model have been evaluated.
4) Production matching and integrative reservoir characterization: New ways to match production, decline curves, and reservoir pressure with 3D imaging have been developed in shale reservoirs in order to optimize the well placement and completion/hydraulic fracturing plans.

5.3 Data for Two Distinct Types of Sweet Spot Identification Workflows

5.3.1 Workflow 5.1: Early-Phase Workflow Elements: Total Petroleum System Approach

Thanks to advances in analytics and the ability to collect, process, and reprocess large data sets, basin analysis can be used in both identifying new plays and optimizing existing ones. The USGS has developed a methodology for a geology-based identification of undiscovered reserves in the world's basins, along with an estimation of technically recoverable reserves, given the level of technology at the time of the study. At the heart of the analysis is the incorporation of a TPS evaluation.

According to McGoon and Schmoker of the USGS (2000), the TPS includes elements and processes along with all genetically related petroleum in all forms generated by an active source rock. Briefly stated, the TPS involves an analysis of source rocks, source rock thermal maturity, hydrocarbon migration, reservoir rocks, traps, and seals. Within each aspect of the TPS, there are numerous subtleties, which allow the prediction of preferential enrichment, which constitutes a sweet spot consisting of technically recoverable reserves. What makes the TPS so powerful as a predictive model is that it maps the flow of petroleum in all its types as it moves through a basin and attempts to capture the distribution of petroleum fluids over time.

Regional geological history: The TPS approach starts with a reconstruction of the regional geological history of major geological provinces to understand the depositional environments, structural movement, heat flow, sedimentation (including erosion and unconformities), upliftment, and burial.

Depositional environment: Modeling of the depositional environment of both source rocks and reservoir rocks is critical, and the information can be used to predict thicknesses, erosional surfaces, facies changes, and relative richness of the organic content.

The TOC of black shales is of particular interest because it is a determinant of the potential for petroleum generation. Further, kerogen typing is extremely useful because it can be a predictor of reservoir quality, and the types of hydrocarbons generated, ranging from methane to petroleum.

Heat flow within the basin is best understood within the context of the geological history, particularly of mantle upwellings that can result in hydrothermal activity, volcanism, metamorphosis, but most importantly, the "cooking" of the kerogen within what is often referred to as the "kitchen."

Hydrodynamics and hydrothermal action: Hydrothermal movement along faults and within fracture networks has been responsible for mineralization, with prominent examples being the

Mississippi Valley-type deposits (including lead-zinc) and the Basin and Range (Nevada gold, silver, etc.). Identifying where the heat flow occurred over time can help identify where thermal maturation took place, with the heating up of the kerogen as well as potential catalysis.

Regional structure: The regional structure is important when considering potential structural traps. An example is salt domes in the Gulf of Mexico. Once a salt dome is identified (usually by seismic), then a number of trapping mechanisms can be postulated, and then prospective areas or "sweet spots" can be identified. The difference between of sweet spot and a prospect is usually one of scale, with the sweet spots being areas of preferential enrichment within a field or structure, and the prospect is one that contains the potential for multiple wells.

Faults, fractures, and migration pathways: Local structure is of particular importance when identifying sweet spots. Knowing where the faults are can allow one to identify where not to drill – where there is likely to be water with very low (or nonexistent) oil saturation. Faults are important also because they often serve as conduits for oil in the migration process. Likewise, fractures are important. Open fractures often served as migration pathways. In addition, they can serve as storage, although there is a chance that the oil and gas have migrated out. Nevertheless, the presence of open fractures is a measure of producibility, especially if the open fractures can be expanded through hydraulic fracturing.

Lithology, rock qualities, geomechanical characterization, and fracability: When identifying sweet spots, lithology and lithological changes are absolutely critical. In highly heterogeneous reservoirs, such as those of shale and mudrocks, it is very important to be able to model the reservoir and to clearly delineate the lithology in terms of its mineralogy, texture, and geochemistry. Obviously, permeability and porosity are important, and on a local level, they need to be described in detail. On the TPS scale, it will not be extremely detailed but will at least indicate the areas of likely reservoirs. Rock strength, brittleness, and ductility are critical in developing a map that details where the prospective reservoir rocks are capable of responding in a positive way to hydraulic fracturing.

The "fracability" is often expressed in Young's Modulus and Poisson's Ratio, which are useful in predicting frac heights as well as the extent of induced fractures. Understanding the geomechanical characteristics is also very useful for predicting on a nano-scale when and how methane adsorption takes place as well as how it is expelled during generation and moves through fracture networks, with a subsequent impact on reservoir pressure. Modeling the reservoir stress fields can also help in predicting the movement of the gas during the enrichment stage, and then later, in production.

Traps and preferential enrichment: The TPS focuses a great deal on traps and trapping mechanisms, and it uses geological history to help predict structural traps (anticlines and fault traps) as well as erosional surfaces that could result in contact with an impermeable seal. In shale reservoirs, what was once simply the source and the seal is now the reservoir as well. So, it is essentially a self-source and self-sealed reservoir, thanks to very low milli- and nano-darcy-level permeability.

5.3.2 Workflow 5.2: Smaller-Scale Field-Level Tools and Techniques

To determine the areas of preferential enrichment and the presence of technically producible hydrocarbons, it is necessary to use additional information to create a set of reservoir models and studies that provide a probabilistic assessment of the reserves. That said, it is important to maintain the regional TPS so that regional facies trends come together with local thickness and thermal maturity information. This approach has been applied with success in the VacaMuerta in Argentina by Chevron (Reijenstein et al. 2015).

Reservoir quality parameters: Porosity, permeability, pay zone thickness, and TOC are the same as in regional studies, but are performed with more detail.

3D modeling brings together seismic data, well logs with geomechanical modeling. According to Rahman and Gui (2016), a full geomechanical model should include pore pressure, vertical and horizontal stress (minimum and maximum), Poisson's ratio, Young's modulus, and fracture toughness (fracability).

Passive seismic-frac height and fracture characterization (induced fractures): Reviewing the results of microseismic to have an idea of the extent of hydraulic fracturing and where it might be possible to extend the fracture networks is important. It is also important to examine the proppants and to see if proppant embedment is a problem. Yuan and Zhang (2016) have developed a model to use existing sonic logs and microseismic monitoring arrays which register responses to fracturing to build a 3D model based on velocities (V_p and V_s) that infers Poisson's ratio and fracability. The areas within the 3D model that are more fracable are assumed to be sweet spots.

History matching: Reviewing the producing wells with their production history, particularly the decline curves in horizontal wells, is important. If a well has a very steep decline curve in relation to others in the same formation, it is good to find out if the well went out of zone, or if there are completion issues, or problems with the reservoir fluid that resulted in clay mineral swelling, extreme emulsification, sanding, or formation damage. Integrating subsurface data with production data was especially important in the Vaca Muerta in Argentina (Licitra et al. 2015).

Reservoir characterization: Bringing together petrophysics, well log data, and any core data in an organized and repeatable way is important. Reservoir characterization that is inconsistent can lead to very expensive mistakes and less-than-ideal decisions.

Fracture characterization: The characterization of natural fractures is absolutely vital in determining the storage as well as the producibility of a reservoir. Natural fractures can be open or closed. Open fractures are more easily expanded and also it is more likely to have better drainage with a combination of induced and natural open fractures. Fracture characterization needs to be done in conjunction with rock strength, fracability, and stress field studies. Fracture conductivity is often determined using a whole geomechanical model (Rahman and Gui 2016).

Pore/grain characterization: In addition to characterizing fractures, it is important to model the pore spaces as well as grains. Modeling the shape of the pores, as well as detecting any overgrowths, will give a good idea of the effect of different completion strategies. If the grains have dolomite or chert overgrowths, that will affect the mechanical strength of the rock. Further, when developing a strategy for the frac fluid and any stimulation, knowing the extent and nature of diagenesis (including overgrowths and dissolution) will affect the kind of fluids to use. For example, an acid frac may be necessary to dissolve calcitic overgrowths, but if the overgrowths are chert, the acid frac may do more harm than good.

Geochemical fingerprinting: Natural gas often contains characteristic and unique combinations of isotopes which are useful as "fingerprinting" and will help one determine the migration pathways.

Chemostratigraphy, which includes isotopes as well as major and trace elemental data to help pinpoint the depositional environmental reasons for heterogeneity within a producing formation has been used with success in the Eagle Ford (Tinnin et al. 2015). They developed a model that incorporated petrophysical, geomechanical, and handheld XRF information. The information included gamma ray, multimin petrophysical model, XRF (major elemental data and molybdenum), resisitivty, porosity, velocity, density, Young's modulus, Poisson's ratio, shear modulus, and bulk modulus (Tinnin et al. 2015).

A well log-only approach to sweet spot determination was developed by Kormaksson et al. (2015). They incorporated information from millions of wells in order to develop a statistically robust method called "functional principal components analysis (fPCA)" (Kormaksson et al. 2015). The workflow starts with well logs from vertical wells and production information from horizontal wells, then the analysis brings in the fPCA, which is then smoothed with Kriging and a multiple linear regression to result in color prediction maps.

Induced and accelerated adsorption/stimulation: Some long-term stimulation plans (in enhanced oil recovery) are designed to deliver chemicals such as nano-surfactants whose behavior changes over time, in response to temperature and pressure changes, or the slow action of dissolution. In those cases, the adsorption of methane is accelerated so that it can be moved toward the production pipe, particularly in the case of a waterflood.

Advanced analytics that incorporates Big Data and artificial intelligence has been developed by BP for identifying sweet spots in the Eagle Ford (Chen et al. 2015). The sweet spots were identified by derived net porosity and thickness, which were calculated by integrating seismic and integrated petrophysical information.

Nuclear magnetic resonance (NMR) petrophysics is being used to identify the properties of porous fluid-saturated media and thus can be used to generate maps that include porosity, fluid saturation, thickness, and permeability. It can also be used to detect and classify facies changes (Jiang et al. 2015). NMR petrophysics is very useful in conjunction with geomechanical characterization and production data. Because of the nature of the data, it is often useful to use distributed computing, especially for real-time logging while drilling.

5.4 Results: Two Integrative Workflows

5.4.1 Early-Phase Exploration Workflow

First step: Identify the basin, then the TPS elements, with an emphasis on source and reservoir rocks.

Second step: Follow with geological history (with heat flow history and depositional environments), which incorporates well log data for source rock TOC, thickness, and regional structure.

Third step: Integrate information to posit migration pathways.

Fourth step: Focus on the reservoir rocks and map thickness, structure, and fluid saturation (where possible).

Fifth step: Make time slices and determine where the fluids migrated, and where they accumulated.

5.4.2 Later Phase Developmental, Including Refracing Workflow

First, reservoir modeling: Focus on the reservoir and migration pathways; emphasize heterogeneity and reservoir storage.

Second, total geomechanical model: Focus on fracability and producibility.

The sweet spots are in the areas of an optimal combination of storage capability, fracability, hydrocarbon saturation, and producibility (tie to pressure as well as rock properties).

5.5 Case Studies

5.5.1 Woodford Shale: Emphasis on Chemostratigraphy

Use chemical fingerprints (usually stable isotopes) for stratigraphic correlation. It can be very useful in developing an idea of (i) which layers are preferentially enriched, and where they are, especially if they are laterally discontinuous; (ii) where the gas or oil may have migrated; (iii) where ideal depositional conditions existed for optimization of total organic carbon (TOC) and hydrocarbon generation. Chemical fingerprinting using stable isotopes can be correlated to depositional conditions. The ideal depositional environments can be correlated by characteristic suites of trace minerals. For example, in the case of the highly productive Woodford Shale in Oklahoma, the most enriched sections came from anoxic conditions where redox-sensitive trace metals (Mo, Ni, and Cu) could be found, along with pyrite and high TOC (Turner and Slatt 2016).

To find potentially overlooked sweet spots by means of chemostratigraphy, researchers start by examining core samples. They are cleaned and then scanned using XRF in order to identify trace elements. The data is then analyzed using hierarchical clustering analysis (HCA) (Turner and Slatt 2016). Both TOC and trace element content are analyzed, and the end results are chemofacies profiles which correspond to ideal depositional conditions for petroleum generation.

The drawback to using chemostratigraphic sweet spot hunting is that it can be difficult to determine the direction of the depositional environment by means of cores and core analysis alone.

However, core analyses can be used to find places where the formation was overlooked, especially in the days when the Woodford (or another shale) was considered an unproducible source rock. Chemostratigraphic analyses can also provide a great deal of value for determining the finer points of a petroleum system and could be key to step-outs, extensions, and analogues.

5.5.2 Barnett Shale: Emphasis on Seismic Attributes

Earlier unconventional plays such as the Barnett Shale in the Fort Worth Basin have been extensively studied to determine if rock properties that have been determined by means of the integration of seismic and well log data. Using a combination of information along with seismic attributes can help develop a model in which the ideal "signature" or suite of rock properties can be identified and then correlated (Aliouane and Ouadfeul 2014).

In the Fort Worth Basin, the Barnett Shale "sweet spots" were ones which exhibited high TOC, high maturity (vitrinite reflectance), low water saturation, and high adsorbed gas. Geomechanical properties measured by Young's modulus and Poisson's ratio were found to be useful for determining brittleness and fracability (Aliouane and Ouadfeul 2014).

The workflow starts with determining TOC by means of well logs when core data is not available. The Passey method can be used to calculate TOC, although there have been instances in which the results are not correct, particularly when the reservoir is not 100% self-sourced or self-sealed.

In order to understand the lithology and rock properties between the wells, seismic data is processed using AVO inversion, which is then tied to core data. The resulting 3D map can help develop a sense of where to drill and how thick the sediments may be.

Brittleness and ductility are calculated using Poisson's ratio from seismic data (V_P/V_S, calculated on *P*- and *S*-impedance). Young's modulus requires a density calculation, also from seismic, with a combination of *P*- and *S*-impedance geostatistically determined and compared with cores for appropriate calibration.

In the Barnett, many operators take the seismic attribute-based sweet spot determination a step further and look for fracture networks that can be an excellent indicator of oil storage. Unfortunately, however, oil-impregnated closed fractures may not be producible, meaning that it is necessary to understand the networks of open fractures, both natural and induced (through hydraulic fracturing). In these cases, microseismic is often used to determine the extent of induced fractures and the potential for optimal recovery.

5.5.3 Eagle Ford Shale: Pattern Recognition/Deep Learning

Gherabati et al. (2016) describe a workflow as an analytics-centered approach to finding sweet spots. Their model was developed for the Eagle Ford that takes into consideration production history as well as petrophysics, seismic, and reservoir fluids.

- The first data set incorporated well logs and seismic to develop thickness, porosity, permeability, and water saturation.
- The second data set included reservoir fluid characterization in order to differentiate between gas, condensate, and oil.
- Finally, production data was incorporated as production history in order to qualitatively differentiate regions based on decline curves.

The result was the pinpointing of areas within the study region where it would be most possible to encounter oil (rather than condensate or gas) that would maintain production levels better than the rest.

This approach has certain limitations because it assumes that all the drilling, completion, and stimulation approaches were the same. Nevertheless, it is a method of finding sweet spots that are grounded on actual production rather than a proxy for production, or what is assumed to be a good environment for producing areas of relative enrichment (sweet spots).

5.6 Conclusion

Because sweet spots change over time, due to geomechanical influences and volume/pressure changes, it is important to develop a continuously updating geomechanical model for the field. Log-based stress calculations, production history, and detail anisotropy for the natural fracture networks. Some fractures will heal and some will change based on stress fields; any refracs should seek to increase fracture complexity (Rahman and Gui 2016).

References

Aliouane, L. and Ouadfeul, S.-A. (2014). Sweet spots discrimination in shale gas reservoirs using seismic and well-logs data. A case study from the worth basin in the Barnett Shale. *Energy Procedia* 59 (2014): 22–27.

Chen, B., Kumar, D., Uerling, A. et al. (2015). Integrated petrophysical and geophysical analysis on identifying Eagle Ford sweet spots. *Unconventional Resources Technology Conference,* San Antonio, Texas (20 July 2015).

Dubey, S., Chakraborty, D., Mishra, S., and Fredd, C. (2017). Systematic evaluation of shale play by introducing integration of inversion geophysics, petroleum system and reservoir simulation workflows. *Society of Petroleum Engineers*.doi:https://doi.org/10.2118/183981-MS.

Gherabati, S.A., Browning, J., Male, F. et al. (2016). The impact of pressure and fluid property variation on well performance of liquid-rich Eagle Ford shale. *Journal of Natural Gas Science and Engineering* 33 (2016): 1056–1068.

Jiang, T., Jain, V., Belotserkovskaya, A., Nwosu, N. K., & Ahmad, S. (2015). Evaluating producible hydrocarbons and reservoir quality in organic shale reservoirs using Nuclear Magnetic Resonance (NMR) factor analysis. *Society of Petroleum Engineers.* doi:https://doi.org/10.2118/175893-MS.

Kormaksson, M., Vieira, M. R., &Zadrozny, B. (2015) A data driven method for sweet spot identification in shale plays using well log data. *Society of Petroleum Engineers.* doi:https://doi.org/10.2118/173455-MS.

Licitra, D., Lovrincevich, E., Vittore, F. et al. (2015). Sweet spots in VacaMuerta: Integration of subsurface and production data in loma campana shale development, Argentina. *Unconventional Resources Technology Conference.* doi:https://doi.org/10.15530/URTEC-2015-2153944.

Magoon, L.B. and Schmoker, J.W. (2000). Chapter PS.The total petroleum system – the natural fluid network that constrains the assessment unit. US Geological survey Digital Data Series 60. https://certmapper.cr.usgs.gov/data/PubArchives/WEcont/chaps/PS.pdf.

Rahman, K., &Gui, F. (2016). Geomechanical sweet spot identification in unconventional resources development. *Society of Petroleum Engineers.* doi:https://doi.org/10.2118/182247-MS.

Reijenstein, H. M., Christopher, L., Manuel, F. et al.(2015). Where is the VacaMuerta sweet spot? The importance of regional facies trends, thickness, and maturity in generating play concepts. *Unconventional Resources Technology Conference.* San Antonio, Texas (21 July 2015).

Tinnin, B., McChesney, M. D., and Bello, H. (2015). Multi-source data integration: eagle ford shale sweet spot mapping. *Unconventional Resources Technology Conference.* doi:https://doi.org/10.15530/URTEC-2015-21545.

Turner, B.W. and Slatt, R.M. (2016). Assessing bottom water anoxia within the late devonian woodford shale in the arkoma basin, southern Oklahoma. *Marine and Petroleum Geology* 78 (2016): 536–546.

Yuan, C. and Zhang, J. (2016). 3D microseismic imaging for identifying shale sweet spot. *Society of Exploration Geophysicists* 144–145.

6

Surfactants in Shale Reservoirs

Susan Smith Nash

American Association of Petroleum Geologists, Tulsa, OK, USA

KEYPOINTS

- Surfactants are effective in hydraulic fracturing in shale wells to improve the dispersion and delivery of chemicals
- Surfactants in EOR mobilize and produce capillary oil by lowering interfacial tension between water and oil
- New "smart" surfactants can accelerate the adsorption of methane molecules away from microfractures to stimulate flow.

6.1 Introduction

Surfactants have been used in shale reservoirs in many different ways. First, they are used in drilling fluids to improve the dispersion and delivery of chemicals. At the same time, they have been used during hydraulic fracturing to create a shortlived microemulsion that transports the oil and gas molecules so they can be produced.

A surfactant triggers a physical-chemical response between two substances, usually two fluids, or a fluid and a solid surface. There is preferential adsorption, which results in a change in interfacial tension (IFT), and it usually lowers capillary pressure and surface tension.

A surfactant molecule has a head and a tail. The head is polar and hydrophilic. The tail is nonpolar and hydrophobic. The tail is often a hydrocarbon.

6.2 Function of Surfactants

Surfactants reduce the IFT between two substances. It can be between two liquids or a liquid and a solid, with the result of a reduction in capillary pressure. There is movement along the surface because the tension has been reduced. The surfactant molecules start to move together and create

Unconventional Hydrocarbon Resources: Prediction and Modeling Using Artificial Intelligence Approaches, First Edition. Edited by Sid-Ali Ouadfeul.

micelles. The micelles often are attracted to a charged particle and enwrap it, creating an emulsion or microemulsion. The emulsion is the vector for transport, and it is what moves the particle (which can be oil molecules in the case of the oil industry).

There are important considerations, however. In low-permeability reservoirs, the microemulsion that was so useful in a conventional reservoir can now block the pore spaces. Further, adsorption of the surfactant could mean that there are chemical losses in the formation. So, a unique surfactant plan must be made for each type of reservoir. Not only it is necessary to take into consideration the reactions of the surfactant with the two substances but it also is important to consider the chemical reactions that can (and will) take place between all the different liquids and solids introduced into the formation, as well as the mineralogy of the formation as well as the reservoir fluids.

There are often complaints that surfactants do not work as promised (or as they seemed to work in the laboratory), and then, later there are surprisingly positive results. Many surfactants are introduced when all hope is gone, or when the field has stopped producing, and thus there is less to lose when introducing a witches' brew of chemicals that may require at least six months to start to see increases in hydrocarbon production, or, in a worst-case scenario, can seem to enhance the production of brine (which must then be disposed of) and inhibit the production of hydrocarbons (Negin et al. 2016, 2017).

The surfactants have different purposes at different points:

6.2.1 Drilling

Surfactants can be used to create desirable emulsions that help reduce borehole instability.

6.2.2 Completion (Hydraulic Fracturing)

Surfactants can be as a "force multiplier" to change the capillary pressure, and then to create shortlived microemulsions that surge out of the formation and are essentially finished after 30–45 days (in the case of extensive multistaged horizontal completions).

6.3 Materials and Methods

In shale reservoirs, methane is found in two places. First, it is in the void space of pores and fractures. Some of the fractures are not open (whether they be natural or induced). So, the gas is not recoverable, unless over time additional fractures or channels develop, which would create pathways. The other place the gas is found is as adsorbed molecules on the surface of the rock (or on the surfaces of the pores). While the adsorption of the surfactant is not good, the adsorption of in situ methane molecules is good news. It means that the molecules, once liberated from the surfaces, can flow, and hence be produced. Thus, anything that will reduce the surface tension (reducing the capillary pressure) can in theory cause the adsorbed molecules to move into the open fractures and thus flow through the formation in pathways that go directly to the production string.

6.4 Characteristics of Shale Reservoirs

Several properties of shale reservoirs must be carefully considered when selecting surfactants.

6.4.1 High Clay Mineral Content

Clay minerals such as bentonite can swell when they have contact with water. Surfactants have to be selected so that their hydrophobic and hydrophilic properties do not cause the shales to swell and thus create formation damage.

6.4.2 Nano-Sized Pores

Surfactant micelles can become too large, and also emulsions can form that clog the pores, thus contributing to a steep decline curve.

6.4.3 Mixed-Wettability Behavior

A shale reservoir may be very heterogeneous, resulting in a wide range of wettability behaviors.

6.4.4 High Capillary Pressures

Capillary pressures are affected by IFT and contact angle in reservoir fluid-rock systems.

Li et al. (2016) conducted a number of simulations and laboratory experiments that demonstrated that pressure has a dramatic effect on methane adsorption. The pressure changes do not have to be dramatic; in fact, they can be very subtle, yet still, have a great impact on methane adsorption (Li et al. 2016).

As pressures increase, adsorption accelerates and the density of methane molecules on the surfaces increases. This means, in essence, that pressure has increased the adsorption to the point that the methane molecules reach capacity. In narrow pores, there is more wall interaction (which is to be expected) (Lu et al. 2016). Now, to move the adsorbed methane and to produce it, it is necessary to reduce the IFT, which can be done via the delivery of surfactant. One can readily see that a massive surfactant flood could potentially be effective, but it may run the risk of creating emulsions that are too large to move in nano- and micropores and fractures (Buijse et al. 2013). While emulsions may be the ideal way to move oil molecules, methane gas molecules are another matter. In their case, a pressure change along with the pinpoint delivery of surfactant could be the best approach. Even small increases in temperature can positively affect pressure, and some temperature changes can be chemically induced.

Kang et al. (2011) examined the interaction between adsorbed layers of shale gas that move along the solid surface of the shale matrix. The adsorbed layer can move by means of slip flow at the wall-gas interface (Li et al. 2016).

6.5 The Klinkenberg Correction

Permeability is affected by the fact that not all porosity is actually available. Some are bound up in closed fractures or blind pore spaces, and even though the pores have gas storage potential, there is no way to get the gas out of the pore space. In addition, permeability measurements are not as they seem they might be due to the effect of pressure within the pores, and the slippage along the pore walls due to pressure changes. The gas slippage at the pore wall results in higher gas permeability. The resulting "Klinkenberg Correction" can be quite dramatic in low-permeability rocks such as shales and tight sandstones.

6.5.1 Klinkenberg Gas Slippage Measurement

In the 1940s, L. Klinkenberg published his findings in *Drilling and Production Practice* (1941). He found that there is "gas slippage" along the walls of the pores. He also described how the pore structures change due to geomechanical forces (stress) during production, and that the pore geometry change results in a change in effective porosity and permeability (Letham and Bustin 2015).

Klinkenberg gas slippage and the resulting pore deformation are extremely important in surfactant program design because it is necessary to re-envision the pore architecture to recalculate the effective surface area, as well as the pore size. Failing to take into consideration the Klinkenberg effect could result in the wrong concentration of surfactant, and the formation of a microemulsion that is too thick and viscous. It could also result in moving too quickly and not instituting a chemical program after 90 days of production in order to take advantage of the positive attributes of gas slippage.

6.6 Completion Chemicals to Consider in Addition to the Surfactant

Formulations often include nonionic surfactants to avoid interactions. However, they may not have as much energy in the reservoir, and there may be interactions with other fluids found in drilling, stimulation, and completion:

Clay stabilizer
Biocide
Friction reducers

6.6.1 Enhanced Oil Recovery (EOR)

Reservoir fluids combined with ultra-diluted surfactants can be injected in a mature field that has lost its drive and where there may be issues of channeling which resulted in unproduced parts of the reservoir. In this case, the increase in produced oil will not be immediate. It will be necessary to re-establish pressure by managing the sweep of the formation. The injection wells will need to be spaced in an ideal pattern. The oil that is encountered by the surfactant-containing fluids will move forward if there is sufficient porosity and the dilute surfactant does not create a viscous emulsion. The IFT reduction should be accompanied by very low-viscosity fluids, along with sufficient pressure to assure movement through the reservoir.

Surfactants in EOR:

- Mobilize and produce capillary oil (Buijse)
- Lower IFT between water and oil
- A dilute solution in brine in combination with co-solvents (alkali and polymer)
- Energetically change the wettability; thus, primarily anionic surfactants used in EOR
- Manage interactions between oil, surfactant, cosolvent, and brine

6.6.2 Liquids-Rich Shale Plays After Initial Decline

Surfactants can be successfully applied as a part of a refrac program after the initial production has declined. However, it must be kept in mind that they can also be disastrously applied because sometimes the initial application of frac fluid results in formation damage (or what appears to be formation damage, which is actually the loss of chemicals into the formation through adsorption). Thus, there will be the equivalent of gooey emulsion cake coating on some of the surfaces, which also happen to be the migration pathways/major induced fractures. Instead of trying to remove the emulsion layer from the surfaces, it is a better idea to induce new fractures that slightly offset the original one, and to use a different type of surfactant; one that is more dilute, and which may result in a salinity change, with heat and pressure changes, which could stimulate production via "siphon" effect (a phenomenon which is often observed to occur in an unanticipated and undesired way, when the offsetting acreage is fraced and wells a mile away demonstrate that they have been affected).

6.7 Mono-Coating Proppant

6.7.1 Zwitterionic Coating

The coating changes a proppant from being a grain that props open a fracture (and potentially becoming a barrier) to being a flow enhancer and also an adsorption accelerant, with the result of having an increase in methane flow after 18–24 months, with a continuing internal stimulus, continuing to increase the flow in both the rate and also volume.

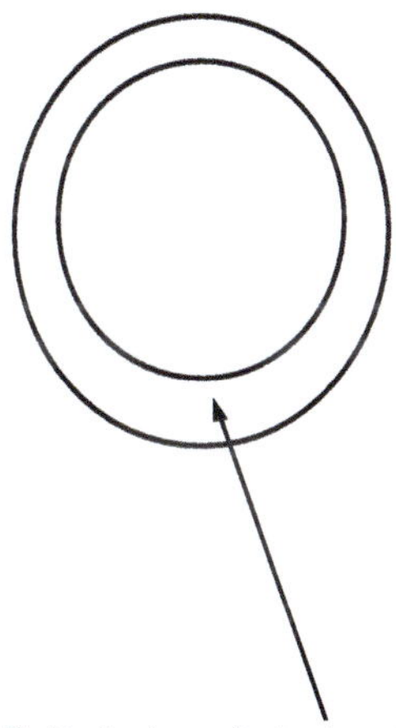

Zwitterionic surfactant coating with both cationic and anionic centers attached to the same molecule. By reducing surface tension on both sides, it is possible to accelerate the process of adsorption of CH_4 molecules, which will increase nanopressure, flow, and production.

6.8 Dual-Coating Proppant

6.8.1 Outside Coating

Zwitterionic

6.8.2 Inner Coating

Inorganic salt that includes magnesium

- Triggers a chemical process
- Increases temperature and pressure with an increase in the expulsion of CH4
- Chemical process creates crystals and thus surface area
- Crystals – more volume for surfactants to work reducing tension
- More adsorption

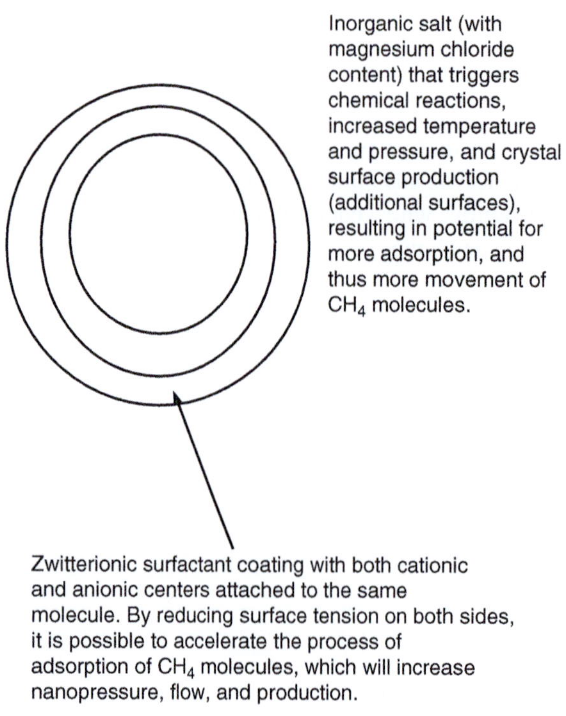

6.9 Dual Coating with Porous Proppant

6.9.1 Zwitterionic Outer Coating; Inorganic Salt Inner Coating, Porous Core

The proppant is a delivery system as well as a physical proppant for induced fractures. The porous proppant grain is impregnated with a liquid surfactant (or potentially a catalyst) to accelerate adsorption as well as generation of methane, and also to dramatically increase (in the nanoscale) temperature and pressure.

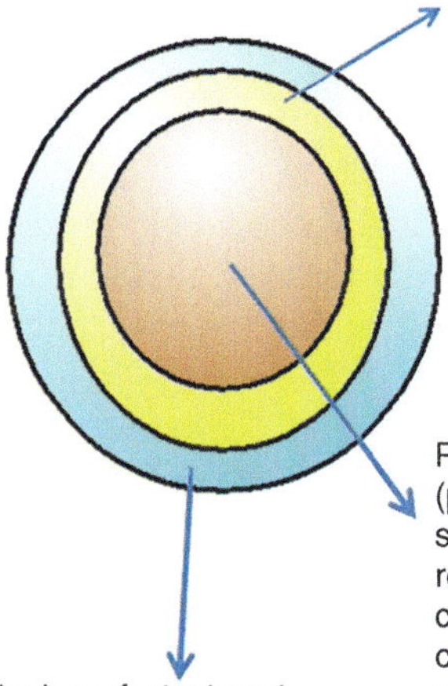

6.10 Data

6.10.1 Types of Surfactants

6.10.1.1 Anionic

Powerful and well-known, including ammonium lauryl sulfate and sodium lauryl sulfate. Anionic surfactants used in the oil industry include the following:

- Alkyl aryl sulfonates
- *N*-ethoxy sulfonate
- Sodium dodecyl sulfate (SDS)
- Alcohol propoxy sulfate (APS)
- Alkyl (or alcohol) ethoxy sulfate (AES)
- Alpha-olefin sulfonate (AOS)
- Alkyl polyalkoxy alkyl sulfonate or alkyl aryl polyalkoxy alkyl sulfonate
- Alpha olefin sulfate
- Branched alkyl benzene sulfonate
- Docusate sodium
- Ethoxy or propoxyglycidyl sulfonates
- Gemini anionic surfactants
- Internal olefin sulfonate (IOS)
- Sulfonated, ethoxylated alcohol, or alkyl phenol
- Sodium petroleum sulfonate
- TDA-9PO-sulfate

6.10.1.2 Cationic

Forms micelles in aqueous solutions. Cationic surfactants used in the oil industry include the following:

- Cetyl trimethyl ammonium bromide (CTAB)
- Coco alkyl trimethyl ammonium chloride
- Stearyl trimethyl ammonium chloride
- Dodecyl trimethyl ammonium bromide (DTAB)
- Ethoxylated alkyl amine

6.10.1.3 Nonionic

Long-chain alcohols may exhibit surfactant properties and are often used as foaming agents. Nonionic surfactants in the oil industry include the following:

- Alkyl ethoxy carboxylate
- Alkyl polyglycoside (APG)
- NEODOL ethoxylate 91-8
- NEODOL 67 propoxylated sulfate (N67-7POS)
- Polyethoxylated alkyl phenols
- Poly-(ethylene/propylene) glycol ether

6.10.1.4 Zwitterionic

Both anionic and cationic centers give them the ability to link to several different types of chemicals at the same time. They are also referred to as amphoteric surfactants. They are used in pharmaceuticals for improved delivery and can also be used as emulsifying agents. They are not considered harsh, and can even be employed for their antiseptic qualities.

6.11 Examples of Surfactants in Shale Plays

6.11.1 Bakken (Wang and Xu 2012)

The most important findings of this study are that the surfactant formulations consistently altered the wetting state of Bakken cores toward water-wet and consistently imbibed to displace significantly more oil than brine alone (Mirchi et al. 2014). Thus, the imbibition of surfactant formulations appears to have a substantial potential to improve oil recovery from the Bakken Formation (He and Xu 2017). Uncertainty exists about recovery factors from the Bakken formation during primary recovery operations (LeFever and Helms 2010), but values of 3–10% OOIP have been suggested. Three of the surfactant-imbibition tests provided EOR values of 6.8–10.2% OOIP, incremental over brine imbibition. Ten surfactant-imbibition tests provided EOR values of 15.7–25.4% OOIP. As a result, surfactant floods are effective in certain circumstances in the Bakken, particularly in cases of ultra-low permeability (Wang et al. 2012; Zhang et al. 2016).

- amphoteric dimethyl amine oxide
- nonionic ethoxylated alcohol
- anionic internal olefin sulfonate
- anionic linear alpha-olefin sulfonate

6.11.2 Eagle Ford (He and Xu 2017)

Cationic (three separate ones, not specified)

The Eagle Ford shale formation exhibits unique responses (He and Xu 2017) which corroborate earlier findings. Because of the electrostatic interaction between the head groups and hydrophobic interaction between hydrocarbon chains, catanionic mixtures present many unique properties, such as low CMC, low surface/interfacial tension, various microstructures (spherical vesicles or rod-like micelles), and slow micelle relaxation (Kim et al. 2016; O'Connor et al. 1997; Patist et al. 2001).

6.11.3 Utica (Shuler et al. 2016)

Surfactant "Huff-n-Puff"

Commercial non-ionic improved oil recovery (IOR) surfactant with dilute KCl solution

6.12 Results

The IFT drops as a microemulsion is formed. This microemulsion is not permanent but is considered a phase, and it allows the oil to mix with the formation fluids and also the surfactant. The microemulsion is what transports the oil out of the reservoir. It is similar to what happens when a detergent surfactant mixes with water and encounters particles it then transports out (as in the case of washing clothes).

The only problem with this approach to enhancing oil production is that the fluids may have to travel too quickly, and there is not enough penetration into the pore spaces and microfractures in a very tight reservoir.

According to work conducted by Lu and Pope (2016), the key to a successful surfactant flood is the velocity at the interface between the microemulsion and the aqueous surfactant solution (Lu and Pope 2016). They used anionic surfactants, cosolvents (isobutyl alcohol and triethylene glycol mono butyl ether), and for laboratory experiments, they used a brine proxy containing sodium chloride and sodium carbonate. For the oil, Lu and Pope (2016) used a blend of three different oils. Lu and Pope were able to demonstrate 99% recovery of oil in a laboratory simulation involving a sandpack. They concluded that their work suggested the potential for effective surfactant floods in deep, high-temperature, high-permeability oil reservoirs that may have residual oil or oil not yet produced after primary production. Whether or not their findings apply to shale reservoirs is a question to ask. It seems clear that some findings, at least, apply to all reservoirs, which is to say that one must be very careful in formulating cosolvents, and also the microemulsion phase duration is critical. One may need to have some kind of artificial lift in order to increase the velocity at the interface.

6.13 Shale Reservoirs, Gas, and Adsorption

Adsorption: undesirable in the surfactant but extremely desirable in methane molecules

Electrostatic interferences: the generation of van der Waals can result in the need to reduce IFT

Surface/liquid interface: too much change in IFT can actually result in diminished efficiency when too many micelles form and adhere to charged particles

The charged surface of the solid/charged head group of surfactant molecule or micelle adsorption occurs in shale formations on kerogen nanopore walls (Ning et al. 2015). It has been demonstrated that there can be a large amount of gas adsorbed on the kerogen nanopore walls, and also that the transport of kerogen occurs through organic porosity. At the same time, the way that adsorbed gas and free gas interact can have a dramatic impact on gas flow behavior (Ning et al. 2015). Interactions between particles may be affected by strong cohesive forces. In these cases, surfactants can reduce the negative impact of the adsorbed and free gas interactions, to either bundle the molecules for easier transport, or to break them apart so that they are able to flow through the smaller nanopores. Most important is the constant generation and regeneration of pore pressure which exerts a force on the methane molecules and allows them to move if their pressure can overcome the chemical forces binding them together or to a surface. In theory, surfactants can be effective, but not if the surfactant molecules are too large (too concentrated) or if the pH is incorrect, which simply introduces charges that cause the molecules to adhere even more tightly.

Models that take into consideration slippage and adsorption in tubes and lattices can be helpful. The fact that natural gas storage capacity is a function of the microscale and nanoscale organic-rich kerogen pores (Ning et al. 2015) limits the number of options for stimulation. At the same time, scanning electron microscopy and pore architecture modeling can begin to simulate the degree of ductility of the pores in the organic-rich kerogen. The more ductile the organic-rich formation (kerogen-rich), the higher the gas slippage.

As a result, strategies that enhance the elasticity of the matrix holding the pore spaces together can potentially intensify the impact of adsorption, with the changes in pore pressure, resulting in a nanoscale "bellows effect" that can periodically flatten, then plump up the pore spaces, changing the pore architecture enough to stimulate adsorption, and then a reduction of surface tension resulting in a movement when the Klinkenberg effect results in a kind of "squeezing" of the lattice (Lohne and Fjelde 2012).

Surface potential measurements by atomic force microscopy conducted by K. Kristiansen et al. (2011) demonstrated that dissolution and surface potential have an impact on the pressure in situ. They suggest that crystal formation is as much a function of pressure as electrochemical forces (Kristiansen et al. 2011).

6.14 Operational Conditions

The goal is to shift the reservoir wettability to make it more water-wet. Often this involves removing alcohol and increasing the microemulsion capacity. It is possible to use anionic surfactants in sandstone reservoirs. In carbonate reservoirs, cationic surfactants are used to move reservoir wettability from oil-wet to water-wet. In shale formations, hybrid surfactants can be used (Morsy and Sheng 2014; Nguyen et al. 2014).

Interfacial tension (IFT) is the force between the molecules between two fluids, or a fluid and a solid. The surfactant works by reducing the force as it is adsorbed, and it thus impacts the capillary force.

Salinity: High salinity can negatively impact the performance of a surfactant. Optimal salinity is the level at which equal quantities of oil and water can be mutually dissolved into a microemulsion (Hirasaki et al).

Surfactant concentration: The concentration of surfacting must be considerably higher than the critical micelle concentration so that micellization can be initiated. A high surfactant concentration is most desirable.

Co-solvent/Co-surfactant: Alcohol is often used as a co-solvent to cause the surfactant film at the oil/water interface to work.

6.15 Conclusions

Too much liquid surfactant can lead to adsorption into the formation of the surfactant, which can cause formation damage. What is more desirable is to achieve adsorption of the gas molecules so that they move into the fractures and migration pathways.

Water-soluble nonionic surfactant creates an oil/water microemulsion, which is very undesirable. What is most desirable is to have the surfactant move oil and gas into the brine. So, the salinity of the brine needs to be known and the pH of the surfactant fluid. Ideally, the surfactant works to break down the IFT to allow micelles to form in the brine and then be stimulated to move in the reservoir by pressure changes (downhole pump). The main issues to avoid include:

- Microemulsions
- Adsorption of the surfactant on the surface of the shale
- Micelle formation
- Maintaining water-wet status

Surfactants are effective in hydraulic fracturing in shale wells because they reduce the IFT between the aqueous and oil phases, alter the wettability of the formation surface to improve hydrocarbon flow, and reduce elasticity and rigidity of the oil-water interface – cause the oil droplets to be deformable and capable of being imbibed.

References

Buijse, M.A., Tandon, K., Jain, S. et al. (2013). Accelerated surfactant selection for EOR using computational methods. *Society of Petroleum Engineers* https://doi.org/10.2118/165268-MS.

Chegenizadeh Negin, Saeedi Ali, Quan Xie, Most common surfactants employed in chemical enhanced oil recovery, *Petroleum*, Volume 3, Issue 2, June 2017, Pages 197–211, https://doi.org/10.1016/j.petlm.2016.11.007.

Negin, C., Ali, S., and Xie, Q. (2016). Application of nanotechnology for enhancing oil recovery – A review. *Petroleum* 2 (4): 324–333. https://doi.org/10.1016/j.petlm.2016.10.002.

He, K. and Xu, L. (2017). Unique mixtures of anionic/cationic surfactants: a new approach to enhance surfactant performance in liquids-rich shale reservoirs. *Society of Petroleum Engineers* https://doi.org/10.2118/184515-MS.

Kang, S.M., Fathi, E., Ambrose, R.J. et al. (2011). Carbon dioxide storage capacity of organic-rich shales. *SPE Journal* 16 (04): 842–855.

Kim, J., Zhang, H., Sun, H. et al. (2016). Choosing surfactants for the Eagle Ford shale formation: guidelines for maximizing flowback and initial oil recovery. *Society of Petroleum Engineers* https://doi.org/10.2118/180227-MS.

Letham, E.A. and Bustin, R.M. (2015). Klinkenberg gas slippage measurements as a means for shale pore structure characterization. *Geofluids* 16 (2): 264–278.

Li, Z.Z., Min, T., Kang, Q. et al. (2016). Investigation of methane adsorption and its effect on gas transport in shale matrix through microscale and mesoscale simulations. *International Journal of Heat and Mass Transfer* 98: 675–686. https://doi.org/10.1016/j.ijheatmasstransfer.2016.03.039.

Lohne, A. and Fjelde, I. (2012). Surfactant flooding in heterogeneous formations. *Society of Petroleum Engineers* https://doi.org/10.2118/154178-MS.

Shuler, P.J., Lu, Z., Ma, Q., and Tang, Y. (2016). Surfactant huff-n-puff application potentials for unconventional reservoirs. *Society of Petroleum Engineers* https://doi.org/10.2118/179667-MS.

Mirchi, V., Saraji, S., Goual, L., and Piri, M. (2014). Experimental investigation of surfactant flooding in shale oil reservoirs: dynamic interfacial tension, adsorption, and wettability. *Unconventional Resources Technology Conference*. https://doi.org/10.15530/URTEC-2014-1913287.

Morsy, S. and Sheng, J.J. (2014). Surfactant preflood to improve waterflooding performance in shale formations. *Society of Petroleum Engineers* https://doi.org/10.2118/169519-MS.

Nguyen, D., Wang, D., Oladapo, A. et al. (2014). Evaluation of surfactants for oil recovery potential in shale reservoirs. *Society of Petroleum Engineers* https://doi.org/10.2118/169085-MS.

Wang, D., Butler, R., Zhang, J., and Seright, R. (2012). Wettability survey in bakken shale with surfactant-formulation imbibition. *Society of Petroleum Engineers* https://doi.org/10.2118/153853-PA.

Yang Ning, Yang Jiang, Honglin Liu, Guan Qin, Numerical modeling of slippage and adsorption effects on gas transport in shale formations using the lattice Boltzmann method, *Journal of Natural Gas Science and Engineering*, Volume 26, September 2015, Pages 345–355.

Zhang, J., Wang, D., and Olatunji, K. (2016). Surfactant adsorption investigation in ultra-lower permeable rocks. *Society of Petroleum Engineers* https://doi.org/10.2118/180214-MS.

7

Neuro-Fuzzy Algorithm Classification of Ordovician Tight Reservoir Facies in Algeria

Mohamed Zinelabidine Doghmane[1], Sid-Ali Ouadfeul[2], and Leila Aliouane[3]

[1] *Department of Geophysics, FSTGAT, University of Science and Technology Houari Boumediene, Algiers, Algeria*
[2] *Algerian Petroleum Institute, Sonatrach, Boumerdes, Algeria*
[3] *LABOPHYT, Faculty of Hydrocarbons and Chemistry, University M'hamed Bougara of Boumerdes, Boumerdes, Algeria*

KEYPOINTS

- This chapter deals with the development of neuro-fuzzy classifier.
- The classifier is used to discriminate the facies in a more detailed way.
- The classified facies belongs to a tight reservoir where the Ordovician is the main evaluated net pay.
- The proposed method has been tested in an Algerian petroleum system.

7.1 Introduction

The Oued-Mya basin, which includes the Haoud-Berkaoui region, is situated in the northern portion of the Algerian Sahara, precisely in the province's center (Figure 7.1). This area was formed by an extended north-east/south-west trending depression that developed during the Paleozoic (Figure 7.2) (Zazoun 2013). Its boundaries are the upper zone of the Cambrian-aged Djamâa-Touggourt to the north, the Talemzane Pier (Hassi R'mel) to the north-west, the Hassi Messaoud to the south-east, and the ridge of El Agreb-El Gassi to the north. Its width varies from 25 to 30 km to the south-west and from 8 to 10 km to the north-east (Doghmane et al. 2019; Eladj et al. 2022a, b, c).

7.2 Neuro-Fuzzy Classification

The method of dividing a characteristic of a species into fuzzy classes is known as neuro-fuzzy classification. In this research paper, we suggest a technique for solving a fuzzy classification problem using ANN backpropagation (Bayram and Barkana, 2010; Doghmane 2019). However, using a backpropagation network, the parameters of each feature in the fuzzy system have been optimized (Imamverdiyev and Sukhostat 2019). The membership and functions specified for each feature (Figure 7.3) are among these parameters (Aïfa et al. 2014; Ehsan and Gu 2020; Mendil et al. 2019).

Unconventional Hydrocarbon Resources: Prediction and Modeling Using Artificial Intelligence Approaches, First Edition. Edited by Sid-Ali Ouadfeul.

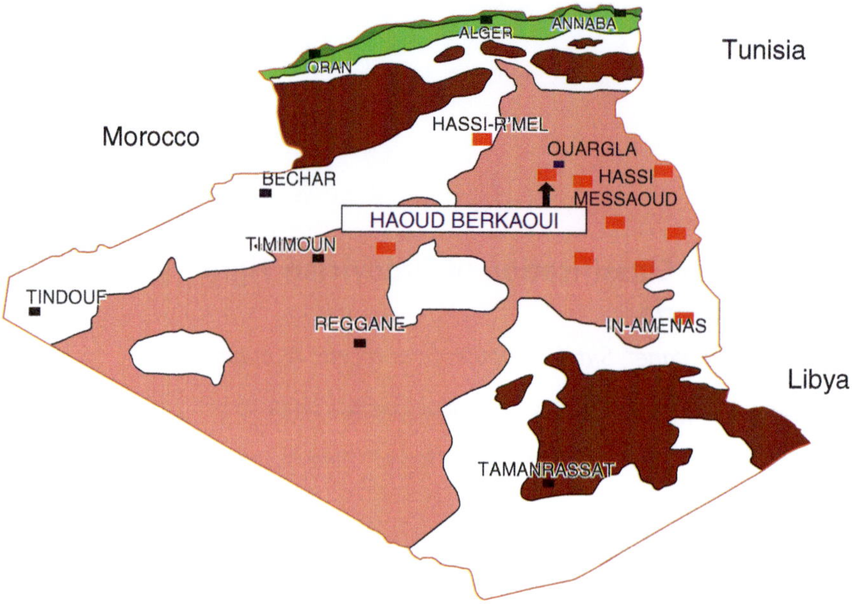

Figure 7.1 Location of the studied field.

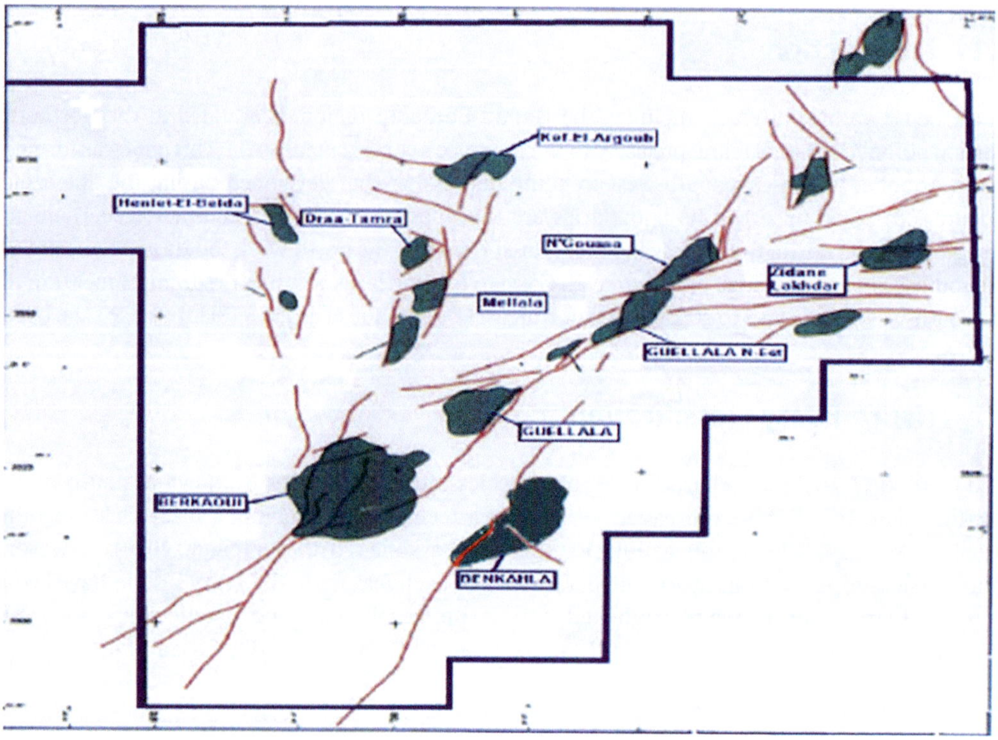

Figure 7.2 Geological structures of the studied field.

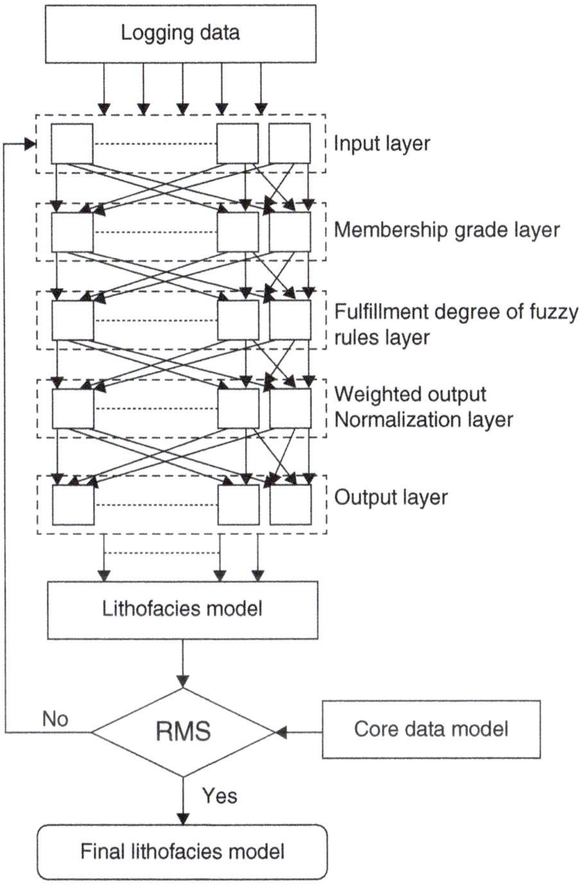

Figure 7.3 A detailed diagram of the proposed neuro-fuzzy lithofacies classifier.

7.3 Results Discussion

The neuro-fuzzy classifier's detailed procedures are shown in Figure 7.3 where three layers of the network are used to discover the best fuzzy system parameters for each class (Doghmane and Belahcene 2019). For each depth point in the logging curves (Ehsan and Gu 2020), the first layer is utilized to determine the membership grade; the second layer is then created to the extent that fuzzy rules are satisfied in order to identify both the dominant and secondary facies. The weighted output reflecting the fuzzy classes of the lithofacies is normalized using layer three. The model must meet a specified root mean square (RMS) error threshold of 10(−2) to be accepted (Doghmane and Belahcene 2019).

Figure 7.4 displays the findings for wells 1, 2, 3, 5, 6, 7, 8, and 9, where facies types 1 through 6 (Sandstone, Clay, Dolomite, Anhydrite, Chlorite, and Salt) predominate. The facies types from 7 to 16 represent the intermediate facies between the dominant facies for each depth interval (Eladj et al. 2020). Facies type 7 is clayey sandstone, facies type 8 is dolomitic clay, facies type 9 is anhydritic dolomite, facies type 10 is chloritic anhydrite, facies type 12 is salty chlorite, facies type 13 is sandstone-clay, and facies type 14 is clayey dolomite. The results of the neuro-classifier for seven of the nine wells that were studied are shown in Figure 7.5. These lithofacies curves demonstrate how the proposed classifier has provided a more precise facies description of the studied reservoir, where the secondary facies has also been taken into account (Cherana et al. 2022a, b). We were able to determine the dominant facies, which has a significant influence on the petrophysical parameters of the reservoir, for a particular depth interval that does not exceed 1 m thickness, as well as the other secondary facies, whose impact cannot be ignored (Bacetti and Doghmane 2020). Due to the fact that the Quanti Elan model was built using these precise data and included extensive descriptions of the rock matrix, a more precise reservoir characterization was obtained.

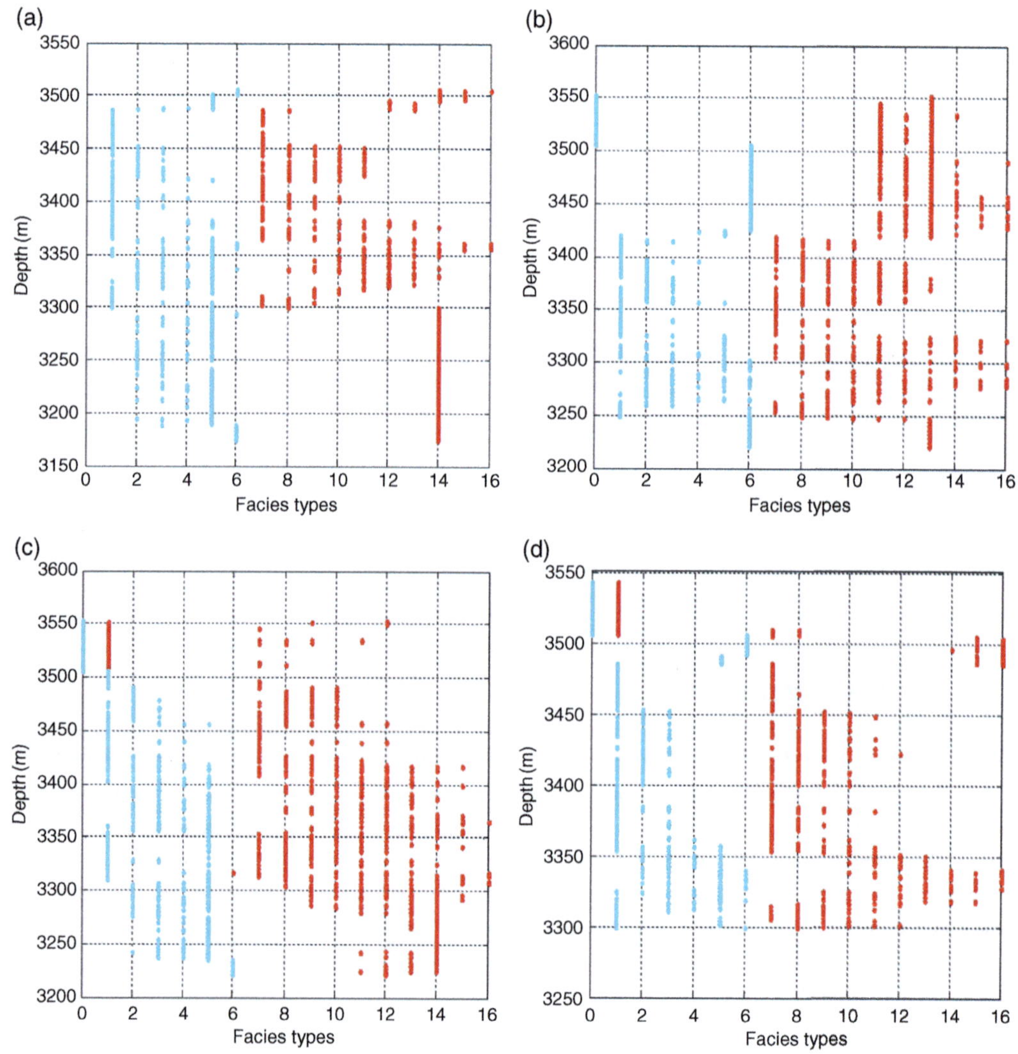

Figure 7.4 Results of neuro-fuzzy classification for: (a) Well-1, (b) Well-2, (c) Well-3, (d) Well-4, (e) Well-5, (f) Well-6, (g) Well-7, (h) Well-8, and (i) Well-9.

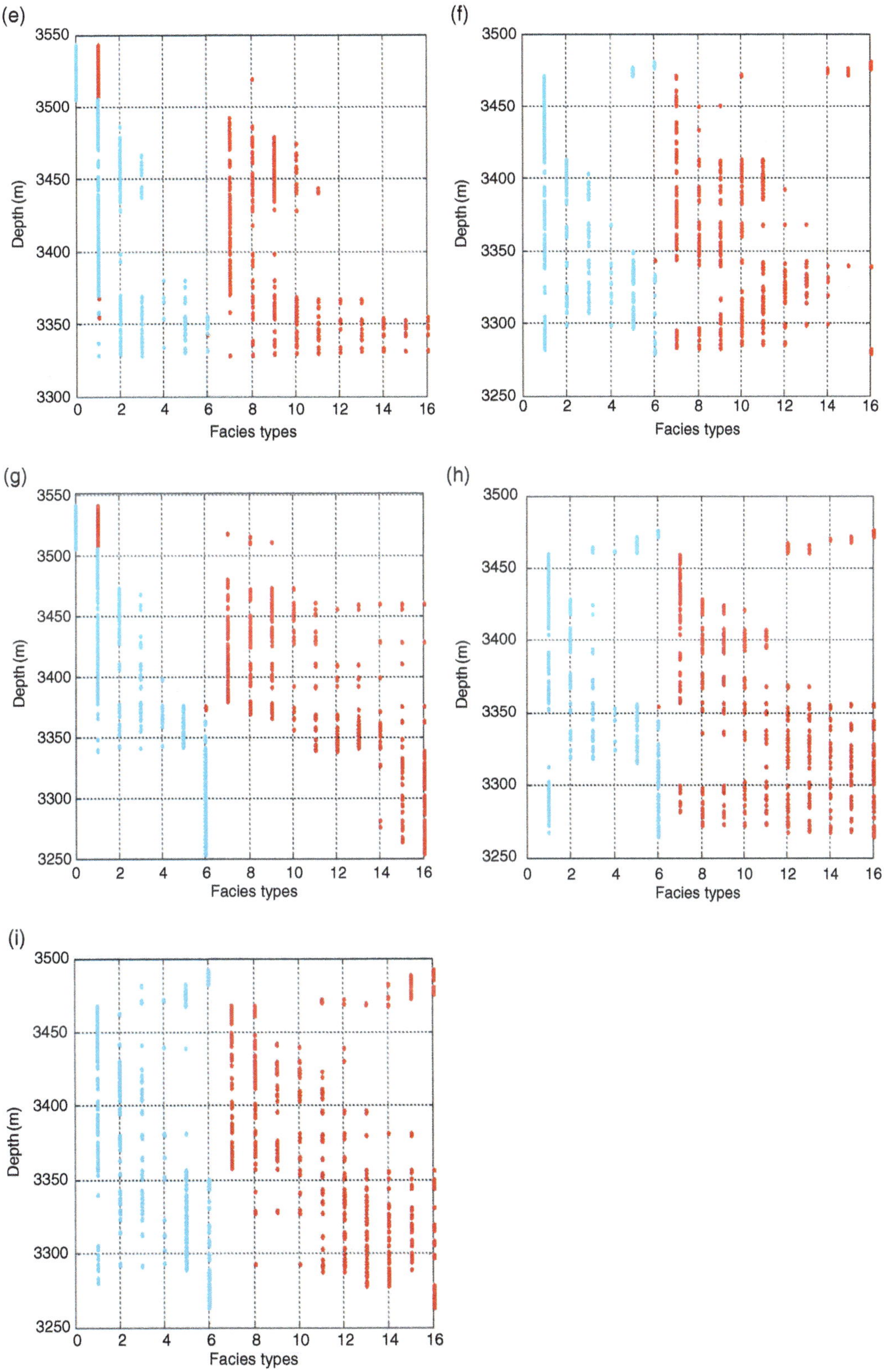

Figure 7.4 (Continued)

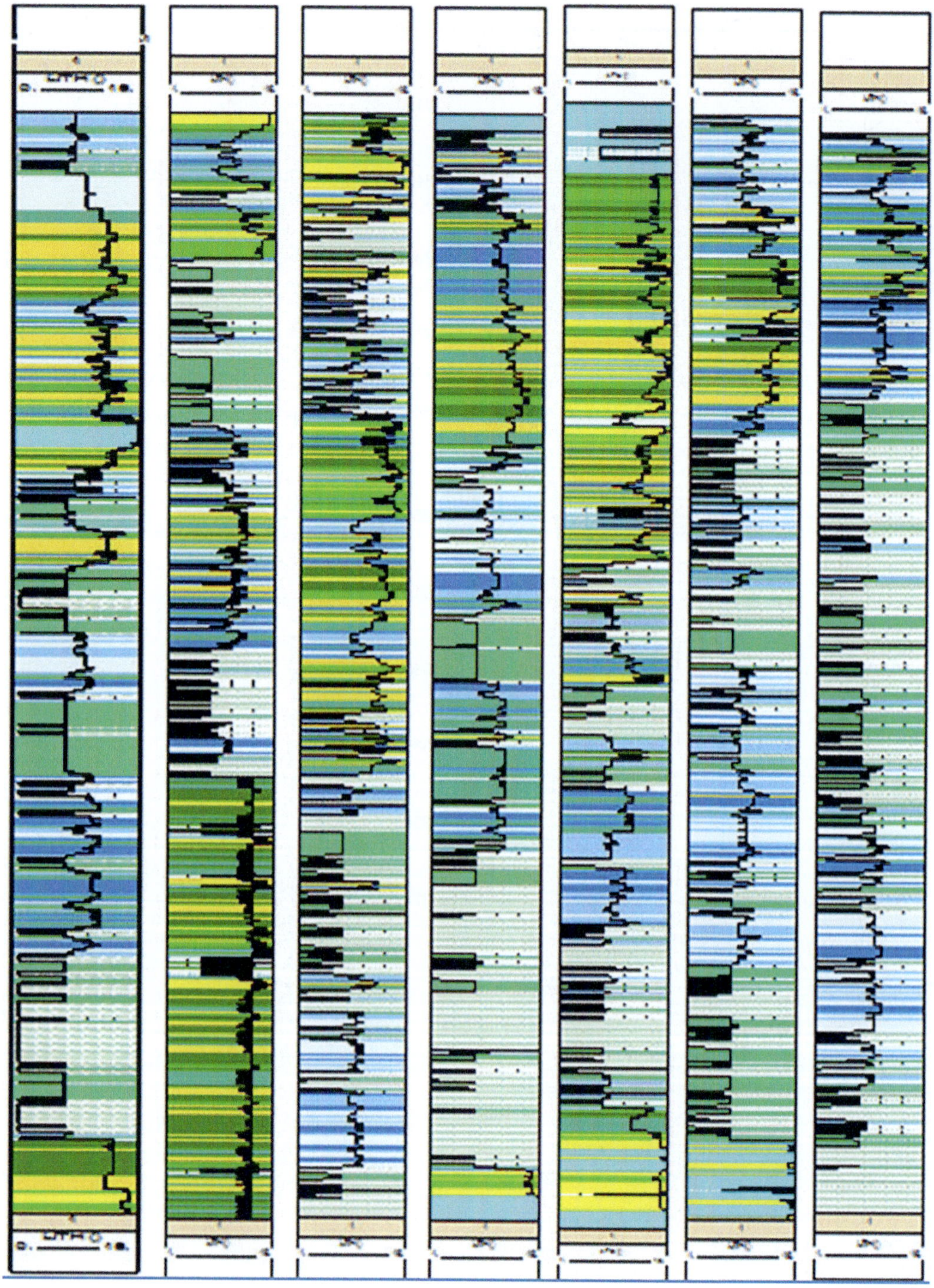

Figure 7.5 Result of neuro-fuzzy classifier for wells 1, 2, 3, 4, 5, 6, and 7 from left to right.

7.4 Conclusion

In this research, we were able to construct a more precise classifier that allowed us to detect the secondary facies in addition to the dominant facies for the same depth intervals by combining the ANN backpropagation technique with a fuzzy clustering algorithm. The classifier is built using cores from lab testing and logging data. It is strongly advised to take into account the proposed classifier for better reservoir classification and evaluation for the examined field as well as for similar fields in the Middle East and North Africa basin based on the results obtained.

References

Aïfa, T., Baouche, R., and Baddari, K. (2014). Neuro-fuzzy system to predict permeability and porosity from well log data: a case study of Hassi R′Mel gas field, Algeria. *Journal of Petroleum Science and Engineering* 123: 217–229.

Bacetti, A. and Doghmane, M.Z. (2020). A practical workflow using seismic attributes to enhance sub seismic geological structures and natural fractures correlation. *Conference Proceedings, First EAGE Digitalization Conference and Exhibition*, vol. 2020, Algeria (30 November–3 December 2020), pp. 1–5. http://doi.org/10.3997/2214-4609.202032062.

Bayram, C.L. and Barkana, A. (2010). Speeding up the scaled conjugate gradient algorithm and its application in neuro-fuzzy classifier training. *Soft Computing* 14: 365–378.

Cherana, A., Aliouane, L., Doghmane, M.Z. et al. (2022a). Lithofacies discrimination of the Ordovician unconventional gas-bearing tight sandstone reservoirs using a subtractive fuzzy clustering algorithm applied on the well log data: Illizi Basin, the Algerian Sahara. *Journal of African Earth Sciences* 196: https://doi.org/10.1016/j.jafrearsci.2022.104732.

Cherana, A., Aliouane, L., Doghmane, M., and Ouadfeul, S.A. (2022b). Fuzzy clustering algorithm for lithofacies classification of ordovician unconventional tight sand reservoir from well-logs data (Algerian Sahara). *Advances in Geophysics, Tectonics and Petroleum Geosciences. CAJG 2019. Advances in Science, Technology & Innovation* (25–28 November 2019). Cham: Springer. https://doi.org/10.1007/978-3-030-73026-0_64.

Doghmane, M.Z. (2019). Conception de commande décentralisée des systèmes complexes en utilisant les stratégies de décomposition et optimisation par BMI. PhD thesis. University M'hamed Bougara of Boumerdes.

Doghmane, M.Z. and Belahcene, B. (2019). Design of new model (ANNSVM) compensator for saturation calculation based on logging curves for low resistivity phenomenon. *Conference Proceeding, EAGE/ALNAFT Geoscience Workshop* 2019: 1–5.

Doghmane, M.Z., Belahcene, B., and Kidouche, M. (2019). Application of improved artificial neural network algorithm in hydrocarbons' reservoir evaluation. *Proceedings of International conference in Artificial Intelligence in Renewable Energetic Systems ICAIRES 2018: Renewable Energy for Smart and Sustainable Cities*, Algeria (24–26 November 2018), pp. 129–138.

Ehsan, M. and Gu, H. (2020). An integrated approach for the identification of lithofacies and clay mineralogy through neuro-fuzzy, cross plot, and statistical analyses, from well log data. *Journal of Earth System Science* 129 (101): 101–110. https://doi.org/10.1007/s12040-020-1365-5.

Eladj, S., Lounissi, T.K., and Doghmane, M.Z. (2020). Lithological characterization by simultaneous seismic inversion in algerian south eastern field. *Engineering, Technology & Applied Science Research* 10 (1): 5251–5258.

Eladj, S., Doghmane, M.Z., Lounissi, T.K. et al. (2022a). 3D Geomechanical model construction for wellbore stability analysis in algerian southeastern petroleum field. *Energies* 15: 7455. https://doi.org/10.3390/en15207455.

Eladj, S., Lounissi, T.K., Doghmane, M.Z., Djeddi, M. (2022b). Wellbore stability analysis based on 3D geo-mechanical model of an algerian southeastern field. *Advances in Geophysics, Tectonics and Petroleum Geosciences. CAJG 2019. Advances in Science, Technology & Innovation* (25–28 November 2019). Cham: Springer. https://doi.org/10.1007/978-3-030-73026-0_136.

Eladj, S., Doghmane, M.Z., and Belahcene, B. (2022c). Design of new model for water saturation based on neural network for low-resistivity phenomenon (algeria). *Advances in Geophysics, Tectonics and Petroleum Geosciences. CAJG 2019. Advances in Science, Technology & Innovation* (25–28 November 2019). Cham: Springer. https://doi.org/10.1007/978-3-030-73026-0_75.

Imamverdiyev, Y. and Sukhostat, L. (2019). Lithological facies classification using deep convolutional neural network. *Journal of Petroleum Science and Engineering* 174: 216–228.

Mendil, C., Kidouche, M., and Doghmane, M.Z. (2019). Automatic control of a heat exchanger in a nuclear power station: the classical and the fuzzy methods. *2019 International Conference on Advanced Electrical Engineering (ICAEE)* (19–21 November 2019), Algiers, Algeria.

Zazoun, S. (2013). Fracture density estimation from core and conventional well logs data using artificial neural networks: the Cambro-Ordovician reservoir of Mesdar oil field, Algeria. *Journal of African Earth Sciences* 83: 55–73.

8

Recognition of Lithology Automatically Utilizing a New Artificial Neural Network Algorithm

Mohamed Zinelabidine Doghmane[1], Sid-Ali Ouadfeul[2], and Leila Aliouane[3]

[1] *Department of Geophysics, FSTGAT, University of Science and Technology Houari Boumediene, Algiers, Algeria*
[2] *Algerian Petroleum Institute, Sonatrach, Boumerdes, Algeria*
[3] *LABOPHYT, Faculty of Hydrocarbons and Chemistry, University M'hamed Bougara of Boumerdes, Boumerdes, Algeria*

KEYPOINTS

- An artificial neural network has been developed for reservoir evaluation in this chapter.
- It has been used for lithofacies classification based on trained data.
- The classification is used to determine dominant and secondary facies.
- A graphical interface has been designed for this algorithm in order to use it as end-user industrial software.

8.1 Introduction

The recording of data from a capture tool is known as well logging. Software based on neural networks has been used to handle issues with classification, feature extraction, diagnostics, function approximation, and optimization. The categorization of lithofacies for well logging is a difficult and nonlinear geophysical problem because of the form of the pores, fluid saturation, grain size, and other variables. The nonlinearity of the physical and statistical solutions, the latter of which is brought on by the noise in geophysical measurements, defines the problem's complexity. Several academics have studied the classification of lithofacies using "cross-plots," a common statistical approach (Eladj et al. 2020). To obtain the lithology curve, two or more logs are cross-plotted (Gassaway et al. 1989; Mavko et al. 2009). The results from the drilling have been examined using multivariate statistical approaches. These methods call for a lot of data, which is expensive and not always available, and are semi-automated. Existing techniques for log data analysis are exceedingly laborious and time-consuming, particularly when dealing with numerous noisy and complex drill readings. The goal of this research is to create an algorithm for the automatic categorization of lithofacies based on an artificial neural network (ANN). We tested the program on data from 11 wells in the Algerian Sahara belonging to Haoud Berkaoui. The automatic recognition using this instrument was compared to manual recognition using IP Log software (used in Sonatrach) positioned at the bottom of the hole, plotted against the depth of the well, to validate the results (Sung and Lee 2001). Searching for regions with recoverable oil is the most typical use of well logging in the Sonatrach oil sector. A thorough description of the geological strata that a

Unconventional Hydrocarbon Resources: Prediction and Modeling Using Artificial Intelligence Approaches, First Edition. Edited by Sid-Ali Ouadfeul.

drill has pierced is helpful. Companies would have a variety of information about geological layers, such as oil content, for the extraction of oil and gas. The sources and sensors are loaded into probes that can be dropped into an existing borehole to measure these parameters (wireline logging) or mounted in a collar behind the drill bit to take measurements as the well is being drilled. Electronic probes and cartridges are chained and reduced in a cased well to a cable that has an electronic signal wire when conducting wireline logging. The electrical resistivity, electron density, sound speed, neutron moderation, radioactivity absorption to thermal neutrons (induced), spectral gamma rays from both natural and artificial sources, Compton scattering, borehole size, and occasionally nuclear magnetic resonance are all measured by the sensors as the chain is raised. Data is captured on a computer at the surface once the data is transmitted across a wire.

8.2 Well-Logging Methods

8.2.1 Nuclear Well Logging

It is important to measure radioactivity when looking for radioactive ores, especially potassium and uranium salts, as well as when looking for clay benches or radioactive beds. It is a measurement of the rocks' inherent radioactivity that a borehole passes through. These observations are helpful for identifying the radioactive minerals U, Th, and K that are present in the formations. Since these elements are particularly connected to clay, the gamma ray was reflected most strongly by Shaliness.

Depending on the amount of clay in the layers, gamma ray is used to define the boundaries of sedimentary formations. Given that the radioactivity of sedimentary rocks is mostly linked to the 40K contained in clay, the quantitative application will occasionally involve the evaluation of the percentage of clay. We can write this condition using Equation (8.1).

$$V_{sh} = \left(\mathrm{GR}_{read} - \mathrm{GR}_{min}\right) / \left(\mathrm{GR}_{max} - \mathrm{GR}_{min}\right) \tag{8.1}$$

8.2.2 Neutron Well Logging

Similar to the previous method, this one measures the density of the formations by bombarding them continuously with electrons that have an energy of 662 kilo, which are released from a source of 137 Cs caesium. This lithology and porosity tool uses gamma rays to determine the rate at which the number of thermal neutrons decreases as a function of distance from the source (Serra 1985), with porosity playing a major role in this rate. When using neutron instruments to determine porosity, it may be required to make certain adjustments to the lithology and hole conditions (Haliburton 2003). The following equation can be used for this method:

$$\varnothing_{Ncorr}\left(\%\right) = \varnothing_{Nread} + 4 - V_{sh} \times \varnothing_{Nsh} \tag{8.2}$$

8.2.3 Sonic Well Logging

Depending on the depth, a sonic log is a continuous record of the sound speed in the formations. It was discovered that porosity affects how far acoustic waves can travel. The signal collected by the receivers, in comparison to that emitted, offers us an idea of the speed of the acoustic wave in the medium, which is related to its compaction and, consequently, related to porosity. The transmitted wave is calibrated in amplitude and frequency. The device measures the acoustic wave's passage time across a one-foot distance (Schlumberger 2000). The transit time, which is measured in (ms/ft), is the amount of time it takes for a sound wave to cross a one-foot shape. Four receivers and two acoustic wave transmitters make up the borehole-compensated (BHC) tool. The measurements and porosity are related by the equation:

$$\varnothing_{\text{scorr}} = \left(\Delta t_f - \Delta t_{\text{read}}\right)/\left(\Delta t_f - \Delta t_{\text{max}}\right) - V_{\text{sh}} \times \left(\left(\Delta t_f - \Delta t_{\text{read}}\right)/\left(\Delta t_f - \Delta t_{\text{max}}\right)\right) \tag{8.3}$$

8.3 Use of ANN in the Oil Industry

Neural max min networks are frequently employed in statistical issues, such as automatic classification and decision making, due to their classification and generalization capabilities. Dreyfus et al. (2004) claim that the advantage of ANN over other methods is their ability to tackle difficult issues as well as their ability to make precise equivalent models with less experimental data and produce more accurate models from the same amount of examples. Several disciplines use this technique. Techniques developed by Rosenberg and Sejnowski (1987) are the most well-known.

Seismic data processing (Debotyam et al. 2014).
Characterization of rocks (Aoun et al. 2022; Irofti et al. 2022).
Interpretation of logs (Rolon et al. 2009).
Diagnostic of the drilling tool (Mendil et al. 2021a).
Control process (Mendil et al. 2021b).
Recognition of lithology and facies analysis (Platon et al. 2003; Doghmane et al. 2022).
Lithological classification of seismic attributes.

A well-known use of the retropropagation algorithm (RPA) shown in Figure 8.1, a supervised learning algorithm, is Rosenberg's presentation from 1987 (Rosenberg and Sejnowski 1987), in which the PRA is utilized in a system that learns to read text. The stock market forecast was another achievement (Zarpanis and Renese 1999).

8.4 Lithofacies Recognition

We concentrated on the technique in this part that was utilized to accomplish the goals of this research; classification systems based on ANNs have nodes in the input layer, one or more hidden layers, and an output layer (Cherana et al. 2022a). The stacking effect is produced because

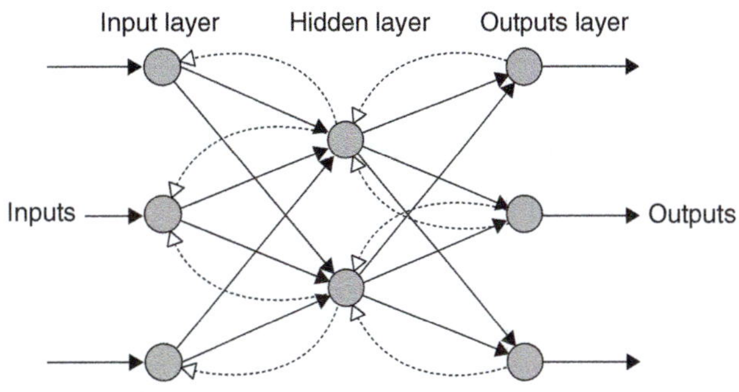

Figure 8.1 A retropropagation model's structure.

each node in one layer corresponds to a node in the following layer. One of the most common ANN structure types is called feed-forward, and it is frequently used to address complicated issues by simulating complex interactions. The following describes the developed algorithm's process:

Step 1: Standardize the data. To modify the interval of defined qualities and prevent neuronal saturation, the obtained data is mapped to the terminal [0, 1]. The normalization of data is determined by

$$I_{i+1} = \left(I_i - I_{\min}\right)/\left(I_{\max} - I_{\min}\right) \times \left(O_{\max} - O_{\min}\right) + O_{\min} \tag{8.4}$$

O_{max} is the maximum value after normalization, O_{min} is the minimum value after normalization, I_{new} is the new value after normalization, and I_{old} is the old value before normalization. I_{min} and I_{max} are the variables' minimum and maximum values, respectively (Masters 1993; Eladj et al. 2022a, b, c).

Step 2: Configure the network; How many levels are hidden? We chose five hidden layers for this study so that the convergence and precision of the findings could be assured. With a depth range of 3300–3500 m, we collected samples for each layer every 0.1054 m. Rate of learning (): In general, too fast or too slow a learning rate is harmful to network convergence (Tabach et al. 2007); for the sake of this study, we have chosen values for testing networks between 0.1 and 1.0. The sigmoid function, with a value in the [0, 1] range, serves as the transfer function.

Step 3: Enter the training example data. The associated input and output values are entered using the algorithm's graphical user interface (GUI).

Step 4: Determine the neuron output signal. The output of each neuron is determined by

$$\text{net}_j = \sum_{i=1}^{l} \omega_{ij} x_i + b_j \tag{8.5}$$

Each neuron in the hidden layers converts its signal using the sigmoid function. The target value of output neuron k and the error of each hidden layer can be used to express the signal from the output layer (Cherana et al. 2022b).

$$\delta_j^l = \sum_{i=1}^{l} \delta_j^{l+1} \omega_{\text{ji}}^l f'\left(\text{net}_j^l\right) \tag{8.6}$$

The first derivative of the sigmoid function is used to determine the progressive change for each weight of each learning interaction.

Step 5: Determine the error values. Repeat steps 3 through 5 until the network converges (Doghmane and Belahcene 2019; Doghmane et al. 2019).

8.5 Log Interpretation

8.5.1 Methodology of Manual Interpretation

We used four curves in this part for human interpretation using the IP Log software: the gamma ray (GR), density (RHOB), neutron (NPHI), and photoelectric factor curves (PE). We used the depths interval in these curves (3300–3500 m). The eleven wells are OKS04, OKS021, OKS023, OKS026, OKS027, OKS047, OKS051, OKS052, OKS053, OKS054, and OKS064 for which the curves of the chosen area (3300–3500 m) provide all the essential information. Following network learning, we applied the sequence to the OKS04 data, then we did the same thing again for ten wells. The progression is as follows:

To identify the clay and nonclay benches in the logs for the chosen location, we use the log gamma (GR).
We may build the lithology curve by calculating the matrix density and fluid density from the graphs.
We manually classified the facies for this well based on the criteria later.

The criteria employed by IP Log for manual categorization include:

If (GR> 70 API), we have clay else
If ((NPHI-PHID) <1%), we have limestone, else
If (1% < (NPHI-PHID) <7.5%), we have sandstone, else
If (7.5% < (NPHI-PHID) <13.5%), we have dolomite, else
If (15% < (NPHI-PHID) <45%), we have anhydrite, else
If ((NPHI-PHID)> 45%), we have salt.

The values used to plot the curve for each well's matrix parameters using the Quick glance method are provided in Table 8.1.

8.5.2 Results of Manual/Automatic Interpretation

In this section, we have established a code source for the automatic recognition of lithofacies. All the goals previously attained using IP Log software are coded using MATLAB, and a GUI is created to make it easier to utilize the suggested technique. Using the MATLAB plot tool, the outcomes of the automatic lithofacies recognition can be displayed as in Figure 8.2.

We observe that the lithology curves are not clearly indicated. For this reason, as well as to enable an effective comparison, we export the MATLAB findings to the IP Log software, where we plot the curves using the same human recognition methods. We compare the outcomes of manual

Table 8.1 Calculated matrix parameters for the different wells.

Well	DeltaMat (us/ft)	Rhomat (g/cm^3)
Well-1	49	2.84
Well-2	51	2.86
Well-3	52	2.78
Well-4	50	2.82
Well-5	55	2.80
Well-6	54	2.82
Well-7	53	2.85
Well-8	53	2.86
Well-9	54	2.79
Well-10	48	2.84
Well-11	56	2.85

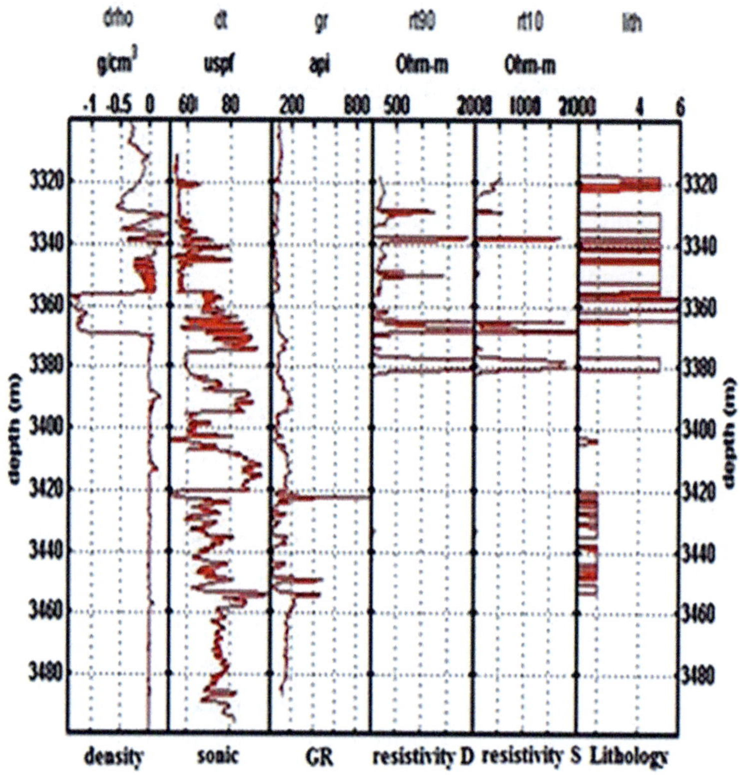

Figure 8.2 The curves displayed by MATLAB algorithm after learning.

interpretation using IP Log software and automatic recognition using the ANN algorithm in this section to validate the use of ANN. We exhibit the outcomes of chosen wells.

Well-1 using an end-user software (Figure 8.3a), we have three clear areas.

Zone 1 (3328–3370 m) contains a sequence of anhydrite and sandstone, and in the location where the two facies converge, there is some limestone present. This is due to the accuracy of the interval for the circumstances laid down.

Zone 2 (3370–3438 m): This region is primarily composed of clay, with traces of anhydrite and sandstone.

A layer of sandstone with some limestone can be seen in Zone 3 (3438–3500 m) in this region; it is buried down through clay.

(Figure 8.3c) There are five regions that we have found to be present, and they are

Zone 1 (3340–3376 m): Anhydrite and sandstone are present in this region, along with some clay.

Zone 2 (3376–3427 m): It is a second clay layer divided by thin sandstone strata.

The existence of limestone in Zone 3 (3427–3474 m), which is composed of strata of limestone separated by a significant layer of clay, is not taken into account.

Zone 4 (3474–3489 m) is a sandstone sequence with a clay layer.

Zone 5 (3474–3500 m): The surface here is covered in clay.

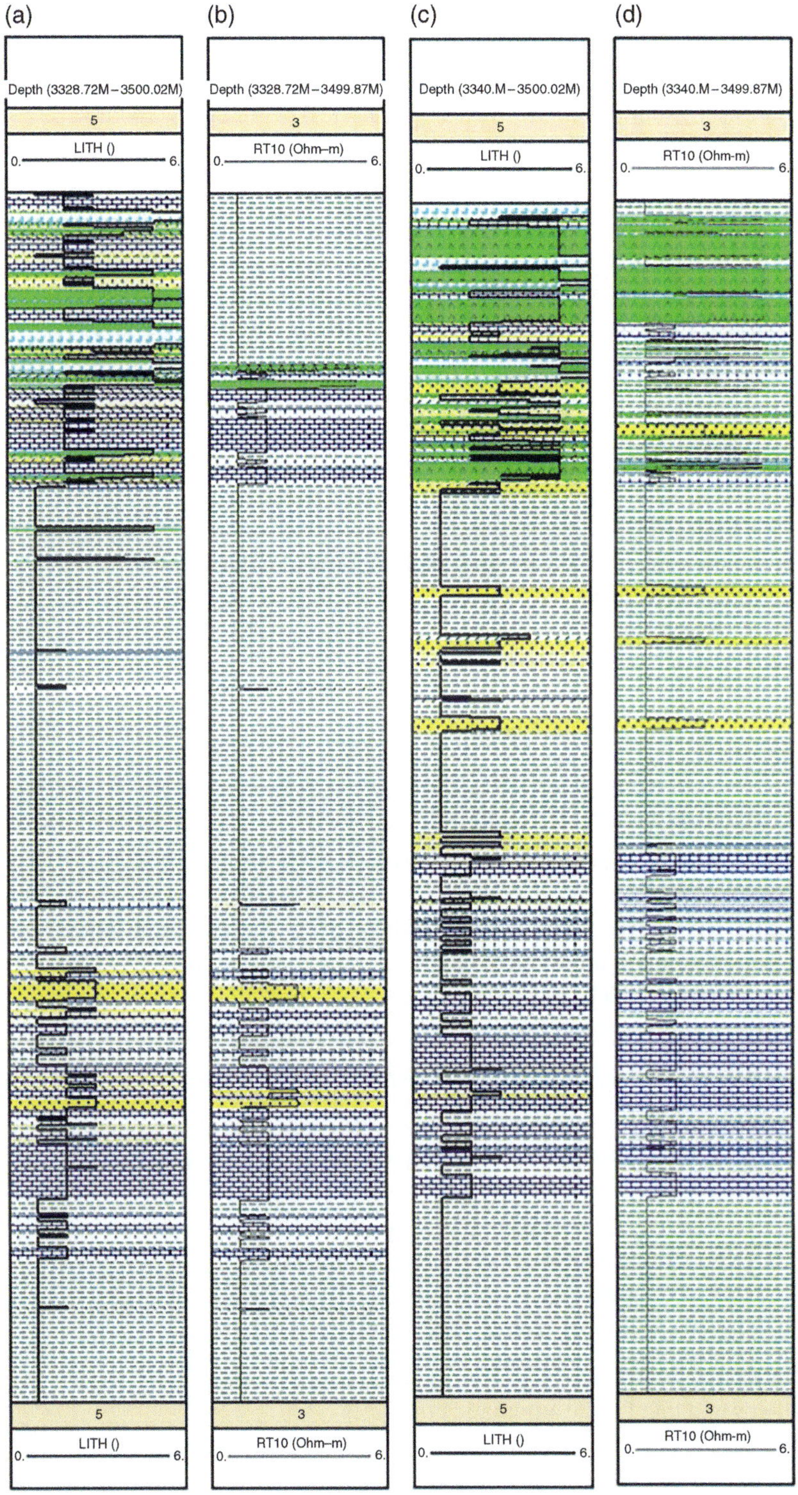

Figure 8.3 Comparison between results of manual and automatic interpretation for: (a) Well-1, manual interpretation. (b) Well-1 ANN interpretation. (c) Well-2 manual interpretation. (d) Well-2 ANN interpretation.

We observe in the data that the IP Log program gives us the opportunity to manually identify facies based on geologist-provided criteria, in addition to giving us the option of employing several curves for log interpretation. Although the presence of salt in this region has no geological explanation, classification accuracy affects the number of distinct zones and the types of facies that are present in each zone. The facies are divided into sub-zones by a more accurate classification as follows. On the other hand, automatic lithofacies recognition enables us to not only identify unique areas facies but also to boost classification accuracy:

Well-2 (Figure 8.3b) with the suggested approach, we can identify five distinct zones.
Zone 1 (3328–3352 m): We observe the presence of a deep clay layer, around 25 m thick.
Sandstone is the predominant facies in Zone 2 (3352–3370 m), while anhydrite and clay are also present in significant amounts.
Another zone of clay coverage, Zone 3 (3370–3440 m), terminates with a thin layer of sandstone in the distance.
Two layers of sandstone and limestone in Zone 4 (3440–3480 m) are separated by a thin layer of clay; limestone is the predominant facies in this region.
Zone 5 (3480–3500 m) has a thick clay layer covering it.
Well-1 (Figure 8.3d): Five places, all of which are present, have been identified by us.
Zone 1 (3300–3340 m): A thick layer of clay cover.
Zone 2 (3340–3377 m): The predominant facies is sandstone, with a thick layer of anhydrite and thin layers of clay and sandstone between them.
Zone 3 (3377–3426 m): A thin layer of limestone separates the massive clay layers.
Zone 4 (3426–3473 m): Sandstones are the predominant facies in a limestone sequence with clay.
Zone 5 (3473–3500 m): It is a clay cover layer.

Well-3 (Figure 8.4a) Four areas that are present have been recognized by us and they are

Zone 1 (3340–3376 m): We observe anhydrite and sandstone layers in this region as well as the presence of some clay layers.
Zone 2 (3376–3427 m) is the second area of clay layers, which are divided by thin sandstone layers.
Zone 3 (3427–3474 m): The existence of limestone is ignored in the third section, which consists of layers of limestone separated by a thick layer of clay.
Zone 4 (3474–3500 m): This area is clay cover layer.

Well-4 (Figure 8.4c): There are six locations that we have found to be present, and they are:

Zone 1 (3316–3356 m): Anhydrite is the predominant facies in this zone, and although we can see a notable salt layer there, the existence of limestone and sandstone is not taken into consideration.
Zone 2 (3356–3467 m): A layer of clay is followed by a thin layer of limestone at the bottom.
Zone 3 (3467–3477 m): A layer of sandstone, then a thin layer of limestone.
Zone 4 (3477–3392 m): A clay cover layer.
Zone 5 (3392–3408 m): A sizeable additional layer of sandstone.
Zone 6 (3408–3450 m): A sandstone layer is surrounded by two clay layers.

However, we have discovered that for the automatic lithofacies
Well-3 (Figure 8.4b): Four regions that are present, which we have discovered, are

Zone 1 (3340–3376 m): Anhydrite and sandstone are found in this region, along with some clay.
Zone 2 (3376–3427 m) is the second area of clay layers, which are divided by thin sandstone layers.

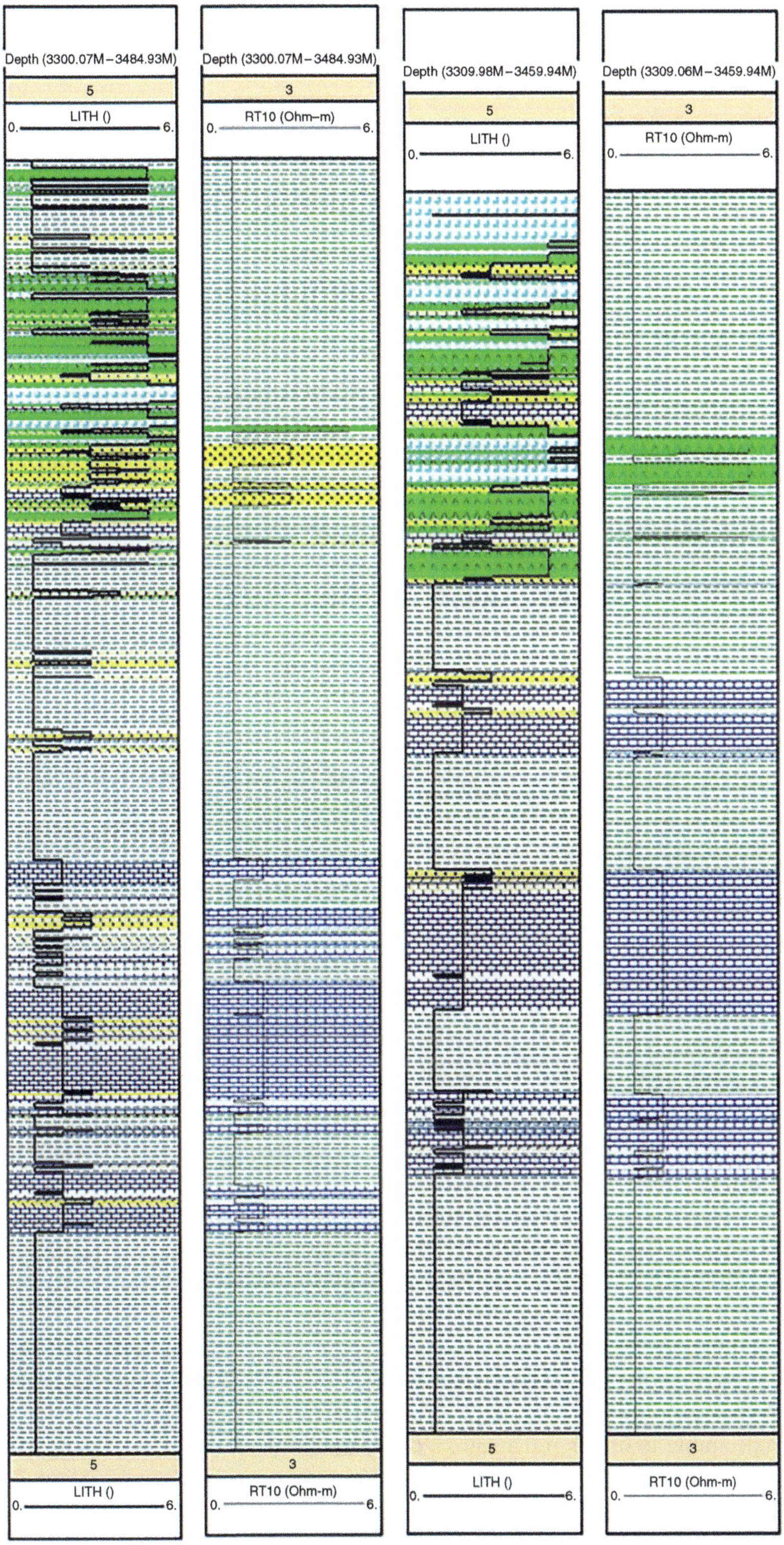

Figure 8.4 Comparison between results of manual and automatic interpretation for Well-3 and Well-4.

Zone 3 (3427–3474 m): The existence of limestone is ignored in the third section, which consists of layers of limestone separated by a thick layer of clay.
Zone 4 (3474–3500 m): This area is a clay cover layer.

Well-4 (Figure 8.4d): We have determined that there are three places, which are:

Zone 1 (3316–3356 m): A thin anhydrite layer separates several thick clay layers.
Zone 2 (3356–3467 m): A clay-sandstone succession, with sandstone as the predominant facies.
Zone 3 (3467–3477 m): A substantial clay layer covering the top layers.

Figures 8.3 and 8.4 illustrate the advantages of utilizing ANN-MATLAB for facial identification over manual facial recognition using log IP software, including:

Reduce the amount of time the interpreter must spend on the job by automating the interpreting process.
The findings obtained by the ANN are more accurate; the algorithm takes into account a region that IP Log software had overlooked.
We can reduce the impact of the precision of the interval on the production of fictitious thin layers that do not exist in the geology of the researched region thanks to the implementation requirements for classification and recognition by this instrument.

8.6 Conclusion

In this study, we set out to create an algorithm for automatic lithofacies recognition based on an ANN, as well as to increase the precision of the techniques used in this sort of work, which are typically carried out by Sonatrach geologists using software called IP Log. To accomplish these goals, we have broken the project down into stages. In the first stage, we used the IP Log software to perform manual classification based on presumptions using the data that was available. This procedure was applied to eleven wells in the Algerian Sahara town of Haoud Berkaoui, and the results were used as comparison samples to validate the results of the neural network classification. The next stage is to train the neural network using the lithology curve from the pilot well. After that, we used five curves and the learned neural network in MATLAB for automatic categorization. Additionally, we have created a GUI that enables users to easily carry out the required steps, thereby streamlining the use of this technique for Sonatrach geologists. The study's final stage compares manual and automatic lithofacies recognition; from this comparison, we came to the conclusion that:

We can improve the lithofacies categorization of Sonatrach exploration data thanks to ANNs.

Sonatrach interpreters' task is made easier and faster by automatic recognition.

Although using IP Log software for manual recognition could result in some ignorance of crucial areas, this instrument gradually realizes the interpretation.

The IP Log software's condition implementation can result in nonexistent layers, especially if the conditions' intervals overlap. But without any geological data, such as the number of lithofacies or the lithology pilot curve, this technique will be useless.

These factors make it desirable to use as much carrot data as possible in order to improve the algorithm's accuracy for distant wells.

References

Aoun, A., Soto, R., Rabiei, M. et al. (2022). Neural network based mechanical earth modelling (MEM): a case study in Hassi Messaoud Field, Algeria. *Journal of Petroleum Science and Engineering* 210: https://doi.org/10.1016/j.petrol.2021.110038.

Cherana, A., Aliouane, L., Doghmane, M.Z. et al. (2022a). Lithofacies discrimination of the Ordovician unconventional gas-bearing tight sandstone reservoirs using a subtractive fuzzy clustering algorithm applied on the well log data: Illizi Basin, the Algerian Sahara. *Journal of African Earth Sciences* 196: https://doi.org/10.1016/j.jafrearsci.2022.104732.

Cherana, A., Aliouane, L., Doghmane, M., and Ouadfeul, S.A. (2022b). Fuzzy clustering algorithm for lithofacies classification of ordovician unconventional tight sand reservoir from well-logs data (algerian sahara). *Advances in Geophysics, Tectonics and Petroleum Geosciences. CAJG 2019. Advances in Science, Technology & Innovation* (25–28 November 2019). Cham: Springer. https://doi.org/10.1007/978-3-030-73026-0_64.

Debotyam, M., Aminzadeh, F., and Karrenbach, M. (2014). Novel hybrid artificial neural network based autopicking workflow for passive seismic data. Vertical seismic profiling and microseismicity frontiers. *Geophysical Prospecting* 62 (4): 834–847. https://doi.org/10.1111/1365-2478.12125.

Doghmane, M.Z. and Belahcene, B. (2019). Design of new model (ANNSVM) compensator for saturation calculation based on logging curves for low resistivity phenomenon. *Conference Proceeding, EAGE/ALNAFT Geoscience Workshop* 2019: 1–5.

Doghmane, M.Z, Belahcene, B., and Kidouche, M. (2019). Application of improved artificial neural network algorithm in hydrocarbons' reservoir evaluation. *Proceedings of International Conference in Artificial Intelligence in Renewable Energetic Systems ICAIRES 2018: Renewable Energy for Smart and Sustainable Cities*, Algeria (24–26 November 2018), pp. 129–138.

Doghmane, M.Z., Ouadfeul, S.A., Benaissa, Z., and Eladj, S. (2022). Classification of ordovician tight reservoir facies in algeria by using neuro-fuzzy algorithm. In: *Artificial Intelligence and Heuristics for Smart Energy Efficiency in Smart Cities. IC-AIRES 2021*, Lecture Notes in Networks and Systems, vol. 361 (ed. M. Hatti). Cham: Springer. https://doi.org/10.1007/978-3-030-92038-8_91.

Martinez, J.-M., Samuelides, M. et al. (2004). *Réseaux de neurones, Méthodologie et applications.* Editions Eyrolles. Dreyfus et al.

Eladj, S., Lounissi, T.K., and Doghmane, M.Z. (2020). Lithological characterization by simultaneous seismic inversion in algerian south eastern field. *Engineering, Technology & Applied Science Research* 10 (1): 5251–5258.

Eladj, S., Doghmane, M.Z., Lounissi, T.K. et al. (2022a). 3D Geomechanical model construction for wellbore stability analysis in algerian southeastern petroleum field. *Energies* 15 (20): 7455. https://doi.org/10.3390/en15207455.

Eladj, S., Doghmane, M.Z., Aliouane, L., and Ouadfeul, S.A. (2022b). Porosity model construction based on ANN and seismic inversion: a case study of saharan field (Algeria). *Advances in Geophysics, Tectonics and Petroleum Geosciences. CAJG 2019. Advances in Science, Technology & Innovation* (25–28 November 2019). Cham: Springer. https://doi.org/10.1007/978-3-030-73026-0_55.

Eladj, S., Doghmane, M.Z., and Belahcene, B. (2022c). Design of new model for water saturation based on neural network for low-resistivity phenomenon (algeria). *Advances in Geophysics, Tectonics and Petroleum Geosciences. CAJG 2019. Advances in Science, Technology & Innovation* (25–28 November 2019). Cham: Springer. https://doi.org/10.1007/978-3-030-73026-0_75.

Gassaway, G.R., Miller, D.R., Benett, L.E. et al. (1989). Amplitude variations with offset: fundamentals and case histories, SEG continuing education course notes.

Haliburton (2003). Logging of Perforating Products and Services, Report.

Masters, T. (1993). *Practical Neural Network Recipes in C++*. San Diego, CA: Academic Press Professional, Inc.

Irofti, D., Ifrene, G.E.H., Pu, H., and Djemai, S. (2022). *A Multiscale Approach to Investigate Hydraulic Attributes of Natural Fracture Networks in Two Tight Sandstone Fields, Ahnet, Algeria, the 56th U.S. Rock Mechanics/Geomechanics Symposium*, Santa Fe, NM, USA (June 2022). https://doi.org/10.56952/ARMA-2022-0450.

Mavko, G., Mukerji, T., and Dvorkin, J. (2009). *The Rock Physics Handbook: Tools for Seismic Analysis of Porous Media*. New York: Cambridge University Press.

Mendil, C., Kidouche, M., Doghmane, M.Z. et al. (2021a). Rock-bit interaction effects on high-frequency stick-slip vibrations severity in rotary drilling systems. *Multidiscipline Modeling in Materials and Structures* 17 (5): 1007–1023. https://doi.org/10.1108/MMMS-10-2020-0256.

Mendil, C., Kidouche, M., and Doghmane, M.Z. (2021b). Hybrid sliding PID controller for torsional vibrations mitigation in rotary drilling systems. *Indonesian Journal of Electrical Engineering and Computer Science* 22 (1): 146–158. https://doi.org/10.1159/ijeecs.v22.i1.pp146-158.

Platon, E., Gary, A., Glenn, J., (2003). Pattern matching in facies analysis from well log data-a hybrid neural network-based application. *AAPG Conference and Exhibition*, Barcelona, Spain (21–24 September 2003).

Rolon, L., Mohaghegh, S.D., Ameri, S. et al. (2009). Using artificial neural networks to generate synthetic well logs. *Journal of Natural Gas Science and Engineering* 1 (4–5): 118–133. https://doi.org/10.1016/j.jngse.2009.08.003.

Rosenberg, C.R. and Sejnowski, T.J. (1987). The spacing effect on NETtalk, a massively-parallel network. *Proceedings of the Eighth Annual Conference of the Cognitive Science Society* (15–17 August 1986). Hillsdale, NJ: Lawrence Erlbaum Associates, pp. 72–89.

Schlumberger (2000). Log Interpretation Charts, the 2000 edition.

Serra, O. (1985). *Diagraphies Différés base de l'interprétation, Tome*, vol. 2. Montrouge: Etudes et productions Schlumberger.

Sung, H. and Lee, D.S. (2001). Neuro-fuzzy recognition system for detecting wave patterns using wavelet coefficients. *IEICE Transactions on Information and Systems* 84-D (8): 2001.

Tabach, E.E., Lancelot, L., Shahrour, I., and Najjar, Y. (2007). Use of artificial network simulation metamodelling to assess groundwater contamination in a road project. *Mathematical and Computer Modelling* 45 (7–8): 766–776.

Zarpanis, A. and Renese, A.P. (1999). *Principles of Neural Model Identification Selection and Adequacy*. London, UK: Centre of Neural Networks, Departement of Mathematics, King's College.

9

Construction of a New Model (ANNSVM) Compensator for the Low Resistivity Phenomena Saturation Computation Based on Logging Curves

Mohamed Zinelabidine Doghmane[1], *Sid-Ali Ouadfeul*[2], *and Leila Aliouane*[3]

[1] *Department of Geophysics, FSTGAT, University of Science and Technology Houari Boumediene, Algiers, Algeria*
[2] *GGR Department, Algerian Petroleum Institute, Boumerdes, Algeria*
[3] *LABOPHYT, Faculty of Hydrocarbons and Chemistry, University M'hamed Bougara of Boumerdes, Boumerdes, Algeria*

KEYPOINTS

- This chapter deals with the computation of water saturation in petroleum field.
- The petroleum field is characterized by the low resistivity values formally known as low resistivity phenomenon.
- The proposed ANN is hybridized with support vector machine approach.
- The main utility of this algorithm is its compensation to the effect of heavy minerals.
- The obtained results have been correlated and compared to water saturation obtained from modular dynamic testing tool.

9.1 Introduction

The world's main energy source is made up of hydrocarbons, especially in Algeria where the exploration of oil reserves continues to be of considerable interest to oil and gas firms (Benedetto 2010). In order to attain high-reliability performance, a variety of exploitation, development, and production methods have been created and are always being improved. As a component of this technology, logging tools play a significant role in the prospecting, assessing, and evaluation of reservoirs (Bacetti and Doghmane 2020). Their primary function is to characterize the petrophysical characteristics of the hydrocarbon reservoirs. However, in many types of reservoirs, including unconventional reservoirs, compact reservoirs, and low resistivity reservoirs, this task is becoming more difficult (Chakraborty and Okaya 1995).

Resistivity logs typically show high values in saturated hydrocarbon reservoirs (Cherana et al. 2022a). However, certain sources that produce hydrocarbons may also have low resistivity (Benedetto 2010; Cherana et al. 2022b). Such an oil and gas reservoir's resistivity (Rt) may only be 2 ohms (Dimri 2005). Algerian clay-sandstone reservoir evaluation presented a substantial problem due to the low resistivity phenomenon (Doghmane et al. 2019). When employing a classical model with conventional resources, the low resistivity phenomena have a significant impact on interpretation (Doghmane et al. 2019). Despite the calculated water saturation (S_w) being higher than 60%, it has been observed that the reservoir under study produced oil with low resistivity

Unconventional Hydrocarbon Resources: Prediction and Modeling Using Artificial Intelligence Approaches, First Edition. Edited by Sid-Ali Ouadfeul.

values (less than 1 ohm.m). A modified version of the Archie equation was later introduced to adjust the influence of clay in resistivity logs independent of its distribution type (Chakraborty and Okaya 1995). The Archie equation was initially devised to compute water saturation for clean sand reservoirs. Later, the Simandoux equation was put forth to take the clay's distribution – laminar, scattered, and structural – into account (Doghmane and Belahcene 2019). However, not all earlier models have taken into account the impact of conductive minerals that might be included in a matrix. On the other hand, a var iety of logging techniques have been developed to measure the water saturation (S_w) values in reservoirs, allowing for the easy computation of hydrocarbon saturation without the need for resistivity (Cherana et al. 2022a; Doghmane et al. 2022). The MDT tool is being utilized in Algeria to address the issue of low resistivity in hydrocarbon reservoirs, where the majority of technical issues in reservoir characterization depend on the functionality and tolerance of laboratory techniques and equipment used in the field and in situ (Eladj et al. 2022a).

Even if their measurements may be accurate, it is not possible to generalize across the entire reservoir section, especially for various hydrocarbon resources (Eladj et al. 2022b). The presence of irreducible water is typically used as an explanation for an increase in conductivity (or decrease in resistivity) in standard log interpretation methods, but this is not always the case (Eladj et al. 2022c). The Archie equation is used to predict the total resistivity in clean sand formation, but it is unreliable for other formations that contain clay minerals and shale deposits. Numerous issues were brought on by the low resistivity phenomenon in Algerian petroleum resources (e.g. the condemnation of the productive reservoir and the underestimation of reserves). These issues can raise a number of queries, such as what causes the low resistivity. How do we identify these kinds of reservoirs? Do there exist any methods to evaluate them?

This study aims to provide answers to these queries. The work is structured as follows. In section two, a brief geological description of the studied field is provided to highlight its mineralogy (Fangyu et al. 2018). The traditional interpretation of the logging data has been provided in section three so that the case of low resistivity reservoirs can better understand its limitations. Section four follows with a thorough explanation of the phenomenon's causes and an illustration of modern tool technology. The proposed ANNSVM model for the case study has been described in depth in section five, and the results have been thoroughly examined. The final section of the report included findings and suggestions for case studies of similar low-resistivity reservoirs in the future.

9.2 Field Geological Description

The field under study, a sizable intra-cratonic basin situated in the northern section of the Saharan platform, is one of Algeria's most productive hydrocarbon basins with a total size of 103 259 km^2 (Figure 9.1). It is the deepest zone of the eastern syncline and a component of the Oriental Erg, where the Paleozoic wall thickness near the basin's center exceeds 6000 m (Doghmane et al. 2019; Eladj et al. 2020). The study area covers a sizable portion of the Eastern Sahara and is surrounded by dunes from the Oriental Erg, which made fieldwork more difficult and expensive and thus decreased the amount of geological data available.

9.2.1 Conventional Interpretation

Log interpretation is very important for evaluating hydrocarbon reservoirs since it makes it possible to identify fluids and estimate reserves (Cherana et al. 2022a). In order to ascertain the mineralogy, lithology, and nature of the fluids from logging data, the conventional evaluation methods of

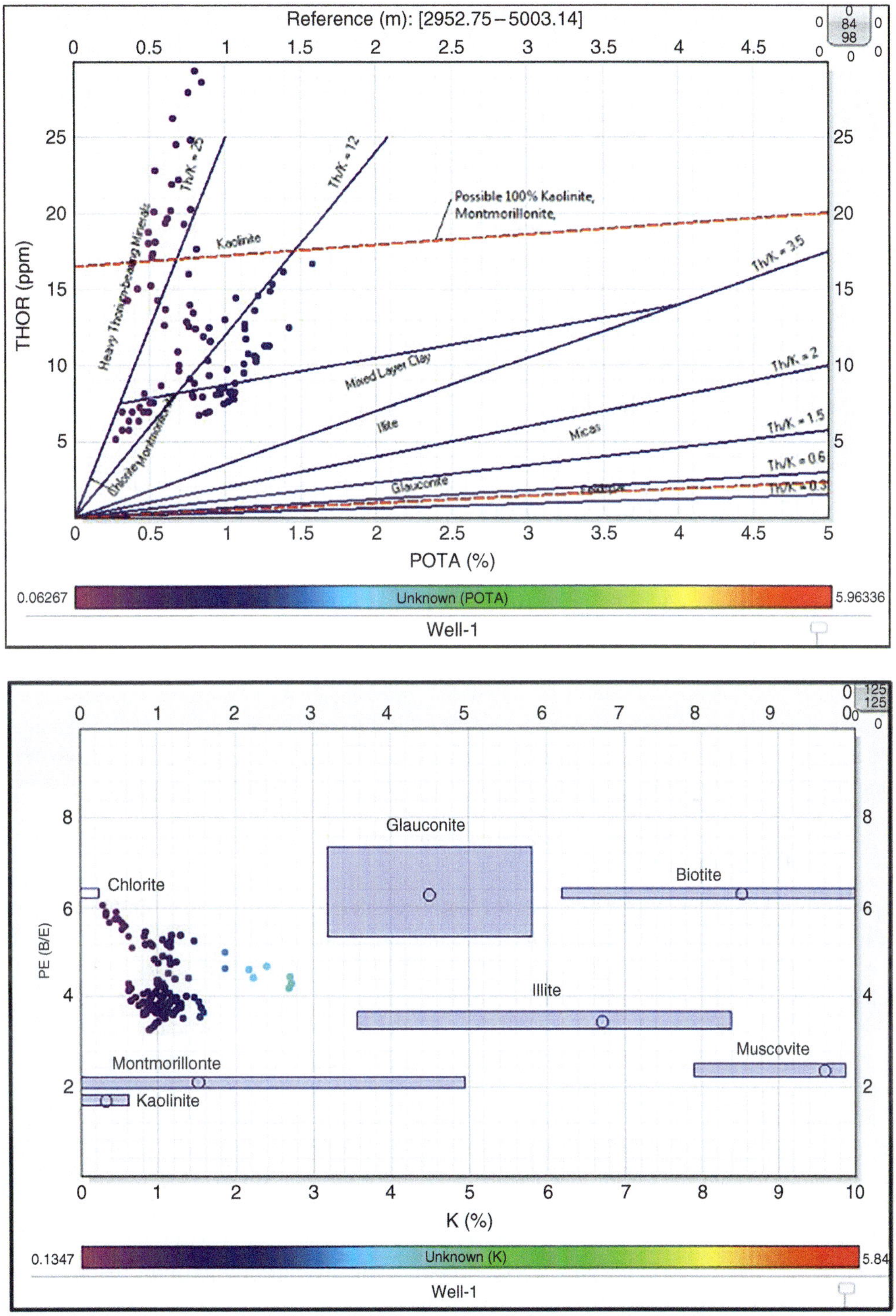

Figure 9.1 Cross-plots for mineralogic study at Well-1.

the examined reservoir's petrophysical parameters are thoroughly explained in this section (Foufoula-Georgiou and Kumar 1994).

9.2.2 Reservoir Mineralogy

The cross-plot of the photoelectric factor as a function of the potassium concentration (PEF vs. K) and the cross-plot of PEF as a function of the thorium-potassium ratio (PEF vs. TH/*K*) have both been used to graphically represent the reservoir mineralogy as shown in Figure 9.1a and b.

These cross-plots indicate that chlorite and montmorillonite are the two main minerals in the reservoir; chlorite is an aluminum-silicate mineral made up of iron and magnesium. Its family's chemical formula is $(F_e, M_g, A_l)_6(S_i, A_l)_4O_{10}(OH)_8$ and it is generally comparable to mica in structure and physicochemical properties. However, it is created by reactions of other minerals, including black mica. It is demonstrated that Well-1 and Well-2 share the same mineralogy since they are part of the same deposit environment by correlating them at the reservoir level.

9.3 Low-Resistivity Phenomenon

Due to the disparity in resistivity values between water and hydrocarbons, the resistivity logs are one of the most crucial markers in petrophysical interpretations (Chakraborty and Okaya 1995). However, if the reservoir exhibits a low resistivity phenomenon, it becomes difficult or perhaps impossible to identify hydrocarbons from these records. The issue is that while hydrocarbons can be generated, the interpretation of resistivity data points to a high water saturation S_w. As a result, this can cause some potential exploratory reservoirs to be missed. Understanding the reasons and origin of this phenomenon is therefore essential. Nuclear magnetic resonance (NMR) and lithology plots are used to analyze the sources of the low resistivity in the studied reservoir at the level of Well-1 (Doghmane et al. 2019).

9.3.1 Cross Plots Interpretation

The potassium-thorium cross-plot in Figure 9.1 has shown that a mineralogical assembly made up of heavy minerals and the minerals chlorite, montmorillonite, and kaolinite exists. The potassium-PEF cross-plot confirms that chlorite is plentiful. The study of the reservoir's cuttings and carrots, which revealed the presence of ferruginous minerals including pyrite and hydrocarbons, further supports the existence of heavy minerals at Well-1 (violet fluorescence). The name "chlorite" no longer refers to a single mineral but rather to the entire category of phyllosilicate minerals. Chlorites are created when ferromagnesian minerals, which are altered minerals including pyroxenes, amphiboles, and biotites, break down. The plagioclases typically undergo a very significant modification along with this chloritization. The most prevalent chlorite in the reservoir under study is chamosite, which is also linked with kaolinite, glauconite (mica), and pyrite (Fangyu et al. 2018). Chamosite is also particularly rich in iron. A typical range of 2–14 meq/100 g, the chamosite envelops the pores and has a limited capacity to exchange cations (CEC). Chlorite can, however, be very conductive in the presence of brine and at high temperatures (more than 105 °C). Additionally, most reservoir sandstones are fine-grained and feature chlorotic grain coatings with high specific surfaces and levels of surface roughness, typically in conjunction with high saturations of irreducible water. Low resistivity responses of logging equipment are also caused by the formation of brine's extremely high salinity (>150 000 ppm), which is produced by the reservoirs of the field under study (Fangyu et al. 2018; Herrera et al. 2013).

9.3.2 NMR Logs Interpretation

Because NMR tool technology can distinguish between irreducible and movable fluids, it provides more accurate porosity logs, which is extremely interesting in the evaluation of reservoirs (Bacetti and Doghmane 2020). The NMR porosity is dependent on continuous fluids in the formation, as opposed to neutron density porosities, and is unaffected by clay minerals. The porosity of Well-1 is 23%, as shown in Figure 9.2, and this value is extremely similar to the predicted effective porosities, supporting the deduced interpretation. Additionally, the NMR logs for Wells 1 and 2 show low permeability values on the order of 6 and 10 m darcy, which suggests the presence of microporosity and a consequent drop in resistivity (Chakraborty and Okaya 1995). The logs also revealed a significant portion of the reservoir's porosity that was occupied by impermeable water that did not circulate, which decreased resistivity readings and obscured the hydrocarbons' signal. In addition to the interpretation of NMR logs, core data, and cross plots, it has been determined that the presence of chlorite, heavy minerals (pyrites, glauconite, and ferruginous minerals), high irreducible water (high wettability), and microporosity are the main causes of the low resistivity phenomenon in this field.

9.3.3 Comparison Between Well-1 and Well-2

Well-2 is situated 1.8 km to the southeast of Well-1, in the eastern portion of the field under study. This well is being drilled to confirm the reservoir's extension to the eastern portion of the area. They share the same log signatures and formation layers due to their proximity and similar log signatures (Figure 9.3).

While Well-1 produced hydrocarbon, Well-2 is entirely filled with water. Resistivity $R_t = R_{xt} = 1$ ohm.m is indicated by the resistivity logs in Well-1 overlapping. However, because Well-2 is an aquifer, the resistivity logs there are separated, and the formation's resistivity there is lower than Well-1's. Despite the fact that the classical interpretation did not suggest it, Well-1 is therefore likely to contain hydrocarbons.

9.3.4 Developed Logging Tools

The most well-known tools are thermal decay time (TDT), gamma spectrometry tool (GST), and ecoscope. New tools' technologies have recently been created to compute the formation water saturation (S_w) without utilizing the resistivity R_t. Unfortunately, they are rarely used in Algerian petroleum exploration activities due to their exorbitant costs.

9.3.5 Proposed ANNSVM Algorithm

Equation (9.1)'s modified Simandoux equation is intended to account for the dispersion of the clay as well as the impact of the matrix's heavy conductive minerals (Doghmane and Belahcene 2019; Huang et al. 2015) (laminated and structured).

$$\frac{1}{R_t} = \frac{\phi_e^m S_w^n}{\left(1 - V_{cl} - C_1\right) a R_w} + \left(\frac{V_{cl}^c}{R_{cl}} + C_2\right) S_w^{n/2} \tag{9.1}$$

Where R_w is water resistivity, ϕ is the rock porosity, a is the tortuosity factor, m is the cementation exponent, n is the saturation exponent, c is the compaction factor, V_{cl} and R_{cl} are the clay volume

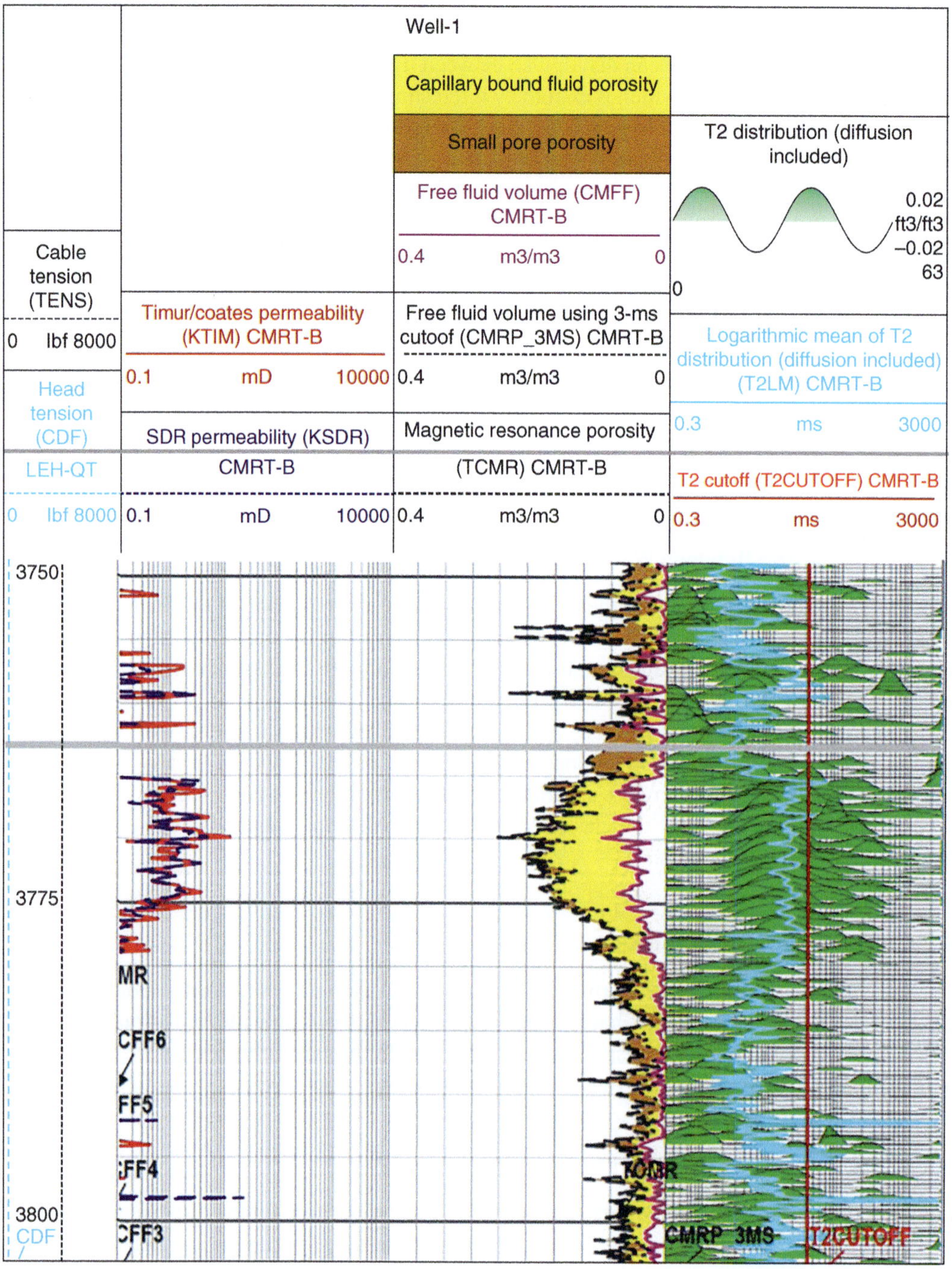

Figure 9.2 NMR tool results for Well-1.

and resistivity each estimated using ANN algorithms, and C 1 and C 2 are compensation values. According to prior fieldwork, the water saturation calculated using such low resistivity values and other equations is not accurate (Dimri 2005; Doghmane and Belahcene 2019).

Because the empirical constants and laboratory data need good association between saturation models, petrophysical and chemical properties of rock, the classical or analytical models are based

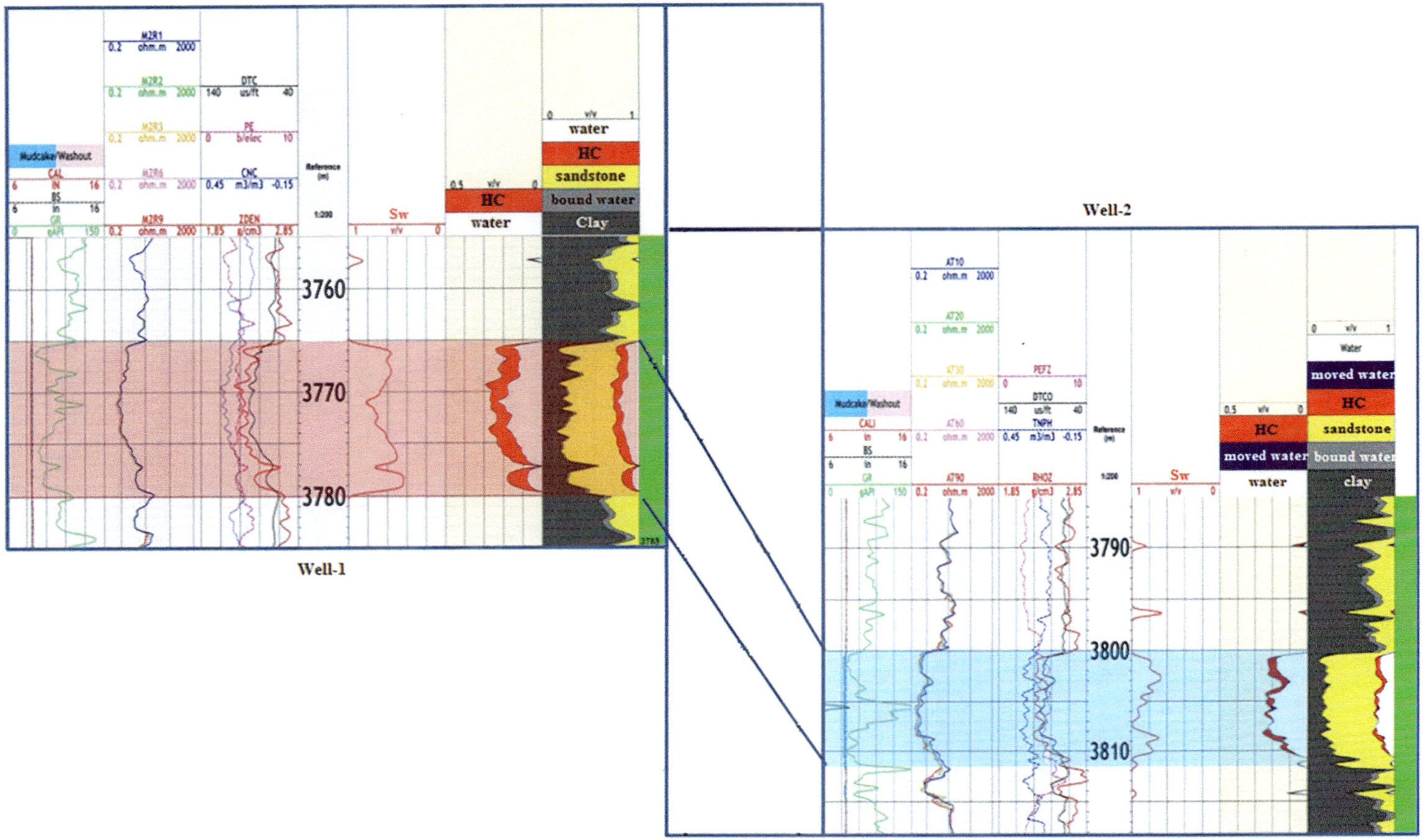

Figure 9.3 Correlation between Well-1 and Well-2 results.

on constrained parameters and are occasionally difficult to apply to other domains. The neural network and support vector machine-based model was created with intelligence, and the MDT assessed results validated its performance (Cherana et al. 2022b; Liu et al. 2016). The fundamental workings of an ANN are based on three variables: weight function (w_i), output results (y_i), and input data or signal (x_i) (Doghmane et al. 2019). After being trained using logging data and MDT samples, ANNSVM was created to be able to forecast the water saturation model (Cherana et al. 2022a; Dimri 2005). The data sets used in this study are from wells in the Algerian field; Well-1's data were used for training and testing, while Well-2's data were used to demonstrate validation.

It has been established that thin beds, salty water, conductive minerals, and a sizeable amount of irreducible water are the primary causes of the low resistivity of oil reservoirs (Serra 1985; Cherana et al. 2022b). But according to a research of more than 40 wells in an oil field in Algeria, the presence of heavy minerals and a significant amount of irreducible water as a result of too much surfactant in the sludge is what causes low resistivity of hydrocarbon reservoirs (Djezzar et al. 2022). Equation (9.1)'s compensation terms (C1) and (C2) have been computed using the trained ANN algorithm (Haliburton 2003; Aoun et al. 2022). According to the traditional interpretation of the Well-1 reservoir results, the porosity is 23.15% and the water saturation is 63%. Gamma ray, neutron density, and PEF have been placed in class 1 (dominant variables), whilst resistivity and sonic are set in class 2. SSVM has been utilized to classify the input log data according to its importance. The ANN training procedure employed the classified data as input, as shown in the algorithm's diagram in Figure 9.4.

Using the Well-1 data displayed, the ANN has been trained, validated, and tested. The results of the training, testing, and validation are shown in Figure 9.5a and b. The tests are carried out using several ANN training topologies and activation functions in order to choose the best

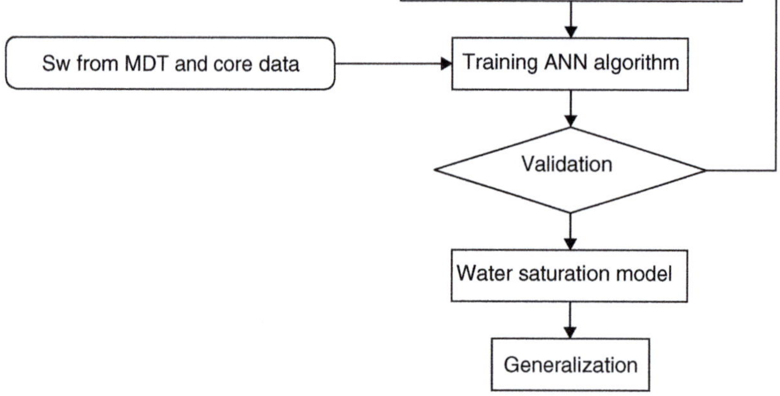

Figure 9.4 Diagram of the proposed algorithm.

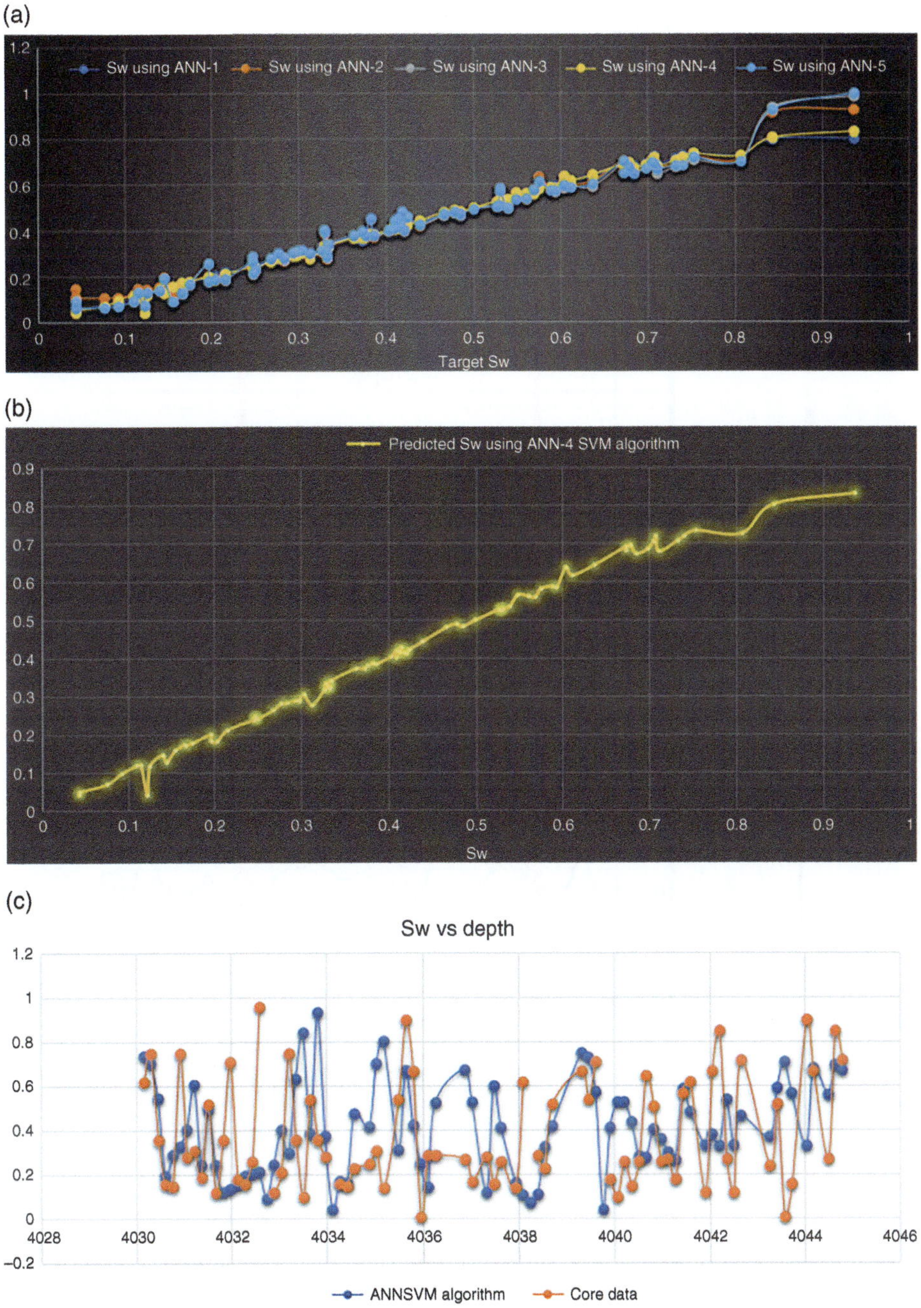

Figure 9.5 Training, testing, and validation of the proposed ANN algorithm: (a) Results for different ANN architecture, (b) Correlation between Sw from ANN and MDT, (c) Prediction of Well-2 Sw.

parameters for the reservoir under study. The correlation matrix is then obtained once the algorithm results and the Well-2 MDT findings have been correlated (Figure 9.6). Additionally, Sw at Well-2 has been predicted using ANNSVM; the enhanced outcomes in the investigated reservoir are shown in Figure 9.5c.

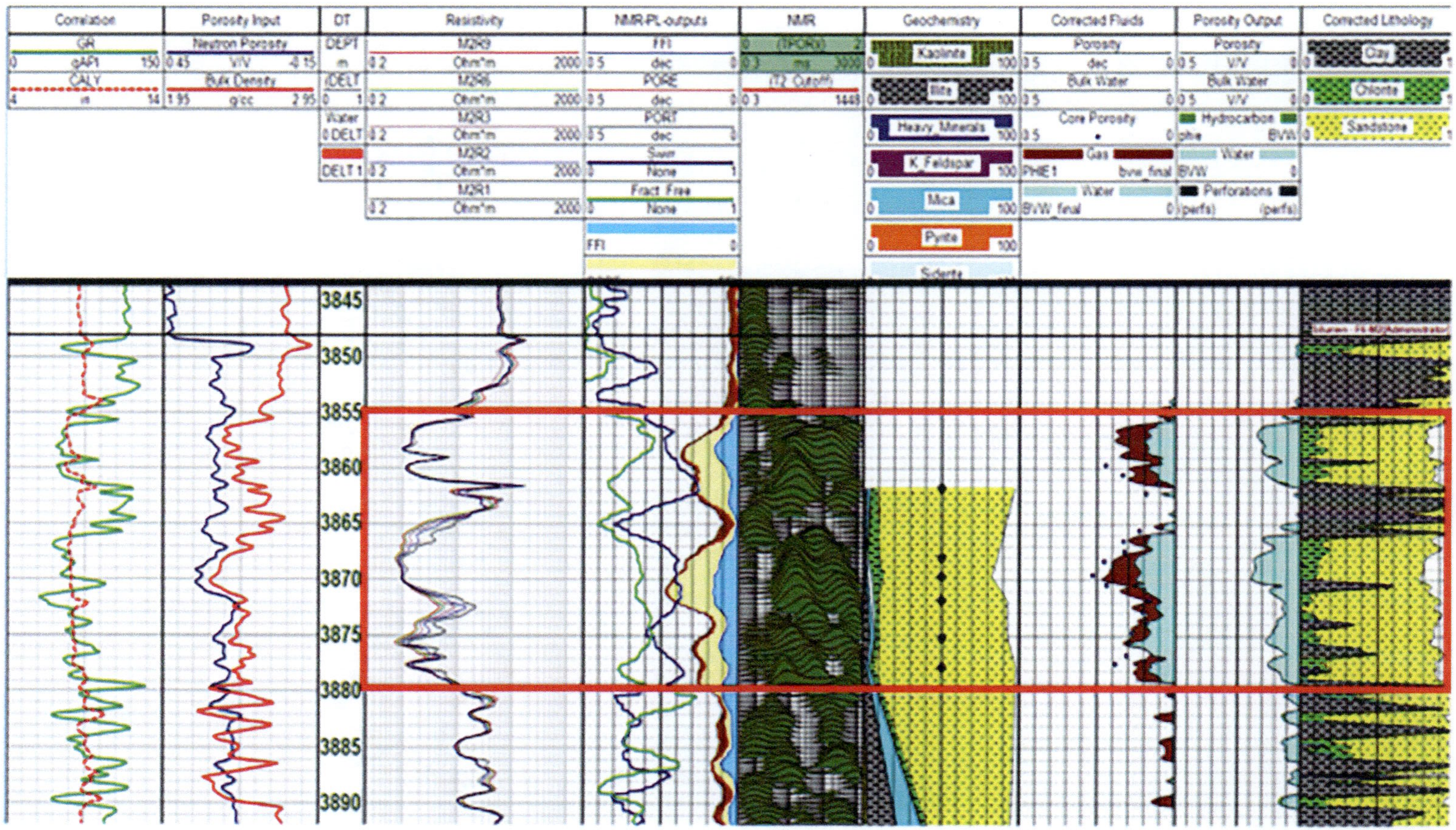

Figure 9.6 Correlation between the predicted water saturation and the measured data.

9.4 Conclusions

In the Algerian hydrocarbon fields, the low resistivity phenomenon is a common and difficult subject. Indeed, the conventional interpretation indicated that the two wells-1 and 2 were aquifers (SW> 60%) because of their low resistivity, which does not surpass (2 ohm.m). However, the pressure readings and samples from the MDT demonstrated that Well-1 can produce oil. This uncertainty is explained by the formation's high levels of irreducible water, microporosity, and conductive clay minerals (chamosite). The fundamental disadvantage of the modified Sinmodoux equation, however, is that it relies on a mineralogy tool to determine the volume of conductive minerals and their distribution, which is not always feasible from an economic standpoint. Even though TDT, GST, and Ecoscope are sophisticated tools, they have never been applied in the field under study because of their exorbitant prices. By applying the traditional methods of interpretation to analyze the petrophysical parameters and determine the reservoir fluids of Well-1, we arrived at an inaccurate estimation of the saturation. The contribution of the MDT tool by measuring the pressure and performing formation tests in front of the reservoir of Well-1 was reliable but insufficient for generalization in the presence of doubts and uncertainties owing to the mineralogical research of clays and cuttings.

The MDT results and the ANNSVM model minimized the uncertainties in the quantitative reservoir evaluation. Additionally, unlike conventional models, the generated model only requires a little amount of laboratory data, and its parameters accurately reflect the characteristics of the local reservoir. As a result, it may be applied to the evaluation of deterministic quantitative reservoirs.

References

Aoun, A., Soto, R., Rabiei, M. et al. (2022). Neural network based Mechanical Earth Modelling (MEM): a case study in Hassi Messaoud Field, Algeria. *Journal of Petroleum Science and Engineering* 210: https://doi.org/10.1016/j.petrol.2021.110038.

Bacetti, A. and Doghmane, M.Z. (2020). A practical workflow using seismic attributes to enhance sub seismic geological structures and natural fractures correlation. *Conference Proceedings, First EAGE Digitalization Conference and Exhibition*, vol. 2020, Algeria, pp. 1–5. http://doi.org/10.3997/2214-4609.202032062.

Benedetto, A. (2010). Water content evaluation in unsaturated soil using GPR signal analysis in the frequency domain. *Journal of Applied Geophysics* 71 (1): 26–35. http://doi.org/10.1016/j.jappgeo.2010.03.001.

Chakraborty, A. and Okaya, D. (1995). Frequency-time decomposition of seismic data using wavelet-based methods. *Geophysics* 60 (6): http://doi.org/10.1190/1.1443922.

Cherana, A., Aliouane, L., Doghmane, M., and Ouadfeul, S.A. (2022a). Fuzzy clustering algorithm for lithofacies classification of ordovician unconventional tight sand reservoir from well-logs data (algerian sahara). *Advances in Geophysics, Tectonics and Petroleum Geosciences. CAJG 2019. Advances in Science, Technology & Innovation*. Cham: Springer. https://doi.org/10.1007/978-3-030-73026-0_64.

Cherana, A., Aliouane, L., Doghmane, M.Z. et al. (2022b). Lithofacies discrimination of the Ordovician unconventional gas-bearing tight sandstone reservoirs using a subtractive fuzzy clustering algorithm applied on the well log data: Illizi Basin, the Algerian Sahara. *Journal of African Earth Sciences* 196: https://doi.org/10.1016/j.jafrearsci.2022.104732.

Dimri, V. (2005). Fractals in geophysics and seismology: an introduction. In: *Fractal Behaviour of the Earth System* (ed. V.P. Dimri). Berlin, Heidelberg: Springer. http://doi.org/10.1007/3-540-26536-8_1.

Djezzar, S., Boualam, A., Ouadi, H. et al. (2022). Geological characterization of lower devonian reservoirs in Reggane Basin, Algeria. *The SPE Annual Technical Conference and Exhibition*, Houston, TX, USA (October 2022). https://doi.org/10.2118/210162-MS.

Doghmane, M.Z. and Belahcene, B. (2019). Design of new model (ANNSVM) compensator for saturation calculation based on logging curves for low resistivity phenomenon. *Conference Proceeding, EAGE/ALNAFT Geoscience Workshop* 2019: 1–5.

Doghmane, M.Z, Belahcene, B., and Kidouche, M. (2019). Application of improved artificial neural network algorithm in hydrocarbons' reservoir evaluation. *Proc. International conference in Artificial Intelligence in Renewable Energetic Systems ICAIRES 2018: Renewable Energy for Smart and Sustainable Cities*, Algeria, pp. 129–138.

Doghmane, M.Z., Ouadfeul, S.A., Benaissa, Z., and Eladj, S. (2022). Classification of ordovician tight reservoir facies in algeria by using neuro-fuzzy algorithm. In: *Artificial Intelligence and Heuristics for Smart Energy Efficiency in Smart Cities. IC-AIRES 2021*, Lecture Notes in Networks and Systems, vol. 361 (ed. M. Hatti). Cham: Springer https://doi.org/10.1007/978-3-030-92038-8_91.

Eladj, S., Lounissi, T.K., and Doghmane, M.Z. (2020). Lithological characterization by simultaneous seismic inversion in algerian south eastern field. *Engineering, Technology & Applied Science Research* 10 (1): 5251–5258.

Eladj, S., Doghmane, M.Z., Aliouane, L., and Ouadfeul, S.A. (2022a). Porosity model construction based on ANN and seismic inversion: a case study of saharan field (algeria). *Advances in Geophysics, Tectonics and Petroleum Geosciences. CAJG 2019. Advances in Science, Technology & Innovation.* Cham: Springer. https://doi.org/10.1007/978-3-030-73026-0_55.

Eladj, S., Doghmane, M.Z., Lounissi, T.K. et al. (2022b). 3D Geomechanical model construction for wellbore stability analysis in algerian southeastern petroleum field. *Energies* 15 (20): 7455. https://doi.org/10.3390/en15207455.

Eladj, S., Doghmane, M.Z., and Belahcene, B. (2022c). Design of new model for water saturation based on neural network for low-resistivity phenomenon (Algeria). *Advances in Geophysics, Tectonics and Petroleum Geosciences. CAJG 2019. Advances in Science, Technology & Innovation.* Cham: Springer. https://doi.org/10.1007/978-3-030-73026-0_75.

Fangyu, L., Bo, Z., Sumit, V., and Marfurt, K.J. (2018). Seismic signal denoising using thresholded variational mode decomposition. *Exploration Geophysics* 49 (4): https://doi.org/10.1071/EG17004.

Foufoula-Georgiou, E. and Kumar, P. (1994). Wavelet analysis in geophysics: an introduction. *Wavelet Analysis and Its Applications* 4: 1–43. https://doi.org/10.1016/B978-0-08-052087-2.50007-4.

Haliburton (2003). Logging of Perforating Products and Services, Report.

Herrera, R.H., Han, J., and van der Baan, M. (2013). Applications of the synchrosqueezing transform in seismic time-frequency analysis. *Geophysics* 79 (3): https://doi.org/10.1190/geo2013-0204.1.

Huang, W., Runqiu, W., Yimin, Y. et al. (2015). Signal extraction using randomized-order multichannel singular spectrum analysis. *Geophysics* 82 (2): https://doi.org/10.1190/geo2015-0708.1.

Liu, W., Cao, S., and Chen, Y. (2016). Seismic time–frequency analysis via empirical wavelet transform. *IEEE Geoscience and Remote Sensing Letters* 13 (1): 28–32. http://doi.org/10.1109/LGRS.2015.2493198.

Serra, O. (1985). *Diagraphies différés base de l'interprétation, Tome 2.* Montrouge: Etudes et productions Schlumbeger.

10

A Practical Workflow for Improving the Correlation of Sub-Seismic Geological Structures and Natural Fractures using Seismic Attributes

Mohamed Zinelabidine Doghmane[1], Sid-Ali Ouadfeul[2], and Leila Aliouane[3]

[1] *Department of Geophysics, FSTGAT, University of Science and Technology Houari Boumediene, Algiers, Algeria*
[2] *Algerian Petroleum Institute, Sonatrach, Boumerdes, Algeria*
[3] *LABOPHYT, Faculty of Hydrocarbons and Chemistry, University M'hamed Bougara of Boumerdes, Boumerdes, Algeria*

KEYPOINTS

- This chapter is dedicated to design a new workflow for seismic attributes.
- This geometry workflow has been applied to 3D seismic data.
- The main contribution of this workflow is that it allows the detection of fractures with dimension less than the seismic resolution.
- This can be a useful tool for optimized exploration and production strategy.

10.1 Introduction

A typical workflow for geometric seismic attributes is proposed in this study to map the structural geology of the tight reservoir of Ain Amenas (Algeria) (Eladj et al. 2020). The proposed attributes allow us to map major faults that appear to have a throw less than the seismic vertical resolution, as well as more subtle folds and flexures and overlying other formations (Satinder and Marfurt 2012; Alridha and Ubaid 2015; Machado et al. 2018). This study develops a useful workflow tool for continuous fracture modeling (CFM) and successfully applies it to several projects (Odling et al. 2001; Doghmane and Belahcene 2019). We used several attributes (curvature, variance, chaos, structural smoothing attribute, edge enhancement, and ant tracking) in this study (Satinder and Marfurt 2012; Alridha and Ubaid 2015). The seismic attributes, which included height, width, and length, provided a plausible characterization of the fracture zone extent (Machado et al. 2018; Doghmane et al. 2022). Despite technological advances in seismic acquisition techniques that provide higher resolution, small faults in deep reservoirs are not always detectable (Robert et al. 1991; Eladj et al. 2022a, b). Furthermore, if vintage seismics are available, expensive high resolution exploration surveys are not always justified. Seismic characteristics have been widely used to solve the aforementioned problem.

Unconventional Hydrocarbon Resources: Prediction and Modeling Using Artificial Intelligence Approaches, First Edition. Edited by Sid-Ali Ouadfeul.

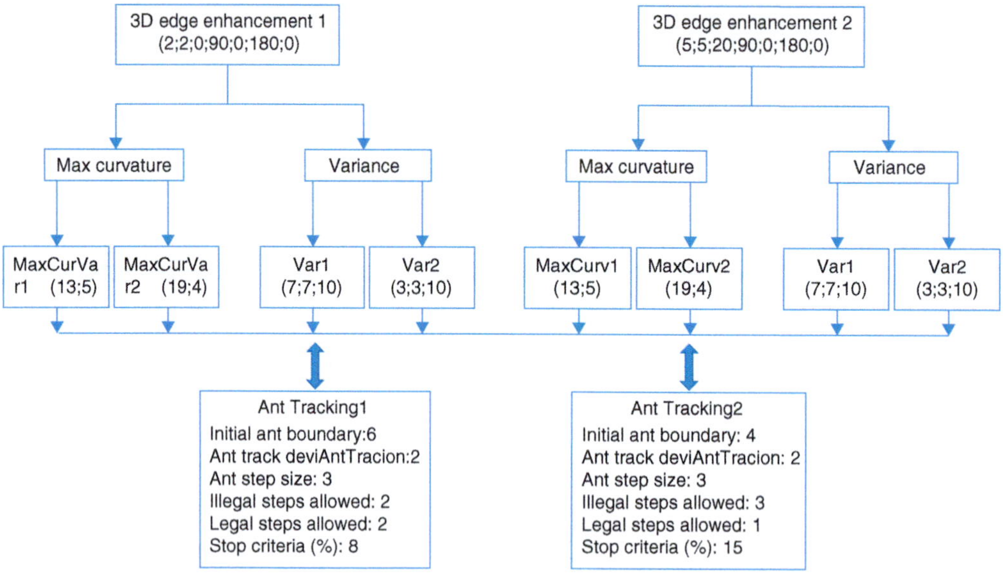

Figure 10.1 Schematic diagram details of the developed workflow.

10.2 Description of the Developed Workflow

Practically, it has been demonstrated that the results obtained from a single attribute are sometimes insufficient, necessitating the use of combined attributes (Machado et al. 2018). Nowadays, geophysical software platforms provide a large number of attributes, which confuses engineers who are faced with the problem of determining the optimal order of combined attributes and their optimal parameters. The workflow described in Figure 10.1 can be regarded as a practical reference solution for geophysicists seeking to uncover the greatest number of hidden geobodies from seismic data.

10.3 Discussion

Many attributes and attribute combinations were employed, with FMI/Core fracture densities correlating to attribute values extracted from wells, as shown in Tables 10.1 and 10.2. Low seismic resolution at the reservoir level and acquisition footprint (Figure 10.2) are thought to be the main reasons for poor correlation (Doghmane et al. 2019). Cropped seismic were processed to reduce the effect of acquisition footprint (Cacas et al. 2001).

Table 10.1 Correlation results FMI/core data.

FMI/Core data $R^2 > 0.2$	Value
Most positive curvature	0.31
3D edge enhancement	0.29
Maximum curvature	0.24
Extreme curvature	0.22

Table 10.2 Correlation results FMI/core data.

Frac. Model $R^2 > 0.2$	Value
Chaos	0.98
Gaussian curvature	0.97
Maximum curvature	0.91
Local flatness	0.84
Mean curvature	0.82
Dip deviation	0.75
Extreme curvature	0.74
Amplitude contrast	0.52
Variance	0.40
Local structure dip	0.36
Minimum curvature	0.33
Most negative curvature	0.27

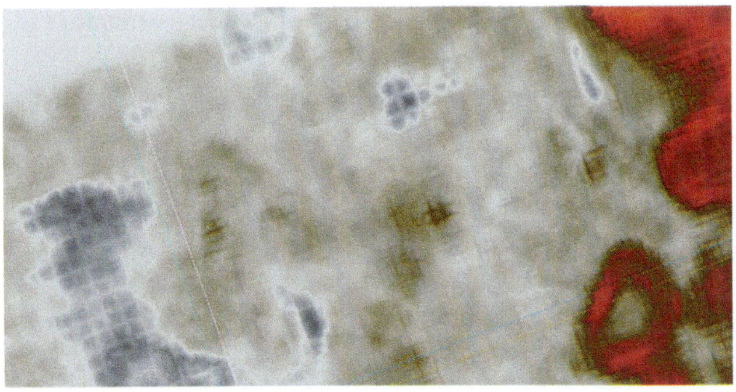

Figure 10.2 Acquisition footprint of the studied field.

10.4 Conclusions

In this paper, a new workflow of geometrical seismic attributes was used for structural geological study, in which small faults with higher throw were detected and naturally fracture models for tight reservoirs were obtained, as shown in Figure 10.3. The fracture model has been correlated with imagery data from field exploration wells (Djezzar et al. 2019). The obtained results were compared to conventional seismic interpretation results, and we show the presence of small faults that are not visible using other methods. Furthermore, the constructed fracture model matched the imagery interpretation results. Finally, in order to be considered for reservoir model construction during the production phase, a relationship between fracture density and the used attributes has been proposed.

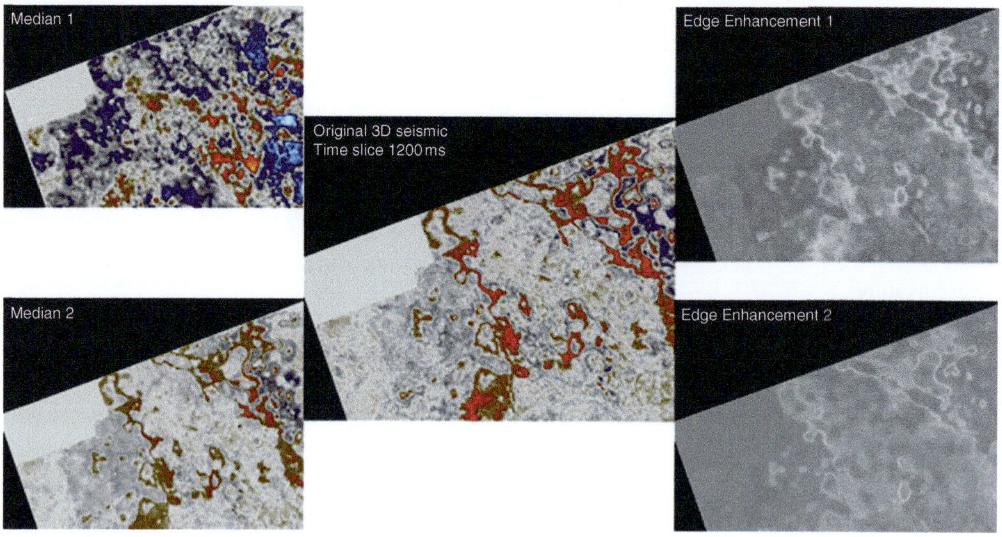

Figure 10.3 Obtained model using the proposed workflow.

References

Alridha, A.N. and Ubaid, M.S. (2015). 3-D seismic attributes analysis in Balad Oil Field – center of Iraq. *Arabian Journal of Geosciences* 8: 2785–2798.

Cacas, M.C., Daniel, J.M., and Letouzey, J. (2001). Nested geological modelling of naturally fractured reservoirs. *Petroleum Geoscience* 7: S43–S52.

Djezzar, S., Rasouli, V., Boualam, A., and Minou, R. (2019). Size scaling and spatial clustering of natural fracture networks using fractal analysis. *The 53rd U.S. Rock Mechanics/Geomechanics Symposium*, New York City, USA, June 23rd, 2019.

Doghmane, M.Z. and Belahcene, B. (2019). Design of new model (ANNSVM) compensator for saturation calculation based on logging curves for low resistivity phenomenon, *Conference Proceedings, EAGE/ALNAFT Geoscience Workshop,* Algiers, Algeria (January 2019), Vol. 2019, pp. 1–5.

Doghmane, M.Z., Belahcene, B., and Kidouche, M. (2019). Application of improved artificial neural network algorithm in hydrocarbons' reservoir evaluation. In: *Renewable Energy for Smart and Sustainable Cities. ICAIRES 2018*, Lecture Notes in Networks and Systems, vol. 62 (ed. M. Hatti). Cham: Springer.

Doghmane, M.Z., Ouadfeul, S.A., Benaissa, Z., and Eladj, S. (2022). Classification of ordovician tight reservoir facies in algeria by using neuro-fuzzy algorithm. In: *Artificial Intelligence and Heuristics for Smart Energy Efficiency in Smart Cities. IC-AIRES 2021*, Lecture Notes in Networks and Systems, vol. 361 (ed. M. Hatti). Cham: Springer https://doi.org/10.1007/978-3-030-92038-8_91.

Eladj, S., Lounissi, T.K., Doghmane, M.Z., and Djeddi, M. (2020). Lithological characterization by simultaneous seismic inversion in algerian south eastern field. *Engineering, Technology & Applied Science Research* 10 (01): 5251–5258.

Eladj, S., Lounissi, T.K., Doghmane, M.Z., and Djeddi, M. (2022a). Wellbore stability analysis based on 3D geo-mechanical model of an algerian southeastern field. *Advances in Geophysics, Tectonics and Petroleum Geosciences. CAJG 2019. Advances in Science, Technology & Innovation* (25–28 November 2019). Cham: Springer. https://doi.org/10.1007/978-3-030-73026-0_136.

Eladj, S., Doghmane, M.Z., Aliouane, L., and Ouadfeul, SA. (2022b). Porosity Model Construction Based on ANN and Seismic Inversion: A Case Study of Saharan Field (Algeria). *Advances in Geophysics, Tectonics and Petroleum Geosciences. CAJG 2019. Advances in Science, Technology & Innovation*. Cham: Springer. https://doi.org/10.1007/978-3-030-73026-0_55.

Machado, G., Marfurt, K.J., and Alali, A. (2018). Attribute-assisted footprint suppression using a 2D continuous wavelet transform. *Interpretation* 6 (2): T457–T470.

Odling, N.E., Bour, O., Davy, A. et al. (2001). Scaling of fracture systems in geological media. *Reviews of Geophysics* 39 (3): 347–383.

Robert, E.G., Heloise, B.L., Bates, C.R. et al. (1991). Detection and analysis of naturally fractured gas reservoirs: multiazimuth seismic surveys in the Wind River basin, Wyoming. *Geophysics* 64 (4): 1277–1292.

Satinder, C. and Marfurt, K. (2012). Curvature attribute applications to 3D surface seismic data, Interpreter's Corner SEG, April 2007. *The Leading Edge* 26 (4): 404–414.

11

Calculation of Petrophysical Parameter Curves for Nonconventional Reservoir Modeling and Characterization

Mohamed Zinelabidine Doghmane[1], Sid-Ali Ouadfeul[2], and Leila Aliouane[3]

[1] *Department of Geophysics, FSTGAT, University of Science and Technology Houari Boumediene, Algiers, Algeria*
[2] *Algerian Petroleum Institute, Sonatrach, Boumerdes, Algeria*
[3] *LABOPHYT, Faculty of Hydrocarbons and Chemistry, University M'hamed Bougara of Boumerdes, Boumerdes, Algeria*

KEYPOINTS

- This chapter deals with the calculation of petrophysicssal parameters, namely lithology factor, cementation exponent, and saturation exponent.
- These parameters are conventionally considered constant over a given net pay since the reservoir is considered homogeneous.
- In case of heterogeneous reservoir, these empirical parameters are not necessarily reliable.
- The obtained curves of this chapter can be considered toward more reliable reservoir modeling of heterogeneous reservoirs.

11.1 Introduction

The Berkine basin is an important region for oil research, with geological and geophysical studies focusing on Triassic Clayey sandstone (TAGI), upper Devonian (Carboniferous), and lower Siegenian rocks (Yisheng et al. 2019). The drilled well is part of the central bloc's delineation; wells in the region are producing oil and gas condensate in the TAGI (Eladj et al. 2022a,b). The basin is of intracratonic type and is located in the northeastern part of the Saharan platform (Doghmane et al. 2019; Doghmane and Belahcene 2019; Eladj et al. 2020). It is defined as a flat basin with infra-Cambrian bedrock that is strongly planned (Figure 11.1).

11.2 Proposed Methods

The proposed method is based on obtaining the tortuosity exponent m, the exponent a, and the saturation exponent n as curves from logging curves (GR, neutron-density, and sonic) (Zhao et al. 2015). This assumption is misleading not only for the Archie equation but also for others like Simandoux, modified Simandous, Waxman, Smith . . . etc. (Mithilesh et al. 2018). In the Archie

Unconventional Hydrocarbon Resources: Prediction and Modeling Using Artificial Intelligence Approaches, First Edition. Edited by Sid-Ali Ouadfeul.

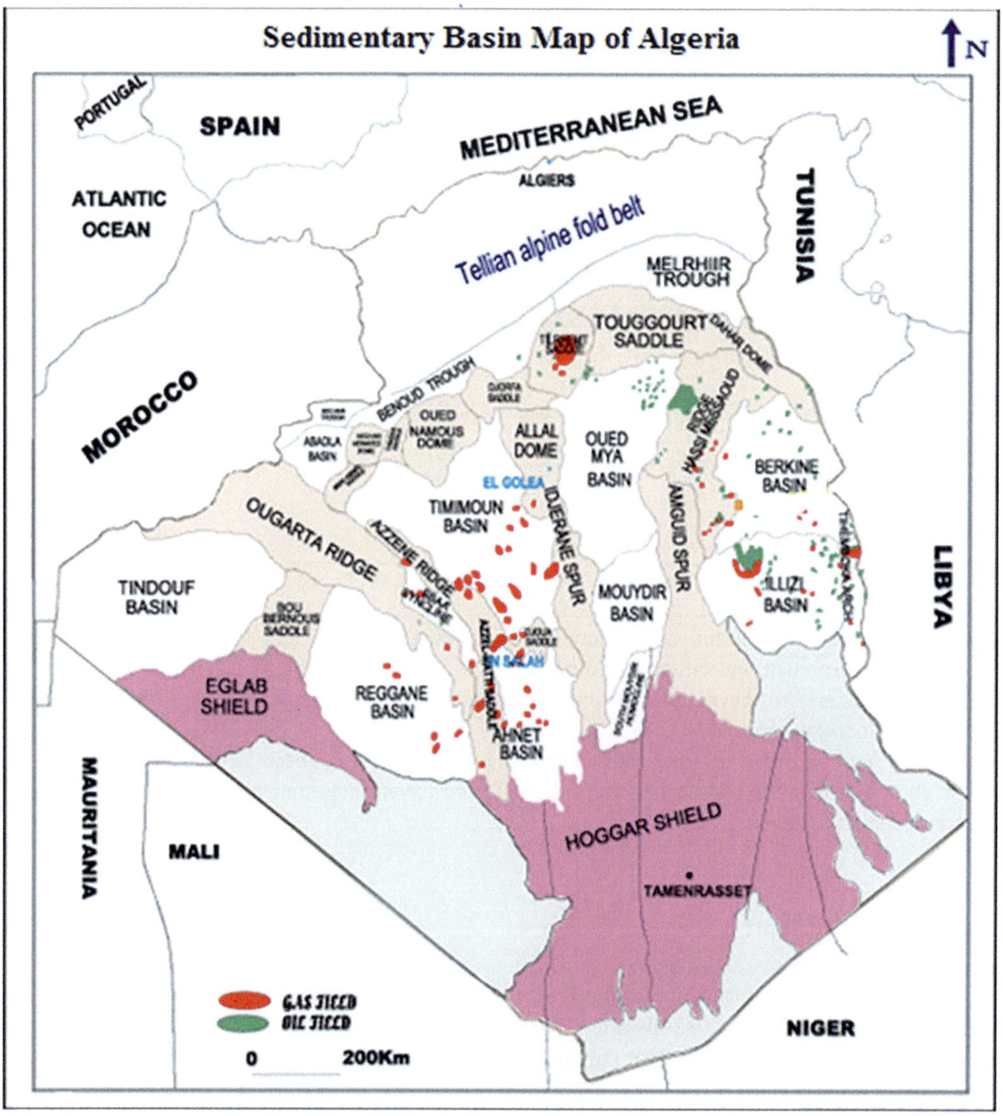

Figure 11.1 Map of sedimentary map of Algeria containing Berkine Basin.

equation, these parameters are considered constant for the entire reservoir section, but it is valid only for homogeneous reservoirs. For unconventional reservoirs with a high degree of heterogeneity, this assumption is misleading not only for the Archie equation but also for others like Simandoux, modified Simandous, Waxman and Smith . . . etc. The coefficient n is also known as the tortuosity factor, the cementation intercept, and the lithology coefficient. It is used to account for compaction, grain size, and pore structure. Because of the strong diagenetic affinity and complex pore structures, considering tortuosity as a constant factor is not a practical assumption in this case, according to the stratigraphic studies (Yisheng et al. 2019).

Many previous studies used an average tortuosity factor value ranging from 1.8 to 2, but due to the need for greater precision, it was necessary to obtain the variation of this parameter throughout the TAGI sections. The proposed equations are based on a relationship between the arrival time of *P* and *S* waves and pore geometry, which can be quantified by tortuosity; tortuosity values greater than one indicate non-regular geometry, while values closer to one indicate regular geometry and homogeneous reservoirs. Based on velocity calculated from sonic logs and adjusted for mud thickness and borehole geometry as described in equation (11.1) (Nordahl et al. 2005).

$$\frac{1}{V} = \frac{\varnothing}{V_f} + \frac{1-\varnothing}{V_{\text{mat}}} \tag{11.1}$$

where V is the sonic velocity calculated by measuring the travel time from the piezoelectric transmitter to the receiver, normally with the units microsecond per foot (a measure of slowness). V_f is the formation fluid velocity, V_{mat} is the matrix velocity, $\varnothing$ is the total porosity obtained from neutron-density logs. Equation (11.2) gives the mathematical equation used to calculate the petrophysical curves (Huawei et al. 2015; Zhao et al. 2015).

$$m = f\left(d, \Delta T_s, \Delta T_p, \varnothing_e\right) \tag{11.2}$$

where d is the depth, ΔT_s is the travel time of S waves, ΔT_p is the travel time of P waves, $\varnothing_e$ is the effective porosity calculated from logging data (Alden et al. 1997). The proposed method has been applied to many wells in the basin, with significant results provided to demonstrate the importance of the approach.

11.3 Results and Discussion

Figure 11.2 depicts the matrix volume calculated using the classical method (on the left) and the matrix volume after tortuosity corrections (on the right) (on the right). When the obtained matrix volumes were compared to laboratory core data, it was clear that the corrections based on variable tortuosity allowed for more precise results in terms of reservoir evaluation, because the volume was much more similar to the one obtained from core data and laboratory tests. Furthermore, the model obtained through the three curves generation was more representative and useful than the model obtained through traditional interpretation techniques.

11.4 Conclusions

The primary goal of this research was to generate the curves tortuosity, cementation, and saturation exponents rather than considering constant in Berkine Basin heterogeneous reservoir. This goal motivates us to propose a mathematical approach that connects logging curves to pore geometry; then, by extracting these curves, the evaluation of thin heterogeneous layers in the reservoir becomes more precise and time efficient. Furthermore, when the obtained matrix volume was compared to that obtained from core data and laboratory tests, it demonstrated the superiority of the proposed method for such nontraditional reservoirs. As a result of the findings, it is highly recommended to consider tortuosity as a variable in evaluating other similar reservoirs in Algerian hydrocarbon fields.

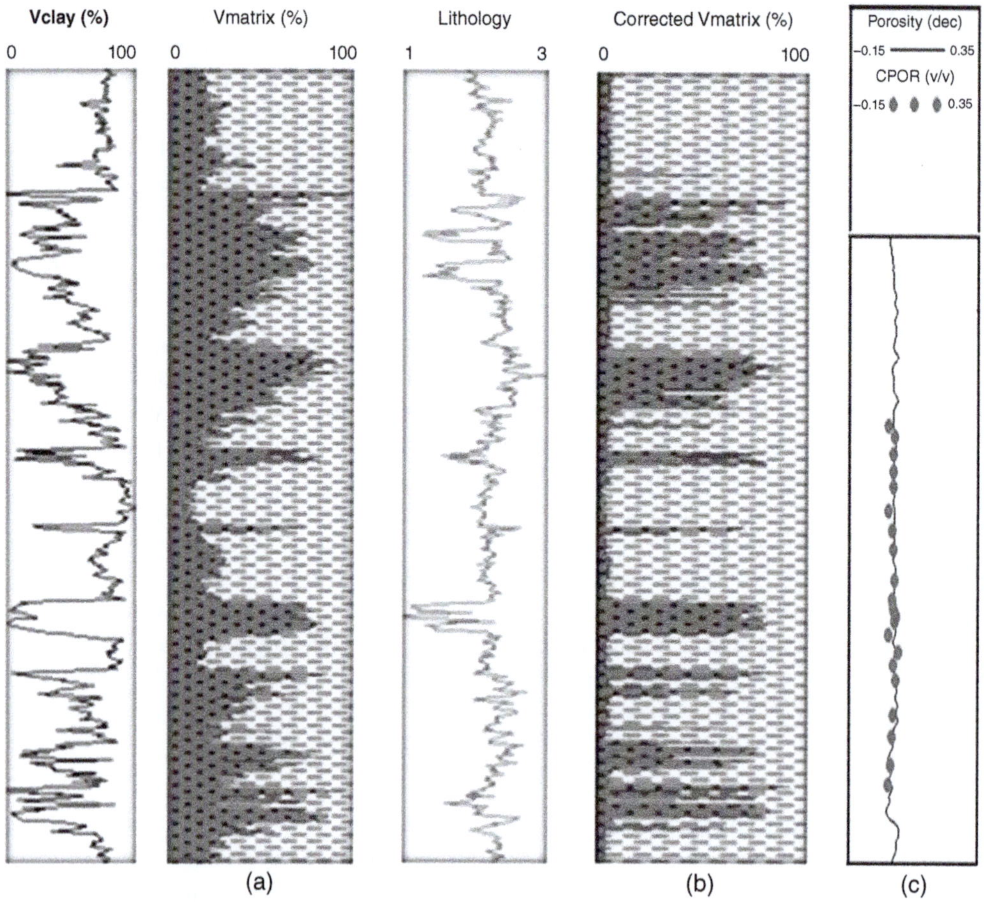

Figure 11.2 Matrix volume: (a) without parameters curves calculation, (b) with and without parameters curves calculation, (c) Porosity estimation with the proposed method.

References

Alden, J.M., Stephen, T.S., and Hartmann, D.J. (1997). Characterization of petrophysical flow units in carbonate reservoirs. *AAPG Bulletin* 81 (5): 734–759.

Doghmane, M.Z. and Belahcene, B. (2019). Design of new model (ANNSVM) compensator for saturation calculation based on logging curves for low resistivity phenomenon. *Conference proceeding, EAGE/ALNAFT Geoscience Workshop* 2019: 1–5.

Doghmane, M.Z., Belahcene, B., and Kidouche, M. (2019). Application of improved artificial neural network algorithm in hydrocarbons' reservoir evaluation. *Proc. International conference in Artificial Intelligence in Renewable Energetic Systems ICAIRES 2018: Renewable Energy for Smart and Sustainable Cities*, 129–138, Tipaza, Algeria (26–28 November 2018).

Eladj, S., Lounissi, T.K., and Doghmane, M.Z. (2020). Lithological characterization by simultaneous seismic inversion in algerian south eastern field. *Engineering, Technology & Applied Science Research* 10 (1): 5251–5258.

Eladj, S., Doghmane, M.Z., Lounissi, T.K. et al. (2022a). 3D geomechanical model construction for wellbore stability analysis in algerian southeastern petroleum field. *Energies 15*: 7455. https://doi.org/10.3390/en15207455.

Eladj, S., Lounissi, T.K., Doghmane, M.Z., and Djeddi, M. (2022b). Wellbore stability analysis based on 3d geo-mechanical model of an algerian southeastern field. *Advances in Geophysics, Tectonics and Petroleum Geosciences. CAJG 2019. Advances in Science, Technology & Innovation* (25–28 November 2019). Springer, Cham. https://doi.org/10.1007/978-3-030-73026-0_136.

Huawei, Z., Zhengfu, N., Qing, W. et al. (2015). Petrophysical characterization of tight oil reservoirs using pressure-controlled porosimetry combined with rate-controlled porosimetry. *Fuel* 154 (15): 233–242.

Mithilesh, K., Dasgupta, R., Singha, D.K., and Singh, N.P. (2018). Petrophysical evaluation of well log data and rock physics modeling for characterization of Eocene reservoir in Chandmari oil field of Assam-Arakan basin, India. *Journal of Petroleum Exploration and Production Technology* 8: 323–340.

Nordahl, K., Ringrose, P.S., and Renjun, W. (2005). Petrophysical characterization of a heterolithic tidal reservoir interval using a process-based modelling tool. *Petroleum Geoscience* 11 (1): 17–28.

Yisheng, L., Yuetian, L., and Qichen, Z. (2019). Petrophysical static rock typing for carbonate reservoirs based on mercury injection capillary pressure curves using principal component analysis. *Journal of Petroleum Science and Engineering* 181: 106175.

Zhao, L., De-hua, H., Qiuliang, Y. et al. (2015). Seismic reflection dispersion due to wave-induced fluid flow in heterogeneous reservoir rocks. *Geophysics* 80 (3): D221–D235. https://doi.org/10.1190/geo2014-0307.1.

12

Fuzzy Logic for Predicting Pore Pressure in Shale Gas Reservoirs With a Barnett Shale Application

Leila Aliouane[1], *Sid-Ali Ouadfeul*[2], *Mohamed Zinelabidine Doghmane*[3], *and Amar Boudella*[3]

[1] *LABOPHYT, Faculty of Hydrocarbons and Chemistry (FHC), University Mhamed Bougara of Boumerdes (UMBB), Boumerdes, Algeria*
[2] *Algerian Petroleum institute, IAP, Boumerdes, Algeria*
[3] *Geophyics Department, FSTGAT, USTHB, Algiers, Algeria*

KEYPOINTS

- Artificial intelligence is used to predict pore pressure in shale gas reservoirs.
- It can be used in case of lack in measurement of the slowness of the *P* wave in horizontal wells.
- Fuzzy logic is used for prediction of pore pressure.
- It can used to replace the Eaton's model

12.1 Introduction

One of the most important tasks in shale gas reservoir characterization is estimating pore pressure from well-log data (Doyen 2007; Sayers 2006; Huffman 2002; Doghmane et al. 2019) for which many models have been proposed in the literature. For example, consider the vertical model, which employs the slowness of the *P* wave and a slowness curve with a constant gradient. Eaton's model (Bowers 1995; Eaton 1975) is another commonly used model for estimating pore pressure; Eaton's (1975) method estimates the vertical component of the effective stress from the seismic velocity v using the following equation:

$$\sigma = \sigma_{\text{Normal}} \left(\frac{V}{V_{\text{Normal}}} \right)^{n} \tag{12.1}$$

σ_{Normal} and V_{Normal} are the vertical effective stress and seismic velocity expected if the sediment is normally pressured, and n is an exponent that describes the sensitivity of velocity to effective stress (Den Boer et al. 2006).

Unconventional Hydrocarbon Resources: Prediction and Modeling Using Artificial Intelligence Approaches, First Edition. Edited by Sid-Ali Ouadfeul.

The pore pressure is then given by:

$$p = S_v - \left(S_v - p_{\text{Normal}}\right) \Big/ \left(\frac{V}{V_{\text{Normal}}}\right)^n \tag{12.2}$$

The deviation of the measured velocity from that of normally pressured sediments (V_{Normal}) as a function of depth is given by Eaton's method as follows:

$$V_{\text{Normal}}(z) = V_0 + k * z \tag{12.3}$$

As previously stated, Eaton's model requires accurate and continuous measurements of the *P* wave slowness (Den Boer et al. 2006). In some cases, and due to acquisition issues such as low compression wave-slowness, this will cause problems with pore pressure estimation. As a result, we propose the use of fuzzy logic for predicting pore pressure from well-log data (Cherana et al. 2022a).

12.2 The Fuzzy Logic

Fuzzy logic is a computing approach that uses "degrees of truth" rather than the traditional "true or false" (1 or 0) Boolean logic on which modern computers are based. Dr. Lotfi Zadeh of the University of California, Berkeley, pioneered the concept of fuzzy logic in the 1960s. Dr. Zadeh was working on a problem involving computer recognition of natural language. Natural language, like most other activities in life, is difficult to translate into absolute terms like 0 and 1. Whether everything is ultimately describable in binary terms is a philosophical question worth exploring, but in practice, much of the data we might want to feed to a computer is in some state in between, and thus they are in general the results of computing (Den Boer et al. 2006).

Fuzzy logic includes not only 0 and 1 as extreme cases of truth (or "the state of matters" or "fact") but also the various states of truth in between so that the result of a comparison between two things could be "0.38 of tallness" rather than "tall" or "short" (Cherana et al. 2022a).

Fuzzy logic appears to be more similar to how our brains work. We aggregate data to form a number of partial truths, which can then be aggregated further to form higher truths. When certain thresholds are exceeded, it can result in additional outcomes such as motor reactions (Cherana et al. 2022a). A process similar to this is used in artificial computer neural networks and expert systems (Doghmane et al. 2022). It may help think of fuzzy logic as the true nature of reasoning, and binary or Boolean logic as a subset of it (Zadeh 1965, 1996).

12.3 Application to Barnett Shale

12.3.1 Geological Context

The Barnett Shale was deposited in North Central Texas today during the late Mississippian Age, during a period of marine transgression caused by the closure of the Iapetus Ocean Basin. The Ouachita Thrust belt began encroaching on the present-day North Texas area by the end of the Pennsylvanian. The thrust belt was formed as a result of the subduction of the South American plate beneath the North American plate. The emergence of the Ouachita Thrust formed

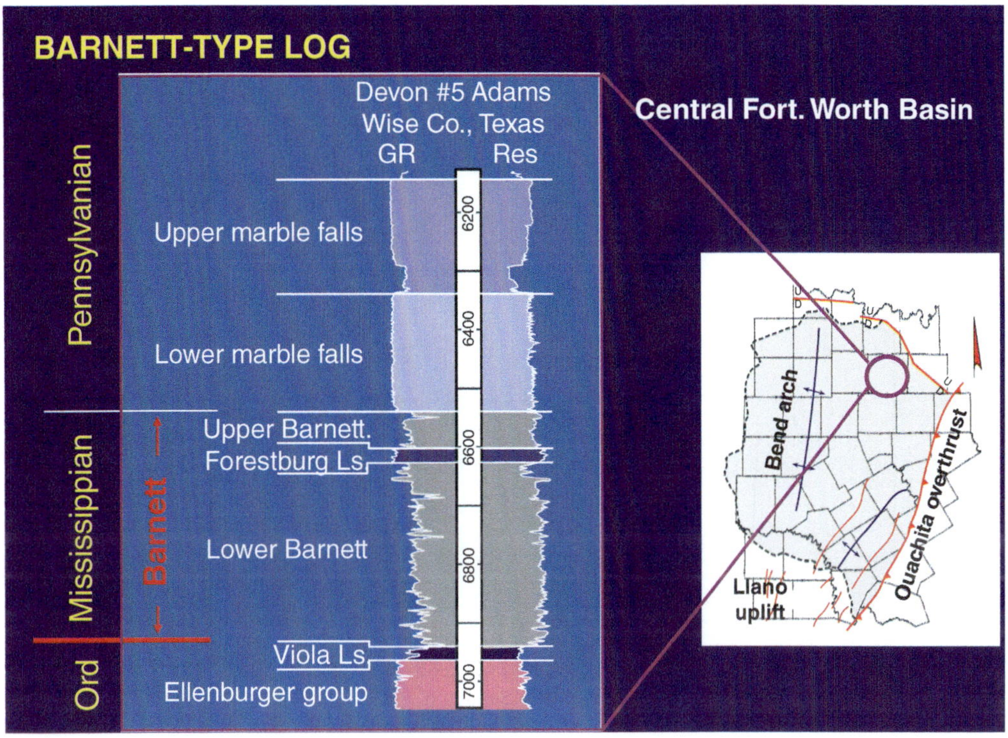

Figure 12.1 Stratigraphic column of Mississippian and the Pennsylvanian ages. *Source:* Sayers et al. (2006).

the foreland basin along the thrust's front. Early studies of the basin attributed Barnett's thermal maturation to burial history and thermal regimes associated with burial depth. Figure 12.1 depicts the stratigraphic column of Mississippian and Pennsylvanian ages; the targeted shale gas reservoir is the lower Barnett, with a top elevation of 6650 ft (Givens and Stewart 2004).

12.3.2 Data Processing

To test the efficiency of artificial intelligence in predicting pore pressure in the absence of measuring the slowness of the *P* wave, well-logs data from two horizontal wells drilled in the Barnett Shale were used (Sayers et al. 2002). Data from the first well H01 were divided into two depth intervals, the first of which is from 6650–7500 ft, and were used for training the fuzzy machine (see Figure 12.2), but data from the second depth interval 7500–8180 ft were used for validation (Figure 12.4). To test the efficiency of artificial intelligence in predicting pore pressure in the absence of measuring the slowness of the *P* wave, well-logs data from two horizontal wells drilled in the Barnett Shale were used. Data from the first well H01 were divided into three depth intervals, the first of which is from 6650–7500 ft, and were used for training the fuzzy machine (Cherana et al. 2022b), but data from the second depth interval 7500–8180 ft were not used. For the generalization, data from the second depth interval, as shown in Figure 12.4, were used. The obtained pore pressure by fuzzy logic compared to the calculated pore pressure by Eaton's model (Figures 12.3 and 12.5) clearly shows the proposed tool's ability to provide pore pressure in the absence of *P* wave slowness measurement.

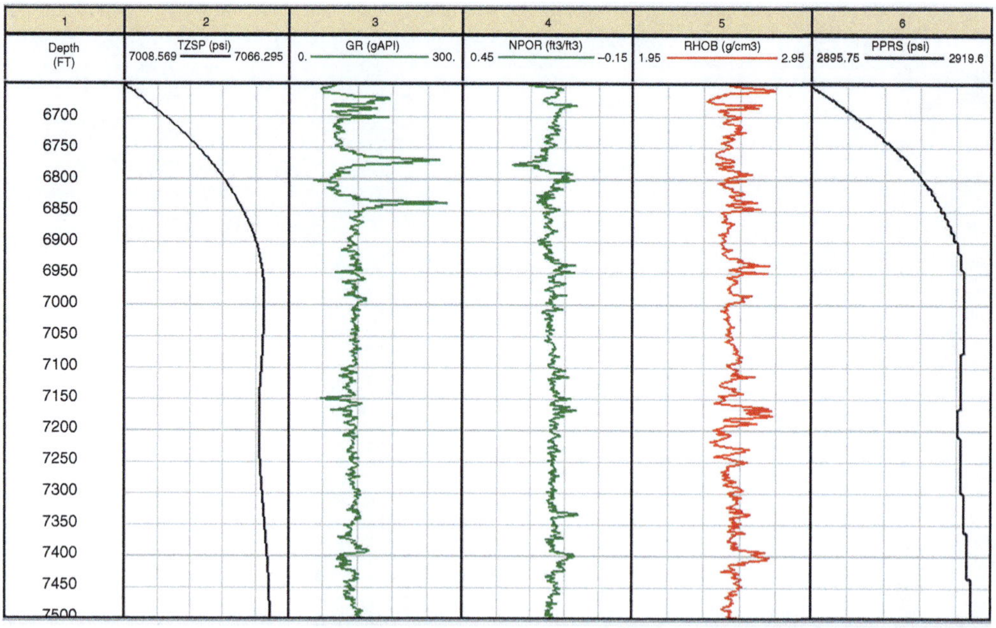

Figure 12.2 Well-logs data of the horizontal well H01 for depth interval 6650–7500 ft: (01) Overburden stress, (02) gamma ray, (03) neutron porosity, (03) bulk density, (04) pore pressure calculated by using the Eaton's model.

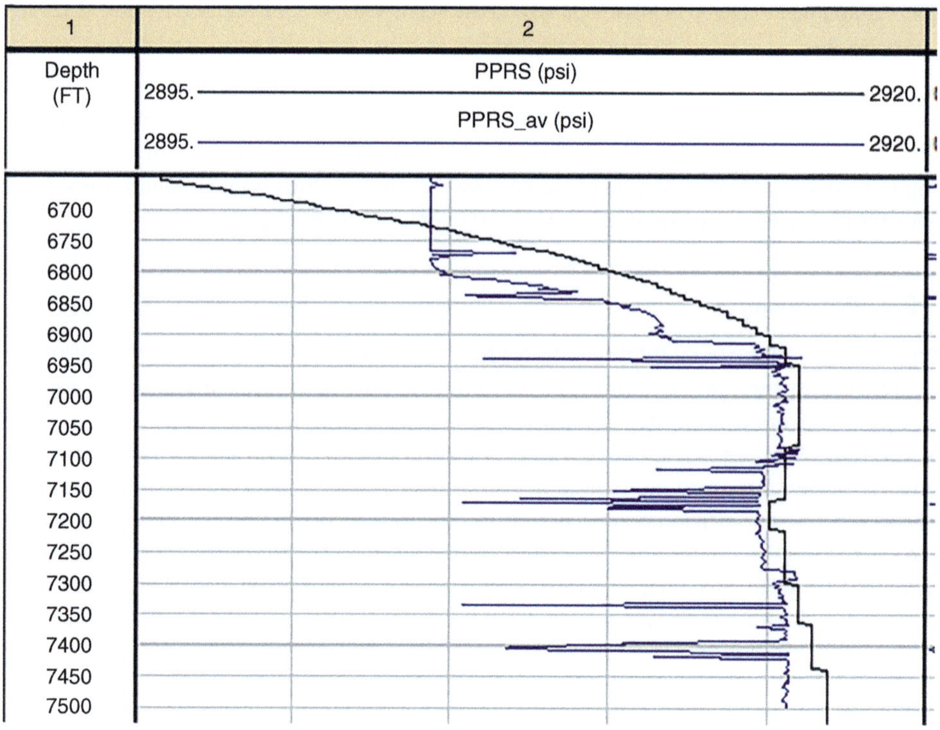

Figure 12.3 Calculated pore pressure by using the Fuzzy logic (blue color) compared with the calculated pore pressure by using Eaton's model for the depth interval of 6500–7500 ft.

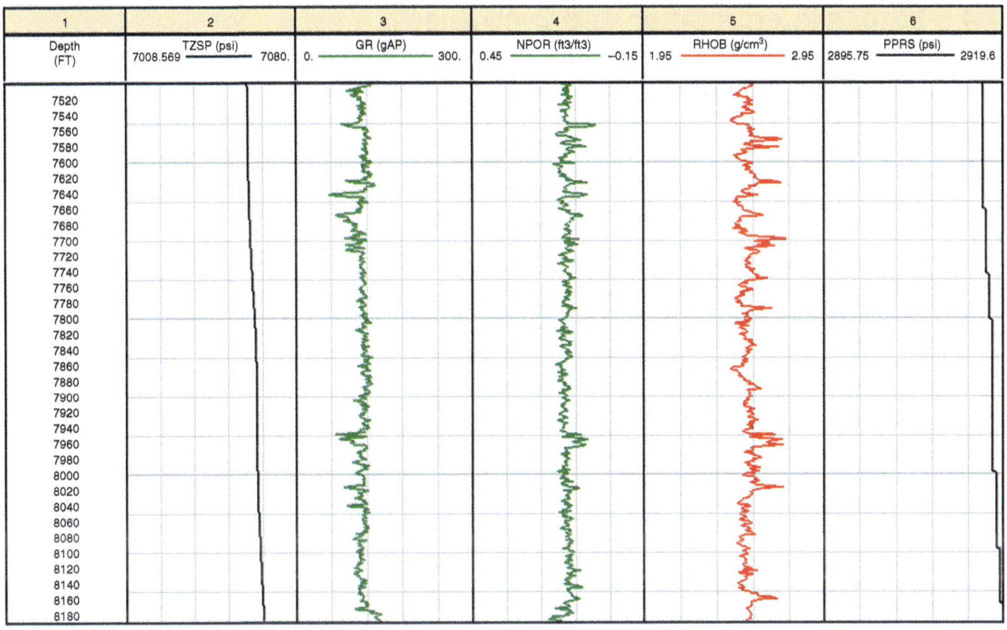

Figure 12.4 Well-logs data of the horizontal well H01 for depth interval 7500–8180 ft: (01) Depth (ft), (02) Overburden stress (Terzaghi 1940), (03) Gamma ray, (04) Neutron porosity, (05) Bulk density, (06) Pore pressure calculated by using Eaton's model.

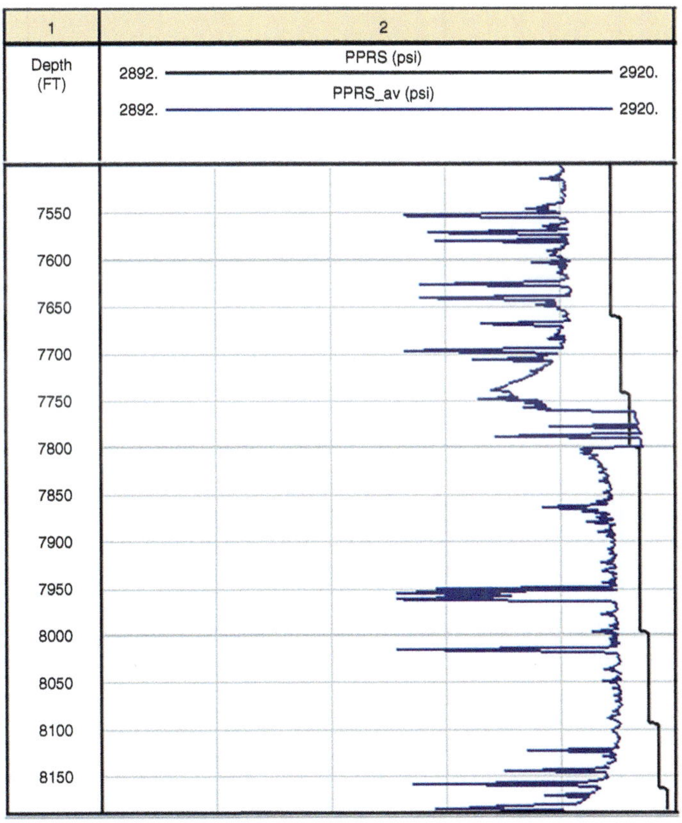

Figure 12.5 Calculated pore pressure by using the Fuzzy logic (blue color) compared with the calculated pore pressure by using Eaton's model for the depth interval 7500–8180 ft.

12.4 Results Interpretation and Conclusions

To test the efficiency of fuzzy logic in predicting pore pressure and replacing Eaton's model in case of a problem in *P* wave slowness measurement, data of the second horizontal well H02 (see Figure 12.6) are propagated through the fuzzy machine. Figure 12.7 shows the predicted pore pressure compared to the estimated pore pressure using Eaton's model; they demonstrate the efficiency of the fuzzy logic method and its power to replace Eaton's model.

Furthermore, the same procedures were carried out with the aid of a multilayer perceptron neural network machine. The obtained results demonstrate the limitations of the multilayer perceptron neural network machine in predicting pore pressure from well-log data when compared to fuzzy logic (Doghmane et al. 2022). As a result, we propose applying the entire process to a large dataset to test the efficiency of the fuzzy logic in pore pressure prediction. It is possible to implement an intelligent machine capable of replacing Eaton's model by implementing our proposed process in this chapter.

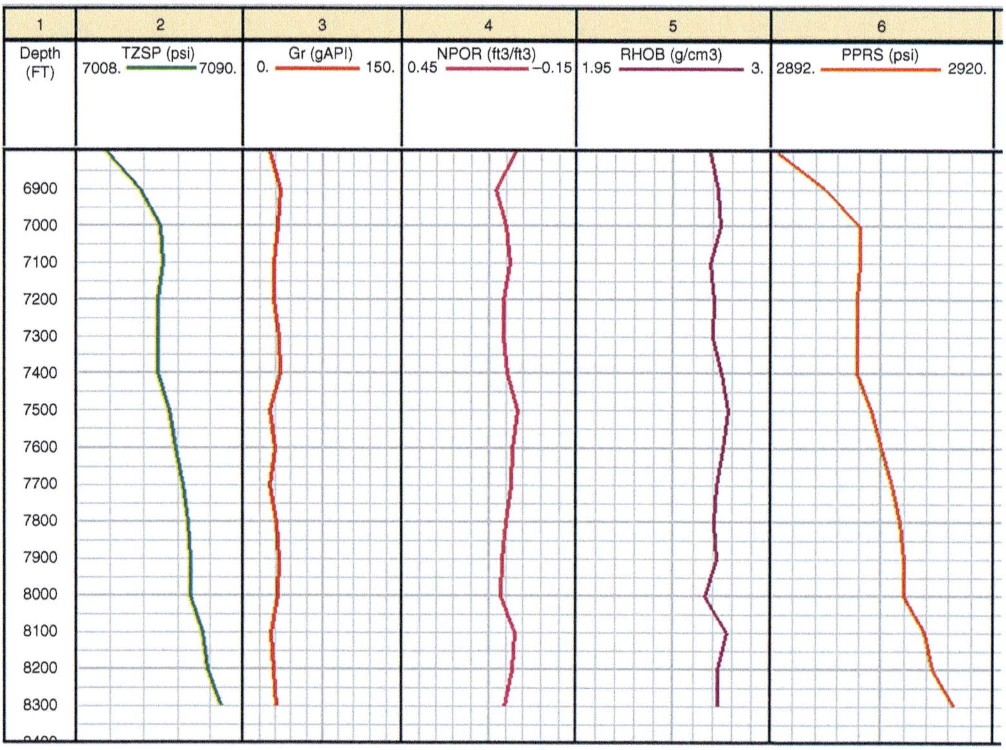

Figure 12.6 Well-logs data of the horizontal well H02: (01) Depth (ft), (02) Overburden stress, (03) Gamma ray, (04) Neutron porosity, (05) Bulk density, (06) Pore pressure calculated by using Eaton's model.

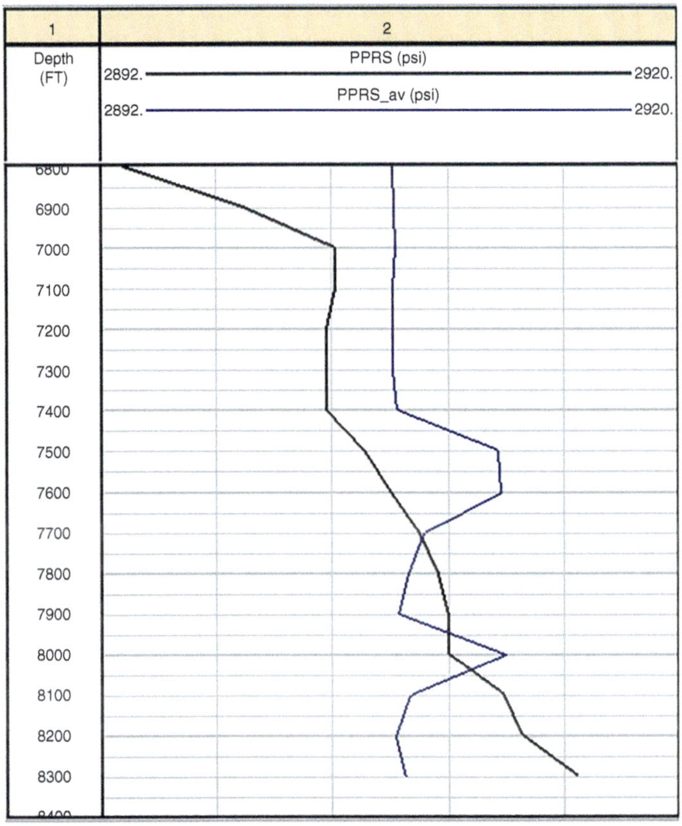

Figure 12.7 Calculated pore pressure using the fuzzy logic (blue color) and Eaton's model (black color) for the horizontal well H02.

References

Bowers, G. I. (1995). Pore pressure estimation from velocity data: accounting for pore pressure mechanisms besides undercompaction. *SPE Drilling and Completion*, 10, pp. 89–95. DOI: https://doi.org/10.2118/27488-PA

Cherana, A., Aliouane, L., Doghmane, M.Z. et al. (2022a). Lithofacies discrimination of the Ordovician unconventional gas-bearing tight sandstone reservoirs using a subtractive fuzzy clustering algorithm applied on the well log data: Illizi Basin, the Algerian Sahara. *Journal of African Earth Sciences* 196: https://doi.org/10.1016/j.jafrearsci.2022.104732.

Cherana, A., Aliouane, L., Doghmane, M., and Ouadfeul, S.A. (2022b). Fuzzy Clustering Algorithm for Lithofacies Classification of Ordovician Unconventional Tight Sand Reservoir from Well-Logs Data (Algerian Sahara). In *Advances in Geophysics, Tectonics and Petroleum Geosciences. CAJG 2019. Advances in Science, Technology & Innovation*. Cham: Springer. https://doi.org/10.1007/978-3-030-73026-0_64.

Den Boer, L. D.; Sayers, C. M.; Nagy, Z. R.; Hooyman, P. J. & Woodward, J. (2006). Pore pressure prediction using well-conditioned seismic velocities. *EAGE Publications*, 24, pp. 43–49. DOI: https://doi.org/10.3997/1365-2397.2006012.

Doghmane, M.Z, Belahcene, B., and Kidouche, M. (2019). Application of improved artificial neural network algorithm in hydrocarbons' reservoir evaluation. *Proc. International conference in Artificial Intelligence in Renewable Energetic Systems ICAIRES 2018: Renewable Energy for Smart and Sustainable Cities*, Algeria (24–26 November 2019), pp. 129–138.

Doghmane, M.Z., Ouadfeul, S.A., Benaissa, Z., Eladj, S. (2022). Classification of ordovician tight reservoir facies in algeria by using neuro-fuzzy algorithm. In: Hatti, M. (eds) *Artificial Intelligence and Heuristics for Smart Energy Efficiency in Smart Cities. IC-AIRES 2021*. Lecture Notes in Networks and Systems, vol 361. Springer, Cham. https://doi.org/10.1007/978-3-030-92038-8_91.

Doyen, P. M. (2007). Seismic reservoir characterization an earth modelling perspective. *EAGE Publications.* DOI: https://doi.org/10.3997/9789462820234

Eaton, B.A. (1975). The equation for geopressure prediction from well logs. *Paper presented at the Fall Meeting of the Society of Petroleum Engineers of AIME*, Dallas, Texas (September 1975). doi: https://doi.org/10.2118/5544-MS.

Givens, Z. and Stewardt, D. (2004). The Barnett shale: not so simple after all. *Publications-West Texas Geological Society* 4: 187.

Huffman, A.R. (2002). The future of pressure prediction using geophysical methods, AAPG Special Volumes, AAPG Memoir 76, Chapter 19, Pages 217–233.

Sayers, C. (2006). Predrill pore-pressure prediction using seismic data. *Geophysics* 67 (4): 1286–1292. https://doi.org/10.1190/1.1500391.

Sayers, C.M., Johnson, G.M., and Denyer, G. (2002). Predrill pore pressure prediction using seismic data. *Geophysics* 67: 1286–1292.

Colin M. Sayers, Lennert D. den Boer, Zsolt R. Nagy, Patrick J. Hooyman; Well-constrained seismic estimation of pore pressure with uncertainty. *The Leading Edge* 2006; 25 (12): 1524–1526. doi: https://doi.org/10.1190/1.2405338.

Terzaghi, R.A.D. (1940). Compaction of lime mud as a cause of secondary structure. *Journal of Sedimentary Research* 10 (2): 78–90. https://doi.org/10.1306/D426908F-2B26-11D7-8648000102C1865D.

Zadeh, L.A. (1965). Fuzzy sets. *Information and Control* 8 (3): 338–353.

Zadeh, L.A. (1996). *Fuzzy Sets, Fuzzy Logic, Fuzzy Systems*. World Scientific Press.

13

Using Well-Log Data, a Hidden Weight Optimization Method Neural Network Can Classify the Lithofacies of a Shale Gas Reservoir: Barnett Shale Application

Leila Aliouane[1], Sid-Ali Ouadfeul[2], Mohamed Z. Doghmane[3], and Ammar Boudella[3]

[1] *LABOPHYT, Faculty of Hydrocarbons and Chemistry, University M'hamed Bougara of Boumerdes, Boumerdes, Algeria*
[2] *Algerian Petroleum Institute, Sonatrach, Boumerdes, Algeria*
[3] *Department of Geophysics, FSTGAT, University of Science and Technology Houari Boumediene, Algiers, Algeria*

13.1 Introduction

Geologists and petrophysicists use their years of knowledge to perform lithologic interpretations of log data. Despite the use of petrophysical calculations, interpretation is mostly relied on qualitative visual data. The interpreters look at curve forms, identifying lithologic boundaries as deflections in the log, and base their interpretation on how those deflections respond in relation to one another. While interpreting a new well, the log analyst may refer to a conceptual model, such as a series of lithologies, as a reference. However, the foundation of this strategy is usually the identification of curve shapes in the well logs utilizing a collection of well logs to interpret the lithologies encountered in the well, frequently in a qualitative manner, typically for stratigraphic purposes and reservoir characterization.

The same method based on well well-log data can be applied to classify the litholology of shale reservoirs whether the well is horizontal or vertical. One of the geophysical issues is finding new shale gas reservoirs since they have poor petrophysical characteristics like low porosity and low permeability (Aliouane and Ouadfeul 2014a; Ouadfeul and Aliouane 2015). Typically, geophysicists search for "sweet spots" in shale gas deposits, or depth intervals with favorable geochemical and geomechanical characteristics (Aliouane and Ouadfeul 2014a). We search for regions with high Young's modulus and low Poisson's ratio in terms of geomechanical characteristics. The rock matrix in geology should have a high proportion of quartz and a low proportion of clay. The pre-identification of lithofacies before drilling is necessary for effective shale gas exploration; it can help save time and prevent drilling in soft rocks (Glaser et al. 2013).

The rapidly evolving fields of applied artificial intelligence and pattern recognition include the building blocks for creating a computer-aided interpreting system and many techniques for pattern recognition include feature extraction, data clustering, and group categorization are used in order to improve the lithofacies classification from well-logs data of sandy shale reservoir. We can cite for example the paper of Ouadfeul and Aliouane (2012), authors of the the paper proposed combining the multilayer perceptron (MLP) and the self-organizing map neural networks Maiti et al. in 2007, suggested neural network model for lithofacies classification and modeling using the Germann

Unconventional Hydrocarbon Resources: Prediction and Modeling Using Artificial Intelligence Approaches, First Edition. Edited by Sid-Ali Ouadfeul.

KTB borehole data. A paper by Kraipeerapun et al. (2007) describes the categorization of lithofacies from well-log data using neural networks, interval neutrosophic sets, and uncertainty quantification. The Appleton Field, an oil field in Escambia County, Alabama, is the site of Chang et al. suggested's. lithofacies detection using multiple adaptive resonance theory neural networks and a group decision expert system.

Zhao et al. (2014) suggested using proximal support vector machines to classify lithofacies in the Barnett Shale. The obtained results show that the PSVM lithofacies classification showed promising results in both seismic and well-log data. Wang et al. (2014) used the support vector machine classifier for identifying organic-rich Marcellus Shale lithofacies in the Appalachian basin. This proved the reliability of the PSVM classifier in the binary categorization of limestone and shale.

In this chapter, the lower Barnett Shale formation's lithofacies have been predicted using an artificial neural network technique. In order to do this, we start the paper by introducing the MLP artificial neural network and the hidden weight optimization (HWO) training technique. The application using recordings of the petrophysical parameters of the Barnett Shale is demonstrated in the second half, which also describes the geological setting of the formation. The classification of lithofacies is implemented using a neural network model. Results analysis and a conclusion are provided, demonstrating how effective artificial intelligence is in characterizing reservoirs.

13.2 Artificial Neural Network

One of the most helpful intelligent systems in geophysics is the artificial neural network (Aliouane et al. 2011, 2012; Ouadfeul and Aliouane 2016a,b, 2017). A neural network is made up of an ensemble of interconnected units, or neurons, and each pair of connected neurons is related to the other by a connection weight. We advise readers to read the book edited by Chi Leung Patrick Hui for additional information about neural networks and their applications. Each neuron in the system receives a signal or information and transforms it using a transfer function (2011).

Pattern recognition, signal processing, remote sensing, and geosciences are just a few of the domains where MLP neural networks have seen extensive use (Aliouane and Ouadfeul 2014b; Aliouane et al. 2013a). A significant class of neural networks is the MLP, also referred to as the universal approximator (Sifaoui et al. 2008). The MLP neural network typically consists of an input layer, an output layer, and numerous hidden layers. The goal is to find the optimal learning method that can produce outcomes that are as close to reality as possible. Many learning algorithms are offered in the literature.

We can list the conjugate gradient and the HWO method as learning algorithms; the reader is invited to read Riedmiller's paper for more information on these learning algorithms (1994).

13.3 Hidden Weight Optimization Algorithm Neural

Typically, the backpropagation technique is used to train MLP neural networks, with linear equations being solved to determine output weights and backpropagation being utilized to determine hidden weights (those which feed into the hidden units). Unfortunately, updating hidden weights through backpropagation is not very successful (Gopalakrishnan et al. 1993; Chen et al. 1999). In this study, we have modified the idea of finding the hidden weights by minimizing a different error function for each hidden unit, and we have called this method HWO (Chen et al. 1999; Aliouane et al. 2013b). The backpropagation algorithm is less efficient than the HWO method. (Sheikhan and Shabani 2009).

13.4 Geological Context of the Barnett Shale

In the Fort Worth basin, there is a peculiar reservoir called the Barnett Shale. The Ouachita Thrust Belt, a structural arch (bend arch), and the Llano Precambrian uplift surrounded the foreland basin, which was formed along the convergent margin of the tectonic plates (See Figure 13.1a). The basin is asymmetrical, with the thrust front's adjacent axis being the structurally deepest. An island-arc system that surrounded the Fort Worth Basin contributed some coarse-grained sediment to the Barnett Shale formation. There are a number of naturally occurring cracks in the Barnett that are more prevalent in interbred limestone, tend to be mineralized, and (Aguilar and Verma 2014). The tectonic movements connected to the Ouachita orogeny created these sets (Bruner and Smosna 2011). Karsts in the underlying Ellenberger cause localized subsidence in the Barnett Shale, high-angle faults, and karst chimneys. These karst failures have less of an impact on the Barnett in our study region (Loucks and Ruppel 2007). The lower Barnett is deposited over this unconformity because the Viola limestone has an erosional surface toward the NE. Additionally, the Forestburg limestone, which thins and vanishes toward the SE, divides the Barnett Shale into two units: the lower and the upper half (see Figure 13.1b). According to Roderick Perez's (2013) analysis of Figure 13.2's mineralogy and total organic carbon, we can see that the lower Barnett is distinguished by high to good TOC. The lower Barnett is characterized by a high Quartz content, which is advantageous for hydraulic fracture, according to the mineralogy tracks. Although the amount of calcite is quite little, the clay concentration is typical. The lower Barnett's briskness score is very high, and all of these characteristics make the Barnett a good shale play (Jarvie et al. 2007).

According to Roderick Perez's (2013) analysis of Figure 13.2's mineralogy and total organic carbon, we can see that the lower Barnett is distinguished by high to good TOC. The lower Barnett is characterized by a high quartz content, which is advantageous for hydraulic fracture, according to the mineralogy tracks. Although the amount of calcite is quite little, the clay concentration is typical. The lower Barnett's briskness score is very high, and all of these characteristics make the Barnett a good shale play (Jarvie et al. 2007).

Wang and Carr (2012) predicted the Marcellus Shale lithofacies by multiclass neural network classification in the Appalachian Basin, and they demonstrated that the predicted shale lithofacies provides an opportunity to construct a three-dimensional shale lithofacies model in sedimentary basins using an abundance of conventional well-logs. Shale gas lithofacies classification using neural networks is becoming an important area of research. The three-dimensional shale lithofacies model is beneficial for planning plans for horizontal drilling and hydraulic fracture when combined with reservoir pressure, maturity, and natural fracture system.

Geologists are constantly looking for new and more effective ways to analyze huge amounts of well data, and as a result, mathematical approaches for correlation and interpretation of well logs are emerging quickly. The majority of log analyzers have been unable to employ many methodologies, mostly due to challenges in getting and working with digital log data. This will change quickly as computer technology progresses, enabling high-speed processing, interactive visuals, and substantial data storage. The percentage of mineral fractions from the lithologic model can be calculated mathematically (Crain and Aharipour 2016). In this instance, the ELAN model (Quirein et al. 1986) To predict lithofacies, two Barnett Shale well recordings of petrophysical data are used. Well H01 and well H02 are these:

Utilizing well-logs from two wells, the multilayer perceptron (MLP) neural network is implemented: Data from well H02 are used for generalization, whereas well H01 is a pilot well and its data are used to train the MLP's design. Figures 13.3 and 13.4 show the raw well-logs data of the two wells. Figure 13.5 shows the three layers that make up this structure: the input layer, which has five neurons that correspond to the five well-logs of data, the hidden layer, and the output

(a)

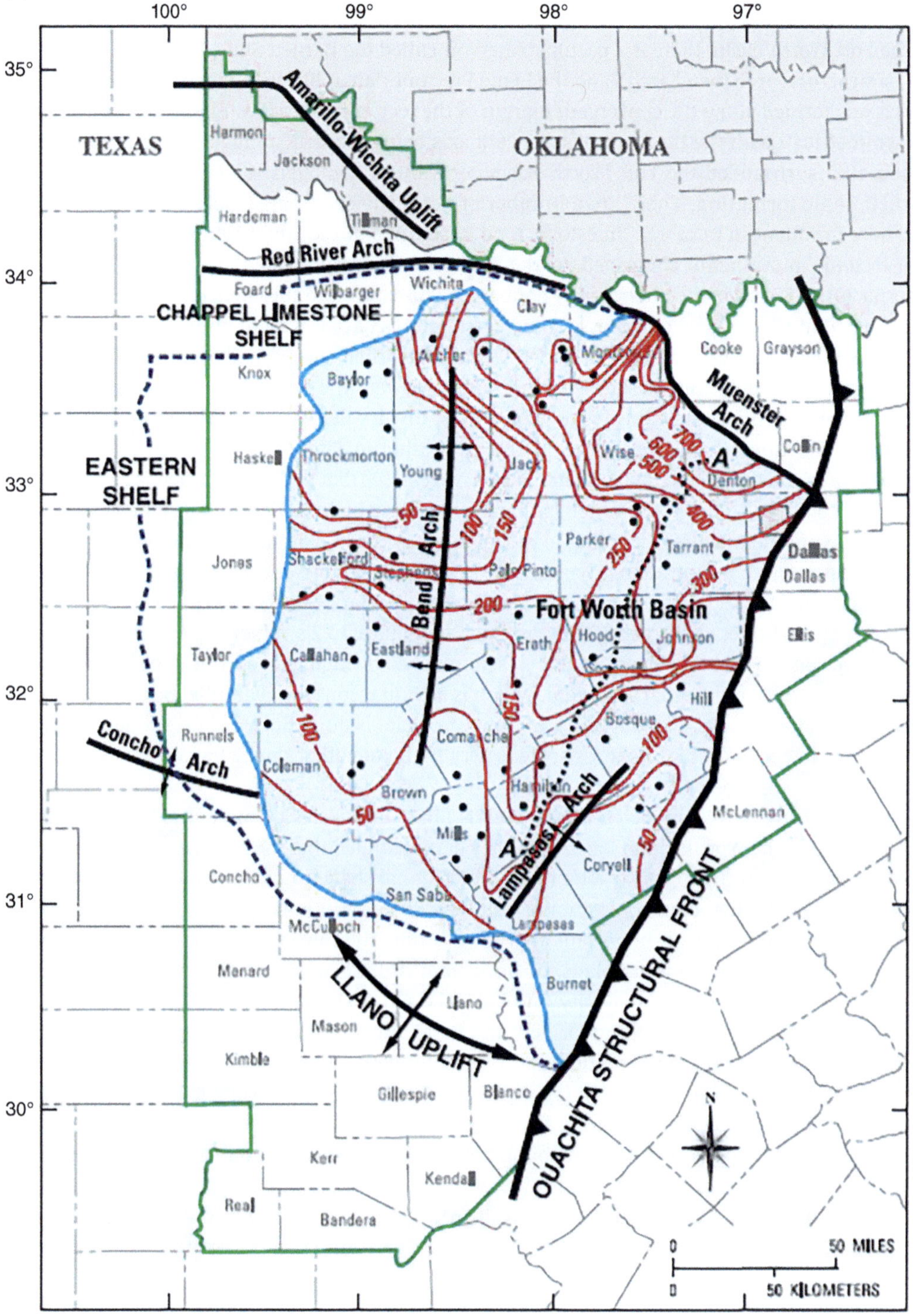

Figure 13.1 (a) Isopach map of the Barnett Shale demonstrating thickness. *Source:* Bruner and Smosna (Bruner and Smosna 2011)/US Department of Energy/Public Domain. (b) Stratigraphy of the Barnett Shale. *Source:* Adapted from Loucks and Ruppel (2007).

(b)

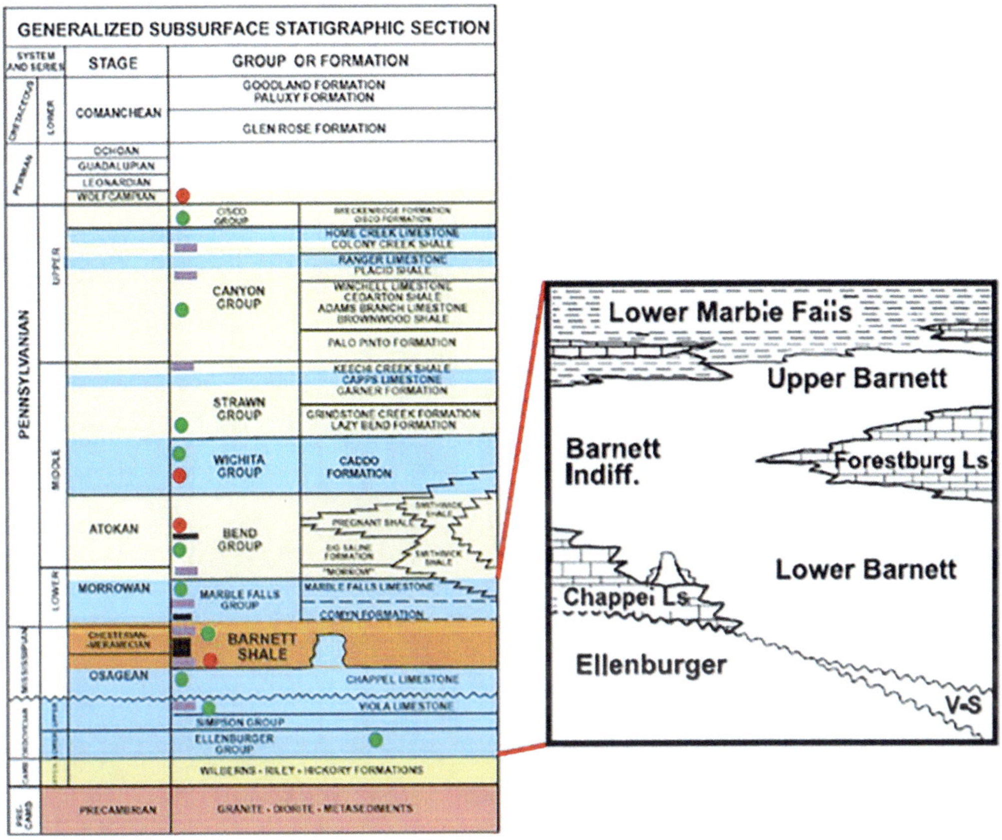

Figure 13.1 (Continued)

layer, which is made up of two neurons that correspond to VCL and VQUA. The neural model was trained using the HWO algorithm through supervised learning utilizing the desired output VCL and VQUA from well H01. Clay and quartz volume for the horizontal well H01 as calculated by the ELAN model is shown in Figure 13.6.

When propagating the connection, the weight of the connection will be derived from the training phase using the desired output (Figure 13.7). A comparison of the predicted and desired outputs is shown in Figure 13.7. This chart makes it quite evident that the training was carried out flawlessly; the root mean square of these outputs is 0.025.

Figures 13.8a and b compare the generated clay volume using the ELAN model to the anticipated clay volume using neural networks (black color). Between the two clay volumes, the root mean square is 0.075.

13.5 Results Interpretation and Conclusions

Figure 13.8b is a cross plot of the predicted clay volume versus the derived volume using the ELAN model, and it shows that the proposed MLP machine is capable of predicting the clay volume with high accuracy. The correlation coefficient between the two volumes is 0.94. Figure 13.8 compares

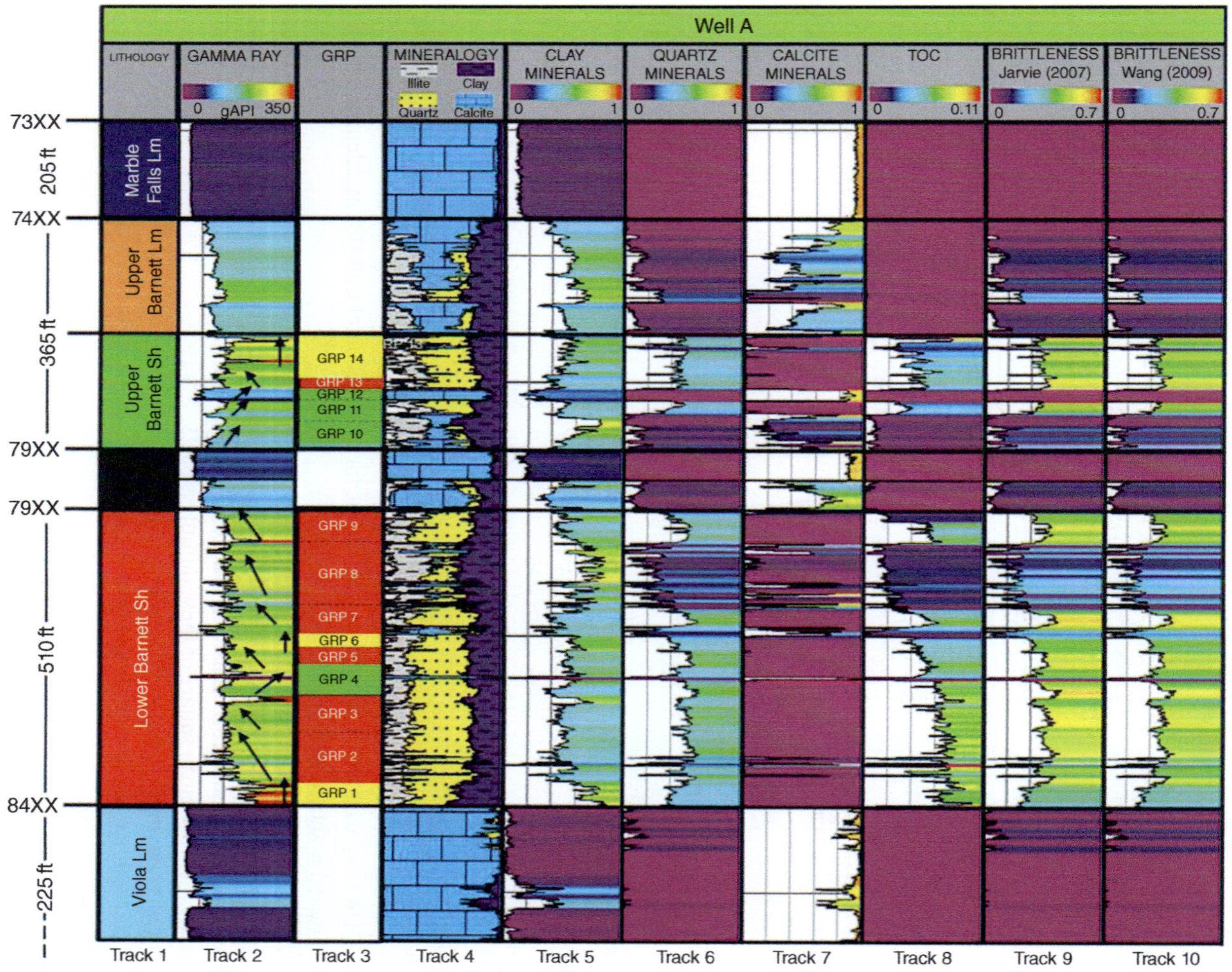

Figure 13.2 Stratigraphic column of the Mississippian, b Mineralogy and TOC content of the lower Barnett after Perez.

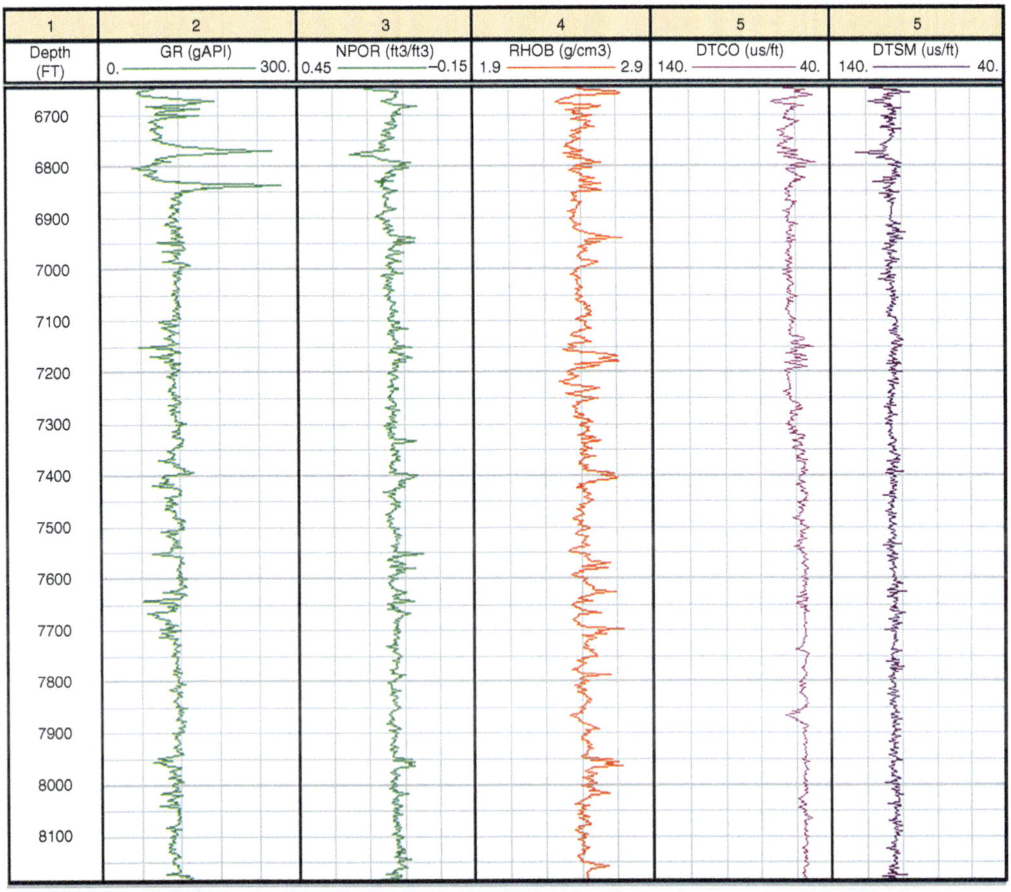

Figure 13.3 Raw well-logs data of a horizontal well drilled in the lower Barnett Shale formation, (02) Gamma ray, (03) Neutron porosity, (04) Bulk density, (05) Slowness of the *P* wave, (06) Slowness of the Shear wave.

the predicted and estimated clay volume using the ELAN model. This outcome demonstrates the MLP neural network's capability to forecast clay volume, a crucial factor in the localization of sweet spots in shale gas reservoirs. In the instance of a lake in elementary analysis, artificial intelligence tools like neural networks can be quite useful for providing information on shale plays (ELAN). For the same well, Figure 13.9 plots the anticipated quartz volume versus depth.

Predicted VCL and VQUA results demonstrate the effectiveness of using an MLP neural network with an HWO approach.

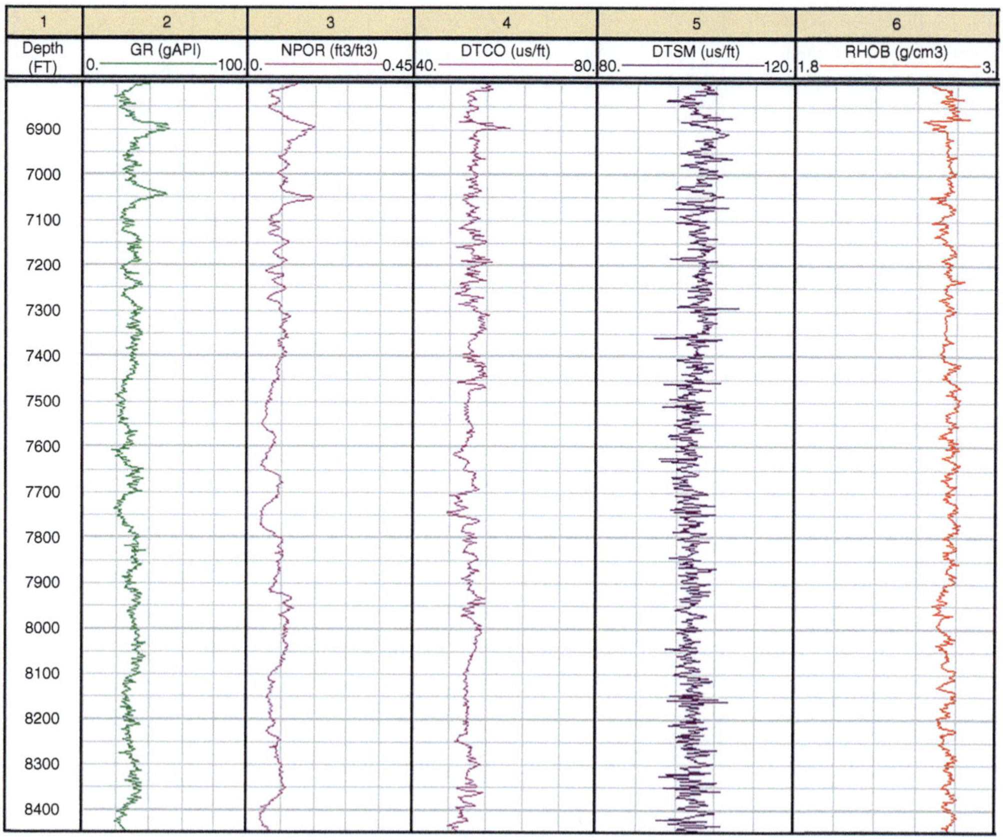

Figure 13.4 Raw well-logs data of a horizontal well drilled in the lower Barnett Shale formation near the pilot well H02. (02) Gamma ray, (03) Neutron porosity, (04) Bulk density, (05) Slowness of the *P* wave, (06) Slowness of the shear wave.

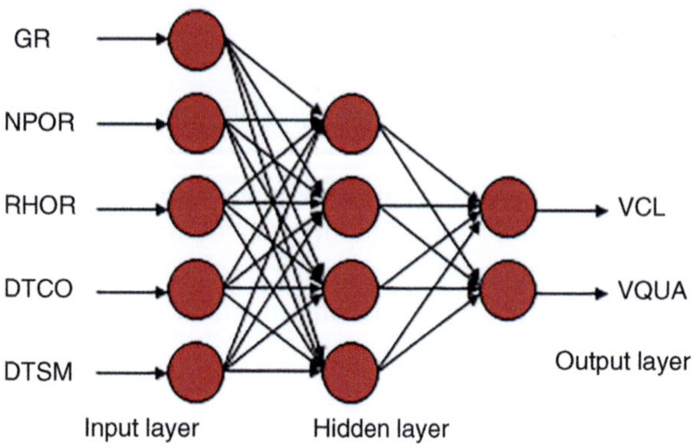

Figure 13.5 MLP neural network architecture.

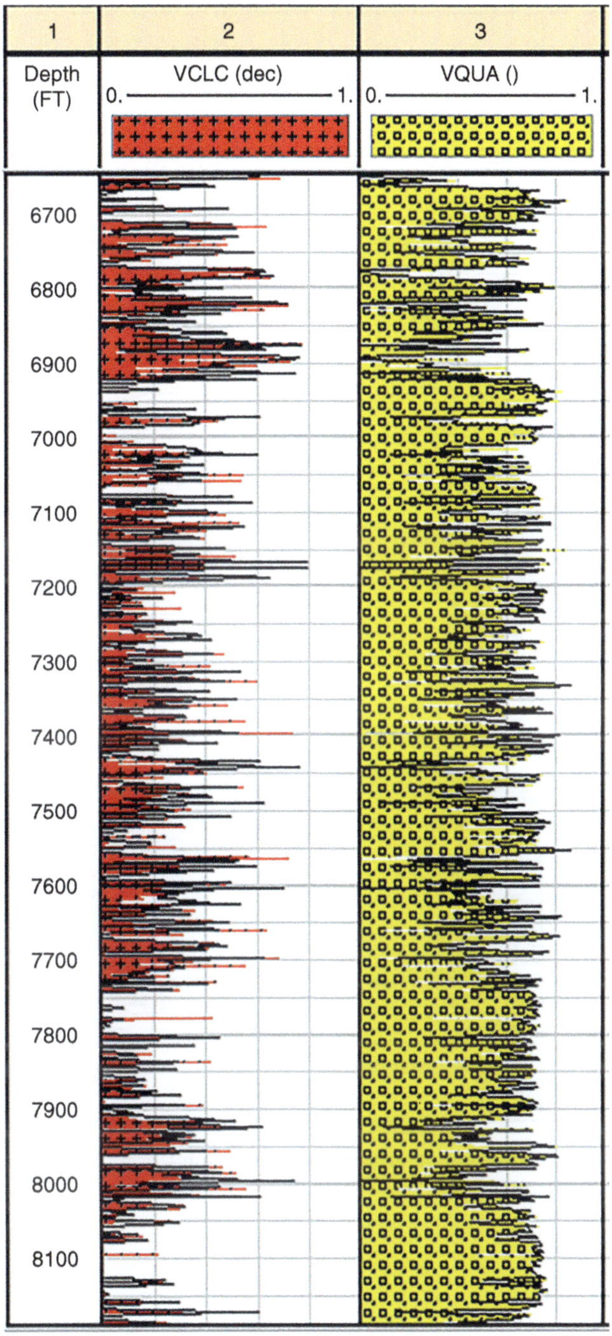

Figure 13.6 Clay and quartz volume for the horizontal well H01 derived using the ELAN model.

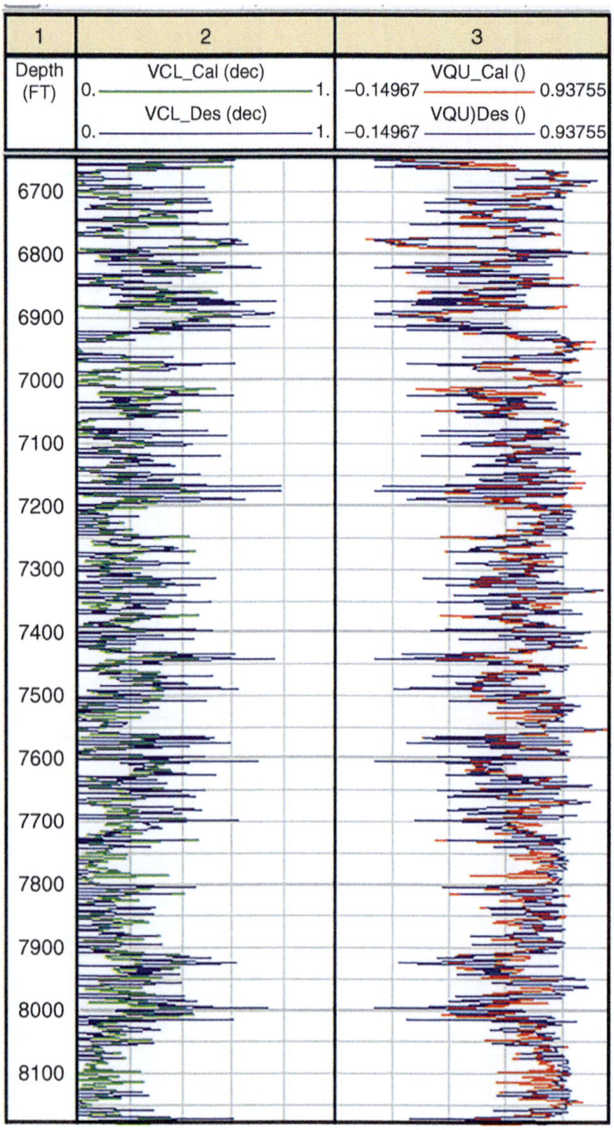

Figure 13.7 Calculated clay and quartz volume using MLP neural network model compared with the desired outputs.

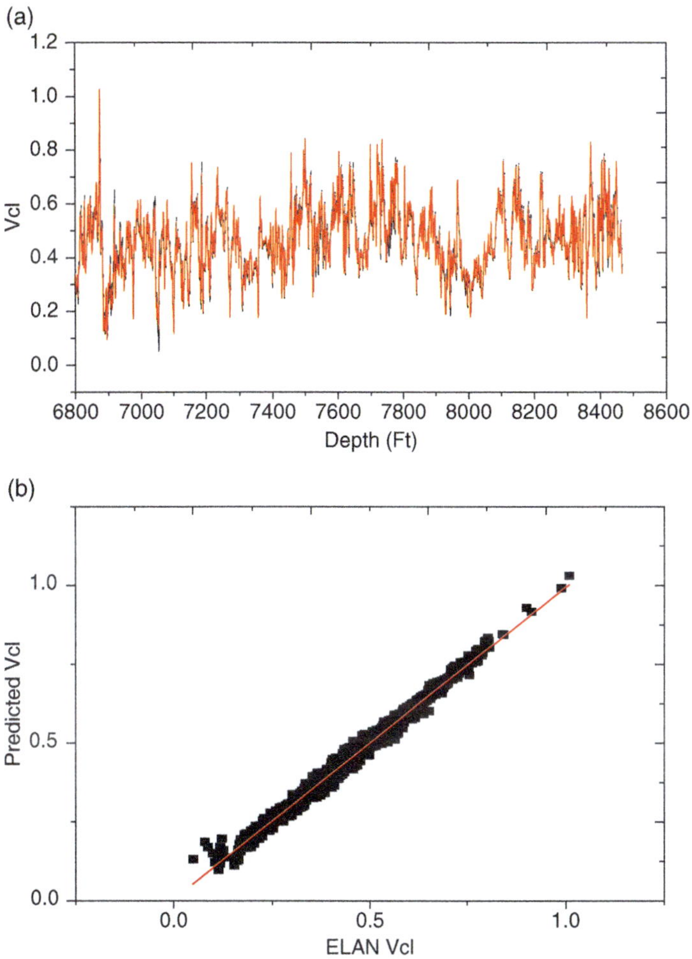

Figure 13.8 (a) Predicted clay volume for the well H02 using MLP neural network (red color) compared with the derived clay volume using the ELAN model (black color). (b) A crossplot of predicted clay volume versus ELAN clay volume.

Figure 13.9 Predicted quartz volume using the MLP neural network for the horizontal well H02.

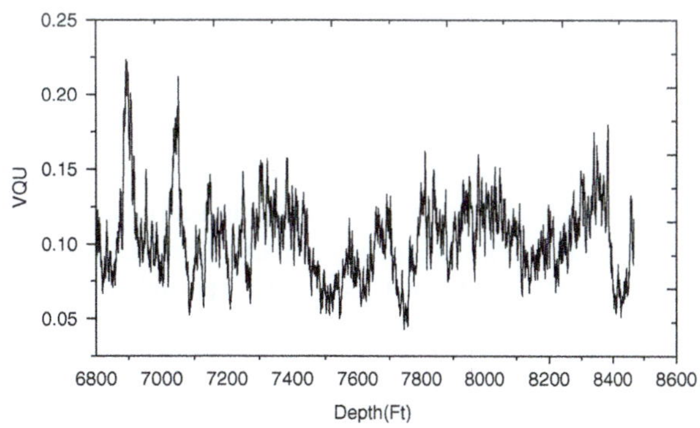

Bibliography

Aguilar, M. and Verma, S. (2014). TOC and fracture characterization of the Barnett Shale with predicted gamma ray and density porosity volumes. *SEG Technical Program Expanded Abstracts* 2662–2666. https://doi.org/10.1190/segam2014-1646.1.

Aliouane, L. and Ouadfeul, S. (2014a). Sweet spots discrimination in shale gas reservoirs using seismic and well-logs data. A case study from the Worth basin in the Barnett shale. *Energy Procedia* 59: 22–27.

Aliouane, L. and Ouadfeul, S. (2014b). *Permeability Prediction Using Artificial Neural, Network. A Comparative Study Between Back propagation and Levenberg-Marquardt*, Lecture Notes in Earth System Sciences, 653–657. Springer.

Aliouane, L. Ouadfeul, S., Djarfour, N., and Boudella, A., fevrier (2013a). Lithofacies prediction from well logs data using different neural network models. *The International Conference on Pattern Recognition: Applications and Methods*, Barcelona, Spain (15–18 February 2013).

Aliouane, L., Ouadfeul, S., and Boudella, A. (2013b). A modified hidden weight optimization algorithm based neural network model for permeability prediction from well-logs data. *Lectures Notes on Computer Sciences-Springer* 8226: 498–502.

Aliouane, L., Ouadfeul, S., Djarfour, N., Boudella, A., 2012, *Petrophysical Parameters Estimation from Well-Logs Data Using Multilayer Perceptron and Radial Basis Function Neural Networks*, Lecture Notes in Computer Science Volume 7667, pp 730–736, Springer. Doi: https://doi.org/10.1007/978-3-642-34500-5_86.

Aliouane, L., Ouadfeul, S., Boudella, A., 2011, Fractal analysis based on the continuous wavelet transform and lithofacies classification from well-logs data using the self-organizing map neural network, *Arabian Journal of geosciences*, doi: https://doi.org/10.1007/s12517-011-0459-4.

Bruner, K.-R. and Smosna, R. (2011). A Comparative Study of the Mississippian Barnett Shale, Fort Worth Basin, and Devonian Marcellus Shale, Appalachian Basin, DOE/NETL-2011/1478.

Crain, E. R. and Aharipour, H. (2016). Characterization of Tight and Shale Unconventional Reservoirs from Logs, SPE-180961-MS.

Chen, H.H., Manry, M.T., and Chandrasekaran, H. (1999). A neural network training algorithm utilizing multiple sets of linear equations. *Neurocomputing*, Elsevier 25: 55–72.

Glaser, K.S., Miller, C.-K., Johnson, G.M., and Toelle, B.-E. (2013). Seeking the sweet spot: reservoir and completion quality in organic shales. *Oilfield Review* 25 (4): 16–29.

Gopalakrishnan, A., Jiang, X., Chen, M.-S. et al. (1993) Constructive proof of efficient pattern storage in the multilayer perceptron. *Conf. Record of the 27th Annual Asilomar Conf. on Signals, Systems, and Computers*, Pacific Grove (1–3 November 1993), vol. 1, pp. 386–390.

Jarvie, D.M., Hill, R.J., Ruble, T.E., and Pollastro, R.M. (2007). Unconventional shale – gas systems: the Mississippian Barnett Shale of north – central Texas as one model for thermogenic shale gas assessment. *AAPG Bulletin* 91: 475–499. https://doi.org/10.1306/12190606068.

Kraipeerapun, P., Fung, C., and Wong, K. (2007). Lithofacies classification from well log data using neural networks, interval neutrosophic sets and quantification of uncertainty. *International Journal of Applied Mathematics and Computer Sciences* 3 (1): 28–33.

Loucks, R.-G., and. Ruppel, S.-C., 2007, Mississippian Barnett shale: lithofacies and depositional setting of a deep-water shale-gas succession in the Fort Worth Basin, Texas, *AAPG Buletin*, v. 91 no. 4 p. 579–601.

Ouadfeul, S. and Aliouane, L. (2017). *Random Noise Attenuation from GPR Data Using Radial Basis Function Neural Network*. SEG Digital library.

Ouadfeul, S. and Aliouane, L. (2016a). Total organic carbon estimation in shale-gas reservoirs using seismic genetic inversion with an example from the Barnett Shale. *The Leading Edge* https://doi.org/10.1190/tle35090936.1.

Ouadfeul, S. and Aliouane, L. (2016b). Shale gas lithofacies classification and gas content prediction using artificial neural network. *2016, ECMOR XV-15th European Conference on the Mathematics of Oil Recovery*, Amsterdam, Netherlands (29 August to 1 September 2016). EAGE Earth doc.

Ouadfeul, S. and Aliouane, L. (2015). Total organic carbon prediction in shale gas reservoirs from well logs data using the multilayer perceptron neural network with Levenberg Marquardt training algorithm: application to Barnett shale. *Arabian Journal for Science and Engineering* 40: 3345–3349. - springer. DOI https://doi.org/10.1007/s13369-015-1685-y.

Ouadfeul, S. and Aliouane, L. (2012). *Lithofacies Classification Using the Multilayer Perceptron and the Self-organizing Neural Networks, Neural Information Processing*, Lecture Notes in Computer Science, vol. 7667 of the series, 737–744. SpringerVerlag.

Perez, R. (2013). Brittleness estimation from seismic measurements in unconventional reservoirs: Application to the Barnett Shale. Ph.D. Dissertation. The University of Oklahoma.

Quirein, J., Kimminau, S., La Vigne, J. et al. (1986). A coherent framework for developing and applying multiple formation evaluation models. *Trans. 21th SPWLA Annual Logging Symposium*, Houston, Texas (9–13 June 1986).

Riedmiller, M. (1994). Advanced supervised learning in multi-layer perceptrons – from backpropagation to adaptive learning algorithms. *Journal of Computer Standards and Interfaces* 16: 265–278.

Sheikhan, M. and Sha'bani, A.A. (2009). Fast neural intrusion detection system based on hidden weight optimization algorithm and feature selection. *World Applied Sciences Journal* 7 (Special Issue of Computer & IT): 45–53.

Sifaoui, A., Abdelkrim, A., and Benrejeb, M. (2008). On the use of neural network as universal approximator. *International Journal of Sciences and Techniques of Automatic control & computer engineering, IJ-STA* 2 (1): 386–399.

Wang G., Carr, T.R., 2012, Marcellus shale lithofacies prediction by multiclass neural network classification in the appalachian basin, *Math Geosci* doi https://doi.org/10.1007/s11004-012-9421-6.

Wang, G., Carr, T.R., Ju, Y., and Li, C. (2014). Identifying organic-rich Marcellus Shale lithofacies by support vector machine classifier in the Appalachian basin. *Computers & Geosciences* 64: 52–60.

Zhao, T., Marfurt, K.J., Zhou, H., 2014, Lithofacies classification in Barnett Shale using proximal support vector machines, *SEG Technical Program Expanded Abstracts* 2014: pp. 1491–1495, doi: https://doi.org/10.1190/segam2014-1210.1.

14

The Use of Pore Effective Compressibility for Quantitative Evaluation of Low Resistive Pays

Mohamed Zinelabidine Doghmane[1], *Sid-Ali Ouadfeul*[2], *and Leila Aliouane*[3]

[1] *Department of Geophysics, FSTGAT, University of Science and Technology Houari Boumediene, Algiers, Algeria*
[2] *Algerian Petroleum Institute, Sonatrach, Boumerdes, Algeria*
[3] *LABOPHYT, Faculty of Hydrocarbons and Chemistry, University M'hamed Bougara of Boumerdes, Boumerdes, Algeria*

KEYPOINTS

- This chapter deals with the petrophysical evaluation of low-resistivity pays.
- The proposed approach is based on extracting the effective pore compressibility from log data.
- The curve of the effective pore compressibility leads to quantitative evaluation of low-resistivity net pays.
- The obtained results can be useful for exploration and production activities.

14.1 Introduction

Given that the oil industry is many countries' primary source of energy, it is critical to adopt new technologies and approaches to increase the productivity of mature reservoirs in tandem with new exploration discoveries. This study is proposed as part of a strategy to increase the production of so-called low-resistivity reservoirs (Davudov et al. 2017) and reduce exploration uncertainties; the positive outcomes of this approach will have an impact on national oil companies' performance. As a result, it will provide new energy resources in the future.

14.2 Low-Resistivity Pays in the Studied Basin

Resistivity logs are one of the main records in conventional reservoirs that allow differentiation between the hydrocarbon zones and the water zone due to the resistivity contrast that exists between these two zones. If the reservoir, on the other hand, has areas of low resistivity with low contrast (LRLC), these logs will be unable to identify these areas and will also be unable to indicate the mobility of produced water (Doghmane et al. 2018). Many potentially productive areas with high irreducible water saturation have been overlooked in recent years due to a lack of adequate technology. As a result, the identification of reservoirs with low resistivity values and weak contrast, the determination of their parameters of interest (porosity, permeability, and hydrocarbon

Unconventional Hydrocarbon Resources: Prediction and Modeling Using Artificial Intelligence Approaches,
First Edition. Edited by Sid-Ali Ouadfeul.

Figure 14.1 Lithology factor visualization at the microscopic scale.

saturation) as shown in Figure 14.1, the calculation of their reserves, and the control of production are major issues in the evaluation of this type of reservoirs that exist in many countries around the world, including Algeria (Davudov and Ghanbarnezhad 2018; Doghmane and Belahcene 2019). The basic resistivity phenomenon has been present in the Berkine basin for a long time, but it was only recently discovered (Eladj et al. 2020). As a result, many studies have been conducted in order to correct the underestimation of hydrocarbon volume calculated using conventional methods, which are empirical (no mathematical basis) and difficult to generalize.

14.3 Water Saturation from Effective Pore Compressibility

To identify the fluids present in the reservoir, new attributes must be introduced, one of which is the reservoir's effective compressibility (Suyang et al. 2018). It gives us a rough estimate of the amount of gas in the reservoir (i.e. the higher the compressibility of the reservoir, the greater the quantity of gas). Because gas is a very compressible fluid, we used this critical physical property to distinguish between gas and oil reservoirs (Eladj et al. 2022a,b). The prediction of fluid pressure gradients exerted within porous and permeable reservoir formations as well as porous and very low permeability clay formations is a major issue in oil exploration. Reliable prediction of these pressure gradients upstream of operations allows for significant cost savings during drilling as well as increased team safety. This represents the pore pressure gradient calculated using seismic inversion attributes (Zp and Zs) as well as other attributes such as the BIOT coefficient, SKEMPTON coefficient, consolidation parameter, and so on... (Zimmerman et al. 1986). The combination of this attribute (i.e. pore pressure gradient) and the reservoir pore compressibility attribute allowed us to detect not only overpressurized areas but also overpressurized areas containing biogenic gas, which is hazardous to the drilling equipment and the crew, even if the quantity of this gas was clearly low (Suyang et al. 2018; Yuzheng et al. 2017).

14.4 Discussion

As shown in Figure 14.2, the obtained petrophysical parameters from pore compressibility have indicated that there are pay intervals with hydrocarbon potential, contrary to what the conventional interpretation indicated (Brown and Korringa 2012). The latter was influenced by the presence of heavy minerals, which resulted in low resistivity values. Furthermore, the good correlation

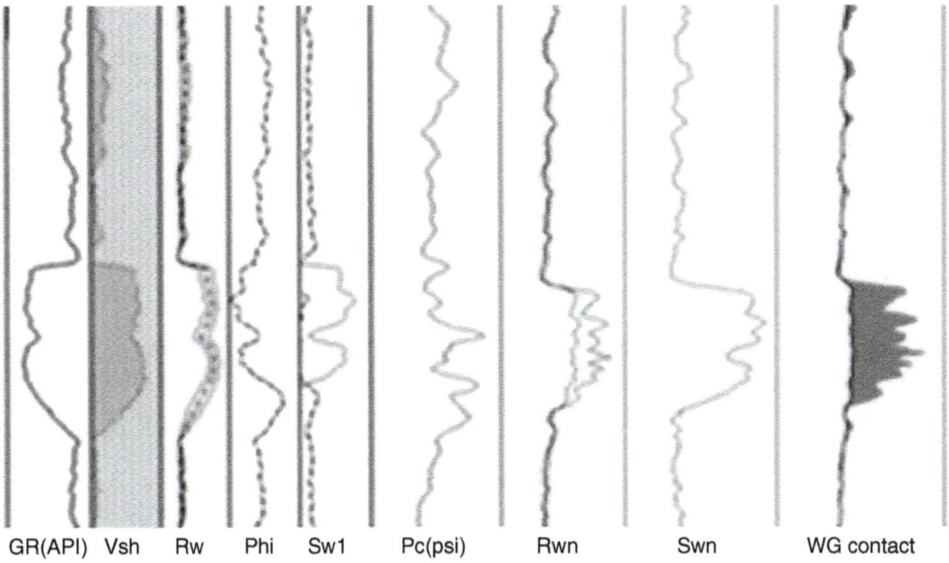

Figure 14.2 Comparison between conventional interpretation and using pore compressibility method.

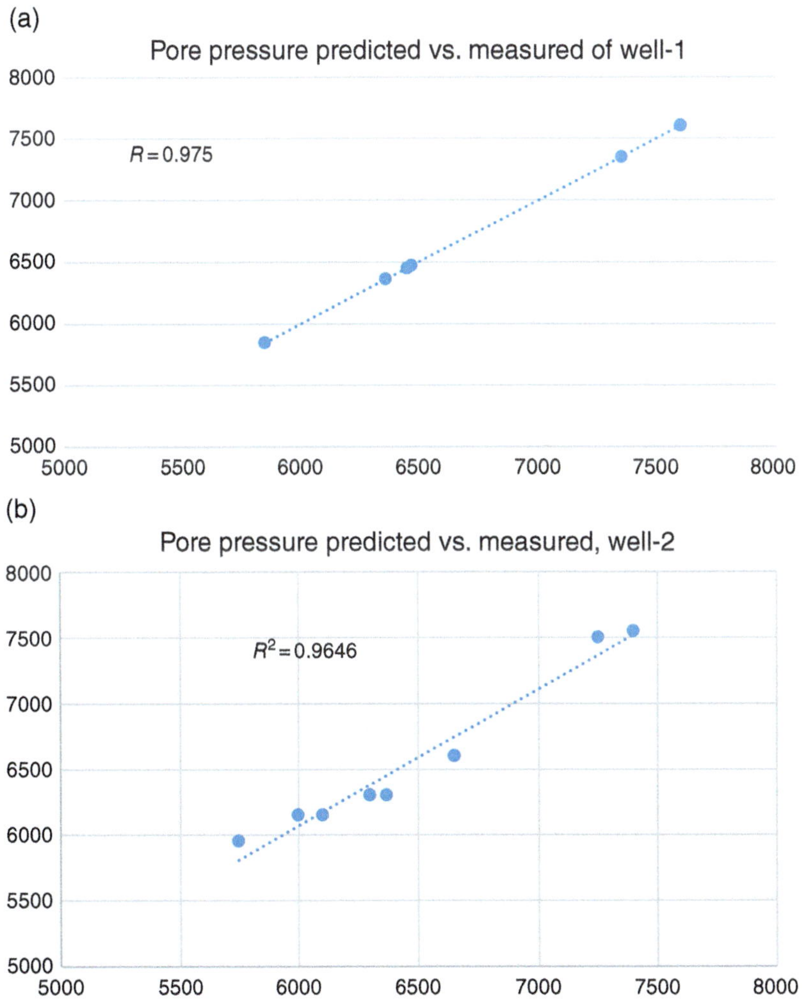

Figure 14.3 Correlation between predicted and measured pore pressures in: (a) Interval Pay 1, (b) Interval Pay 2.

shown in Figure 14.3 confirmed the accuracy of the obtained conclusive pore pressure models (Alexander 2020; Boualam et al. 2020). As a result, the extracted porosity and water saturation are closer to reality than the ones obtained previously because the pore pressure measured from MDT tests in intervals 1 (Figure 14.3a) and 2 (Figure 14.3b) were used as references.

14.5 Conclusions

Based on the findings, we can conclude that pore compressibility extracted from sonic logs can be a useful tool for evaluating LRLC reservoirs because it eliminates the need for resistivity logs to calculate porosity and water saturation. It does not require any additional logging tools, only the sonic logs, making it both economically and technically feasible. As a result, it is highly recommended that this method be considered to re-evaluate the low resistive pays in the studied basin in order to improve reservoir productivity.

Bibliography

Alexander, Y.R. (2020). Effective fluid bulk modulus in the partially saturated rock and the amplitude dispersion effects. *Journal of Geophysical Research-Solid Earth* 125 (3): 1–21.

Boualam, A., Rasouli, V., Dalkhaa, C., and Djezzar, S. (2020). Advanced petrophysical analysis and water saturation prediction in three forks, Williston basin. *The SPWLA 61st Annual Logging Symposium*, Virtual online Webinar, June 2020. https://doi.org/10.30632/SPWLA-5104.

Brown, R.J.S. and Korringa, J. (2012). On the dependence of the elastic properties of a porous rock on the compressibility of the pore fluid. *Geophysics* 40 (4): 608–616.

Davudov, D. and Ghanbarnezhad, R.M. (2018). Impact of pore compressibility and connectivity loss on shale permeability. *International Journal of Coal Geology* 178: 98–113.

Davudov, D., Ghanbarnezhad, R.M., Yuzheng, L., and Dalton, V. (2017). Investigation of shale pore compressibility impact on production with reservoir simulation. *SPE Unconventional Resources Conference* (15–16 February). Calgary, Alberta, Canada, SPE-185059-MS.

Doghmane, M.Z. and Belahcene, B. (2019): Design of new model (ANNSVM) compensator for saturation calculation based on logging curves for low resistivity phenomenon. *Conference proceeding, EAGE/ALNAFT Geoscience Workshop*, Algiers, Algeria (28–29 January 2019). vol.2019, p. 1–5.

Doghmane, M.Z., Belahcene, B., and Kidouche, M. (2018). Application of improved artificial neural network algorithm in hydrocarbons' reservoir evaluation. *Proc. International conference in Artificial Intelligence in Renewable Energetic Systems ICAIRES 2018: Renewable Energy for Smart and Sustainable Cities*, Tipaza, Algeria (26–28 November 2018), pp. 129–138.

Eladj, S., Lounissi, T.K., and Doghmane, M.Z. (2020). Lithological characterization by simultaneous seismic inversion in algerian south eastern field. *Engineering, Technology & Applied Science Research* 10 (1): 5251–5258.

Eladj, S., Doghmane, M.Z., Lounissi, T.K. et al. (2022a). 3D Geomechanical model construction for wellbore stability analysis in algerian southeastern petroleum field. *Energies 15*: 7455. https://doi.org/10.3390/en15207455.

Eladj, S., Lounissi, T.K., Doghmane, M.Z., and Djeddi, M. (2022b). Wellbore stability analysis based on 3D geo-mechanical model of an algerian southeastern field. *Advances in Geophysics, Tectonics and*

Petroleum Geosciences. CAJG 2019. Advances in Science, Technology & Innovation. Cham: Springer. https://doi.org/10.1007/978-3-030-73026-0_136.

Suyang, Z., Zhimin, D., Chuanliang, Y. et al. (2018). An analytical model for pore volume compressibility of reservoir rock. *Fuel* 232: 543–549.

Yuzheng, L., Ghanbarnezhad, R.M., and Davudov, D. (2017). Pore compressibility of shale formations. *SPE Journal 22* (06): 1778–1789.

Zimmerman, R.W., Somerton, W.S., and King, M.S. (1986). Compressibility of porous rocks. *Journal of Geophysical Research* 91 (B12): 12765–12777.

15

The Influence of Pore Levels on Reservoir Quality Based on Rock Typing: A Case Study of Quartzite El Hamra, Algeria

Nettari Ferhat[1], Mohamed Z. Doghmane[2], Sid-Ali Ouadfeul[3], and Leila Aliouane[1]

[1] *LABOPHYT, Faculty of Hydrocarbons and Chemistry, University M'hamed Bougara of Boumerdes, Boumerdes, Algeria*
[2] *Department of Geophysics, FSTGAT, University of Science and Technology Houari Boumediene, Algiers, Algeria*
[3] *Algerian Petroleum Institute, Sonatrach, Boumerdes, Algeria*

KEYPOINTS

- This chapter is focused on study of hydrocarbons reservoir.
- The studied reservoir is Quartzite EL Hamra and it tight.
- It is located in Hassi Messaoud field.
- The chapter deals with rock typing study.
- The FZI has been also identified in this chapter.
- The obtained results have been used for reliable reservoir simulations.

15.1 Introduction

The studied region is in the governorate of Ouargla, nearly 650 km south of Algiers and 44 km south of Hassi Messaoud, and is bounded by the parallels 31° and 32° North and the meridians 6° and 7° East. According to Sonatrach's exploration division, the Hassi Tarfa field is part of block N° 427 (Figure 15.1). Its stratigraphic series is primarily composed of Mesozoic deposits with a thickness of 3118 m that lie unconformably on the Paleozoic, which has a thickness of 407 m. Finally, a 300-m tertiary detrital spreading rests in a discontinuity on the Mesozoic. This research enabled us to describe the reservoir's petrography for the first time (Eladj et al. 2022a,b). We needed a detailed quantitative and qualitative analysis of thin sections at the microscopic level for this (Doghmane et al. 2019; Eladj et al. 2020). This study considered two wells (Well-1 and Well-2) and demonstrated the results of Well-1.

15.2 Quick Scan Method

This study implemented five stages of flow unit identification and characterization based on the research of Gunter et al. (2014) in Gunter et al. 2014: (i) Rock-type identification to illustrate the Winland transverse path of porosity-permeability, (ii) Create a stratigraphic modified Lorenz plot

Unconventional Hydrocarbon Resources: Prediction and Modeling Using Artificial Intelligence Approaches,
First Edition. Edited by Sid-Ali Ouadfeul.

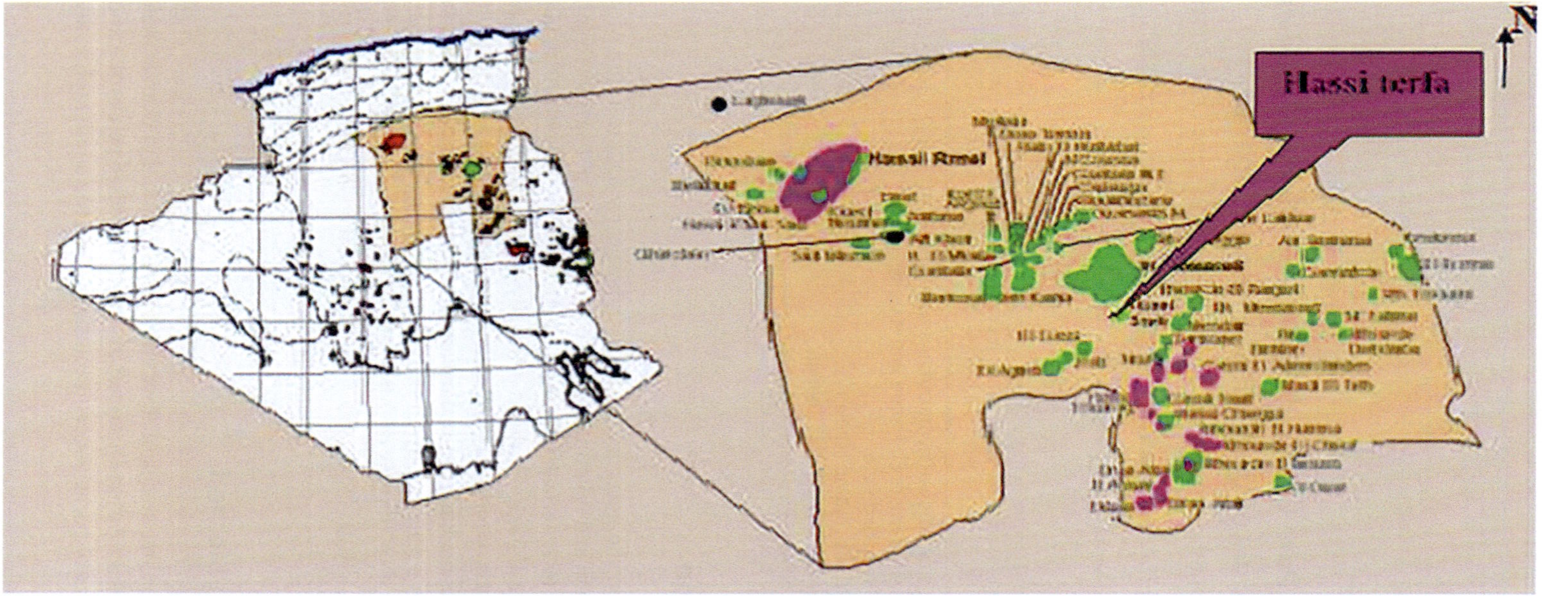

Figure 15.1 Location of the studied hydrocarbon field (studied region).

(SMLP) by calculating the flow capacity percentage (permeability thickness) and storage capacity percentage (porosity thickness), (iii) Flow unit intervals should be chosen based on the SMLP inflection points; these preliminary flow units should be validated using the R35 geological framework (calculated pore radius (um) at 35% mercury saturation) and the K/ø ratio, (iv) Creation of the final stratigraphic flow profile (SFP) using the correlation curve, porosity-permeability (Belhouchet and Benzagouta 2019) *K*/ratio, R35, percent storage, and percent capacity, and (v) Construction of the Modified Lorenz Plot (MLP) by ordering units of final flow (Doghmane and Belahcene 2019). The following is the empirical relation of the Winland method for determining the radius of the channels' connection between the pores:

$$\text{Winland}\log\left(R_{35}\right)=0.732+0.588\times\log\left(K\right)-0.864\times\log\left(\varnothing\right) \tag{15.1}$$

15.3 Results

The microscopic mineralogy of the Quartzite El Hamra reservoir is depicted in Figure 15.2 (Rushing et al. 2008). Instead of HFUs, the Winland equation (R35) has been widely used to determine reservoir rock types. Three types of reservoir rocks were discovered in Well-1, ranging from micro to macro pore type. Six HFUs were determined in Well-1 using the proposed rock typing method. We concluded that approximately 95% of the flow occurs between depths of 3392 and 3407.5 m, and the reservoir is homogeneous in its deepest part (Benayad et al. 2014). Because the HCPV curve is not flat, there are no water zones. The simplest analysis method is graphical grouping using the log-log plot of RQI as a function of *Z*, but it is insufficient to distinguish between different HFUs and estimate their limits. The permeability was calculated using the intercept of FZIMOY values (Doghmane et al. 2022).

15.4 Discussion

Figure 15.3 depicts a linear relationship between R35 and RQI. As a result, the Winland equation is similar to the general mathematical form of FZI. As a result, we concluded that in order to quickly determine the type of petrophysical rock drainage, Winland R35, Pittman, Leverett, RQI,

Figure 15.2 Mineralogy of Quartzite El Hmara in the studied field.

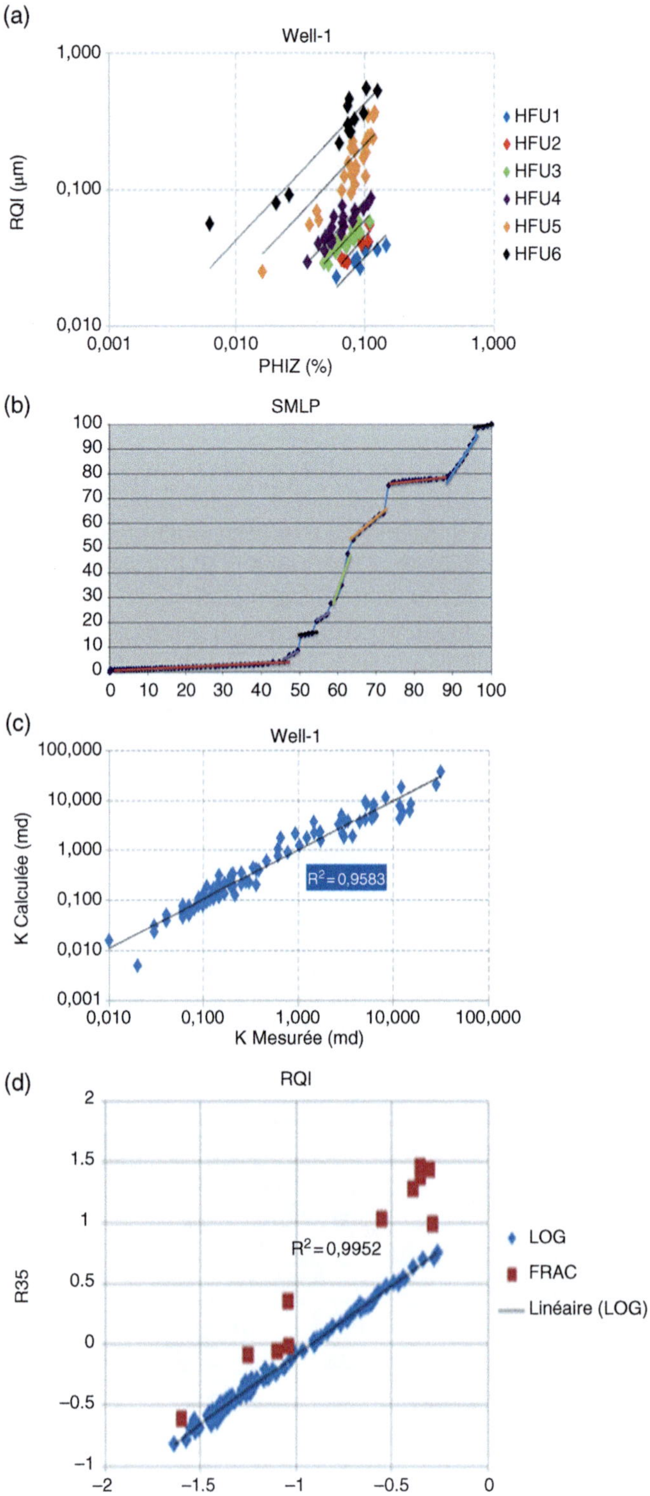

Figure 15.3 Results for Well-1: (a) RQI vs. Porosity, (b) SMLP, (c) Calculated permeability vs. measured permeability, and (d) RQI vs. R35.

or FZI have all produced the same expected results (Sistan et al. 2013; Bacetti and Doghmane 2020). Permeability is the most important factor in determining the reservoir's fluid flow characteristics. The size and geometry of the pore channels, which are influenced by the deposition medium and subsequent diagenetic processes, are the primary determinants of permeability.

15.5 Conclusions

The FZI and Winland methods were used to estimate the absolute permeability from log data from the Quartzites El Hamra. The obtained results proved the existence of six HFUs. The Winland method is very reliable for determining channel radius and providing pore throat. The effective (noncapillary) channel diameter is an important parameter in fluid movement control that takes wettability into account. The radius of the channels, which was determined by R35, controls the water saturation in a reservoir. Lorenz's modified stratigraphic method provided simple and fast graphical discriminations for flow units, as well as a graphical description of the fluid types in producing areas, barriers, and reservoir homogeneity. To compare results and reduce uncertainty, we recommend using the capillary pressure and J function methods for Quartzite El Hamra.

Bibliography

Bacetti, A. and Doghmane, M.Z. (2020). A practical workflow using seismic attributes to enhance sub seismic geological structures and natural fractures correlation. *Conference Proceedings* (30 November–3 December 2020), Vienna, Austria.

Belhouchet, H.E. and Benzagouta, M.E. (2019). Rock typing: Reservoir permeability calculation using Discrete Rock Typing methods (DRT): Case study from the algerian B-H oil field reservoir. In: *Advances in Petroleum Engineering and Petroleum Geochemistry. CAJG 2018. Advances in Science, Technology & Innovation (IEREK Interdisciplinary Series for Sustainable Development)* (ed. S. Banerjee, R. Barati, and S. Patil). Cham: Springer.

Benayad, S., Park, Y., Chaouchi, R. et al. (2014). Parameters controlling the quality of the Hamra Quartzite reservoir, southern Hassi Messaoud, Algeria: insights from a petrographic, geochemical, and provenance study. *Arab. J. Geosci.* 7: 1541–1557.

Doghmane, M.Z. and Belahcene, B. (2019). Design of new model (ANNSVM) compensator for saturation calculation based on logging curves for low resistivity phenomenon. *EAGE/ALNAFT Geoscience Workshop* 2019 (1): 1–5.

Doghmane, M.Z., Belahcene, B., and Kidouche, M. (2019). Application of improved artificial neural network algorithm in hydrocarbons' reservoir evaluation. In: *Renewable Energy for Smart and Sustainable Cities*, ICAIRES 2018. Lecture Notes in Networks and Systems, vol. 62 (ed. M. Hatti). Cham: Springer.

Doghmane, M.Z., Ouadfeul, S.A., Benaissa, Z., and Eladj, S. (2022). Classification of ordovician tight reservoir facies in algeria by using neuro-fuzzy algorithm. In: *Artificial Intelligence and Heuristics for Smart Energy Efficiency in Smart Cities*, IC-AIRES 2021. Lecture Notes in Networks and Systems, vol. 361 (ed. M. Hatti). Cham: Springer.

Eladj, S., Lounissi, T.K., Doghmane, M.Z., and Djeddi, M. (2020). Lithological characterization by simultaneous seismic inversion in algerian south eastern field. *Eng. Technol. Appl. Sci. Res.* 10 (1): 5251–5258.

Eladj, S., Doghmane, M.Z., Lounissi, T.K. et al. (2022a). 3D Geomechanical model construction for wellbore stability analysis in algerian southeastern petroleum field. *Energies* 15 (20): 7455.

Eladj, S., Doghmane, M.Z., Aliouane, L., & Ouadfeul, S.A. (2022b). Porosity model construction based on ANN and seismic inversion: A case study of saharan field (algeria). *Conference of the Arabian Journal of Geosciences Advances in Geophysics, Tectonics and Petroleum Geosciences. CAJG 2019. Advances in Science, Technology & Innovation*. Springer, Cham.

Gunter, G.W., Spain, D.R., Viro, E.J., et al. (2014). Winland pore throat prediction method – a proper retrospect: New examples from carbonates and complex systems. *SPWLA 55th Annual Logging Symposium* (18–22 May), Abu Dhabi, United Arab Emirates.

Rushing, J.A., Newsham, K.E., and Blasingame, T.A. (2008). Rock typing: keys to understanding productivity in tight gas sands. *SPE Unconventional Reservoirs Conference* (10–12 February), Keystone, Colorado, USA.

Sistan, M., Jamshidi, S., and Salehi, S. (2013). Investigation of capability of FZI and Winland methods for rock typing using pore network modeling. *International Journal of Petroleum Engineering* 1 (2): 1–12.

16

An Example from the Algerian Sahara Illustrates the Use of the Hydraulic Flow Unit Technique to Discriminate Fluid Flow Routes in Confined Sand Reservoirs

Abdellah Sokhal[1], Sid-Ali Ouadfeul[2], Mohamed Zinelabidine Doghmane[1], and Leila Aliouane[3]

[1] *Department of Geophysics, FSTGAT, University of Science and Technology HouariBoumediene, Algiers, Algeria*
[2] *Algerian Petroleum Institute, Sonatrach, Boumerdes, Algeria*
[3] *LABOPHYT, Faculty of Hydrocarbons and Chemistry, University M'hamed Bougara of Boumerdes, Boumerdes, Algeria*

KEYPOINTS

- This chapter deals with the flow fluid path identification.
- The flow zone identification has been used to realize the objective of this chapter.
- The hydraulic flow units have been also studied in order to identify net pays.
- A fuzzy system has been constructed to differentiate the transition zones.
- A case study from Hassi Terfa field is demonstrated to highlight the effectiveness of the proposed approach.

16.1 Introduction

Methods for reservoir characterization are useful because they give a more accurate representation of a petroleum reservoir's flow and storage capacity. It is possible to distinguish the layers of the examined reservoirs by their lithofacies, petrofacies, electrofacies, and HFUs, also known as flow units, using techniques called reservoir rock typing. The usual definition of an HFU is the representative volume of total reservoir rock where the geological characteristics that govern fluid flow are internally consistent and predictably distinct from those of other rocks (Ebanks 1987).

An HFU, on the other hand, was described by Amaefule and colleagues in 1993 as a reservoir portion that differs from other regions due to hydraulic properties or traits that influence fluid flow in the reservoir. Therefore, permeability may be calculated with adequate accuracy if the reservoir is divided into several flow units. The flow zone indicator (FZI), like HFUs, is a function of the reservoir quality index (RQI), void ratio, and z (Al-Ajmi and Holdtich et al. 2000).

Before examining the types of fluids, the quantity of fluids, and the rates of fluid flow, it is crucial to have a thorough understanding of these two properties. Only then can fluid recovery estimates be established. So much of the technical literature in the oil business deals with techniques for evaluating porosity and permeability. One of the crucial factors in modeling, calculating, and estimating

Unconventional Hydrocarbon Resources: Prediction and Modeling Using Artificial Intelligence Approaches, First Edition. Edited by Sid-Ali Ouadfeul.

production from reservoirs is permeability. It is determined using a variety of techniques, including core analysis, log analysis, and test data. Since the core is not always available for research in some periods or wells, permeability should be estimated using other sources of information.

From well logs, several authors have attempted to calculate the permeability of rock mass (Soto et al. 2001). It is crucial to be able to predict permeability accurately and affordably. The mathematical model between the input and output data can be represented mathematically to calculate the FZI. In this study, the flow zone indication in uncored wells is determined using fuzzy logic (FL). This chapter introduces the idea of an HFU based on the FZI and demonstrates how it may be applied to a tight sandstone reservoir from the Algerian Sahara's Hassi Terfa field.

16.2 Regional Geologic Setting

It is an oil play with extremely high potential. The examined reservoir (Hamra Quartzites) has been eroded under the Hercynian unconformity and forms the Hassi Messaoud ring (Figure 16.1).

The Hassi Messaoud field's Cambro-Ordovician stratigraphic succession is made up of a siliioclastic series that lies irregularly above an eruptive metamorphic basement. Figure 16.2 depicts the succession in ascending order. Although this sandstone was produced during the Arenig period in a shallow marine environment, its specific sedimentology and sequence structure are still poorly understood at the local scale (KRACHA.N Sonatrach 2011).

The sandstone is heavily bioturbed (skolithospipe rock) and suffered significant quartz cementation during its history of burial, which changed its mechanical characteristics and porosity. The Lower Ordovician reservoir of the Hamra Quartzites is made up of clay-interbedded quartzite and quartzitic sands. Given that quartz arenites make up the majority of its composition, it is regarded as mature. The main diagenetic variables influencing reservoir quality are quartz overgrowths,

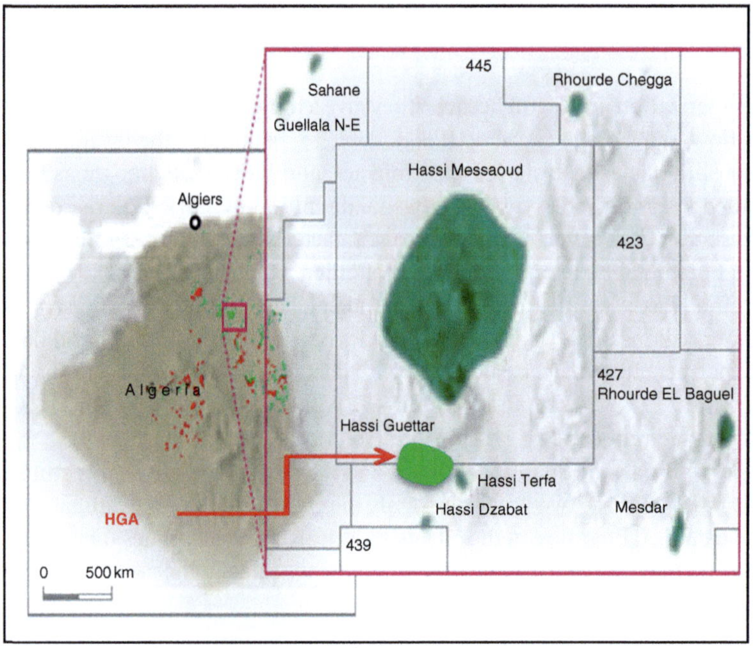

Figure 16.1 Location of the Hassi Messaoud and Hassi Terfafields. *Source:* Tawadros (2011).

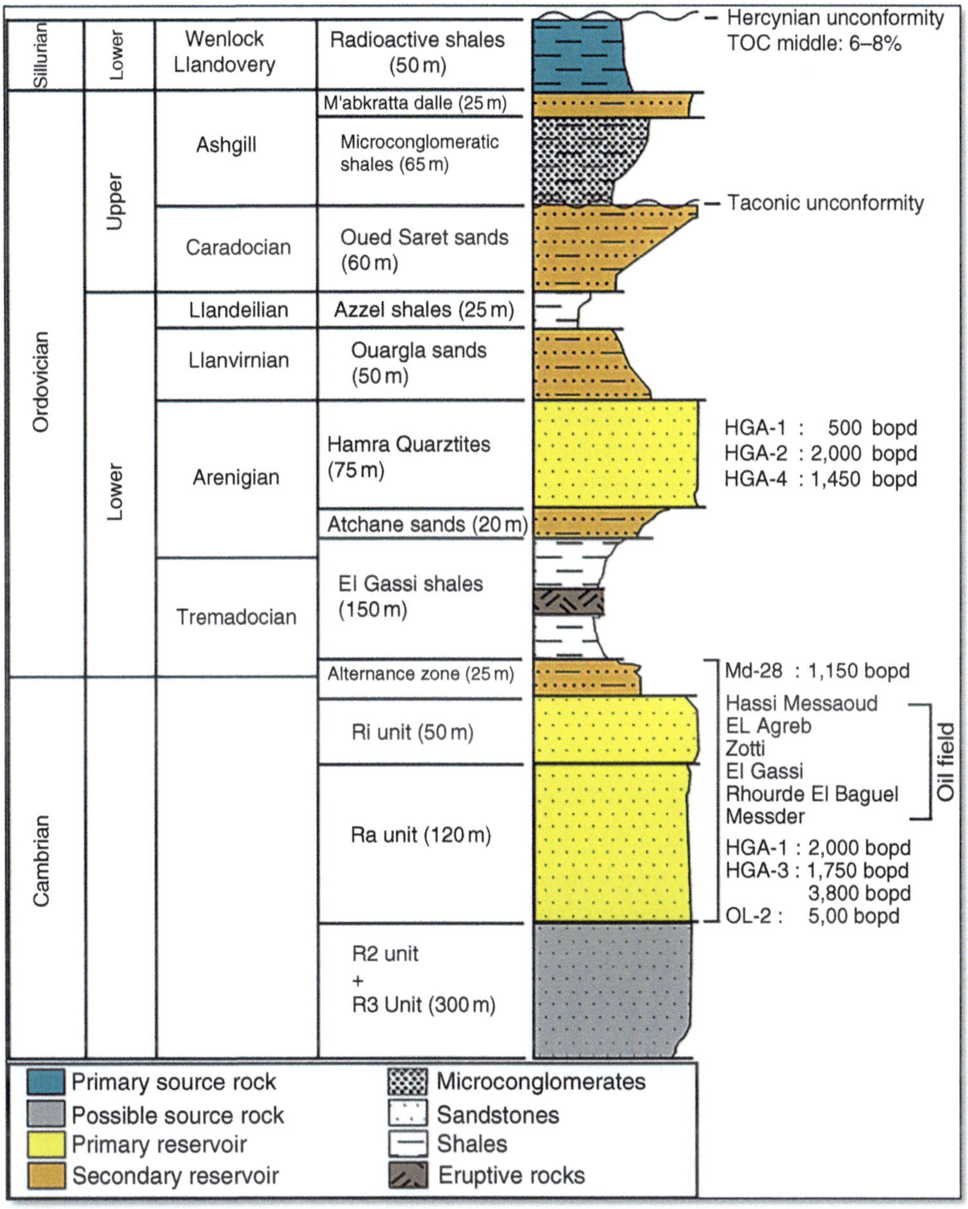

Figure 16.2 Generalized stratigraphic column of Hassi Messaoud field (Tawadros 2011 modified).

authigenic clay development, and carbonate cement precipitation. During various diagenetic processes, the sediments were strongly cemented and silicified in the majority of cases.

A marine geological unit with shallow depths is the Hamra Quartzite formation. Based on environmental deposition, two parasequences have been identified; the bottom unit is made up of coastal plain, backshore, and tidal channel deposits. The higher unit has poor reservoir characteristics because it is made up primarily of shoreface and berm deposits and has more diagenetic cementation and siliciclastic cementation.

16.3 Statement of the Problem

This study's primary goal is to identify the flow units in the Hassi Tarfa field's "Quartzite Hamra" tight sand gas reservoir. The available core plug data measurements and the traditional well logging data serve as the foundation for the flow unit delineation idea.

16.3.1 Concept of HFU

According to Amaefule et al. (1993), core data reveal a variety of depositional and diagenetic influences on pore geometry, and the differences in pore geometry features result in the existence of distinct zones of HFUs with comparable flow parameters. The representative volume of the entire reservoir rock known as HFU is where the geological characteristics that govern fluid flow are internally consistent and predictably distinct from those of other rocks (Ebanks 1987).

However, Amaefule et al. (1993) provided three parameters that can be determined using the following equations to determine the HFU:

$$\mathrm{RQI}(\mu\mathrm{m}) = 0.0314 \times \sqrt{\frac{K}{\phi_e}} \tag{16.1}$$

$$\varphi_z = \frac{\varphi_e}{1-\varphi_e} \tag{16.2}$$

$$\mathrm{FZI} = \frac{1}{\sqrt{F_S \tau S_{gv}}} = \frac{\mathrm{RQI}}{\phi_z} \tag{16.3}$$

where Φe represents effective porosity in fraction, *k* represents permeability in md, RQI represents reservoir quality index in um, Φz represents normalized porosity index, FZI represents flow zone indicator in um, F_s represents shape factor, τ represents tortuosity, and S_{gv} represents surface area per unit grain in um. The following formula can be used to compute the permeability in md for FZI in um and Φ*e* in fraction:

$$K = 1014 \times \mathrm{FZI}_{\mathrm{moy}}^2 \times \frac{\varphi_e^3}{\left(1-\varphi_e\right)^2} \tag{16.4}$$

RQI, FZI, and Φz are used for HFU classification, which will be covered in more detail in the parts that follow.

16.3.2 HFU Zonation Process

The most widely used flow unit determination methods have been taken into account in this study because there are numerous flow unit determination methods described in the literature, and each of them may produce distinct benefits and outcomes. The stratigraphic modified Lorenz plot (SMLP), Winland R35, and flow zone index (FZI) are the most used ones.

The HFUs can be distinguished based on FZI values following the classification of samples by eliminating samples with induced fractures and computing pore-throat associated metrics of RQI and FZI from core data. To employ FZI for zonation, various clustering techniques may be used. They fall into three major categories:

- Log-Log plot of RQI vs Φ_z
- Histogram analysis of FZI distribution
- Probability plot of FZI distribution

The amount of HFU is determined in this chapter by examining the outcomes of the various techniques used. Moreover, Ward's approach confirms the precise HFU number.

16.4 Results and Discussion

Data for this section of the investigation were gathered via HT2 well standard log data and conventional core analysis. All of the samples from the fractured cores were categorized before any data processing. These core samples were rejected if the fractures were caused by core recovery, but not omitted if they were organic open fractures.

16.4.1 FZI Method

Based on the aforementioned calculations applied to the well HT-2 core data, three parameters ϕ_z, RQI, and FZI were determined. Three different clustring approaches have been taken into consideration in order to calculate the number of HFUs, and the results are then compared: graphical clustering utilizing the traditional RQI-Φ_z plot, histogram analysis, and probability plots. These plots ought to be able to differentiate different distributions for each HFU, according to theory. The individual distributions cannot be distinguished on the histogram due to data overlap. For these rock types, the cumulative probability plot enables the selection of at least 4 HFU and the estimation of FZI borders (Figure 16.3).

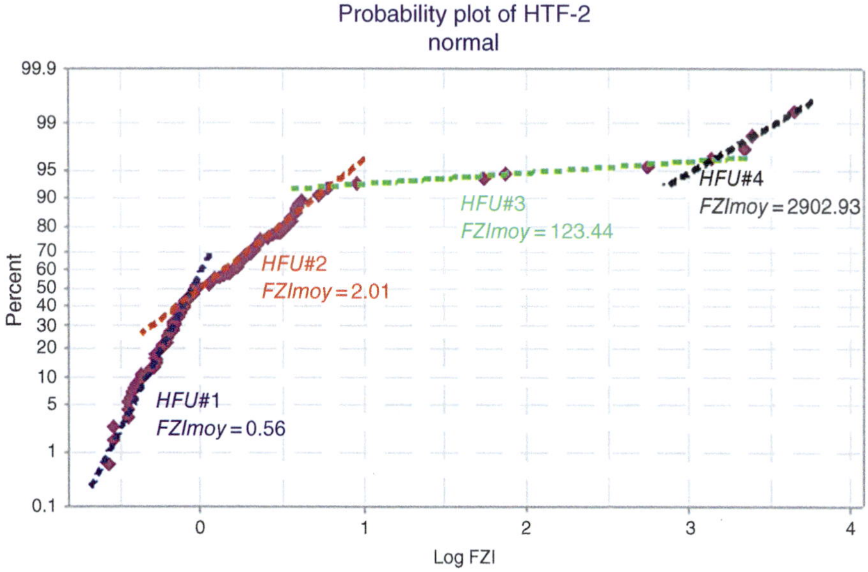

Figure 16.3 Probability plot of FZI for well HT-2.

Finally, the ward's approach was employed to determine the precise number of HFUs. In Figure 16.4, the inaccuracy of the sum squares is displayed versus the quantity of HFUs.

According to Figure 16.4, the ideal number of HFUs is three, which indicates that the reservoir under study contains three different types of rocks. The permeability from the equation is calculated using the intercept (FZI mean) values (5). Figure 16.5 shows a cross-plot of the estimated and observed permeabilities, which shows how accurate the computed permeability is.

This classification approach can carry out a very competent numerical classification among the data, as is seen in Figure 16.6. Each group's depth was highly dispersed.

16.4.2 FZI Method

For a certain collection of experimental sites, this new technique makes an effort at a direct mapping between well log data and core data. The FL approach is the best solution for this kind of issue. The first step is to choose an acceptable model for the well/log data and FZI in a cored well. The method is then expanded to include uncored wells and intervals in order to derive FZI from log data. The graphic in Figure 16.7 summarizes the steps that were taken.

The data set was split into two subsets at the initial stage of this procedure, one for network training and the other for training validation. The FL received the collection of logs NPHI, RHOB, PHIE, and GR as inputs. By deducting each of these records from the mean and dividing the result

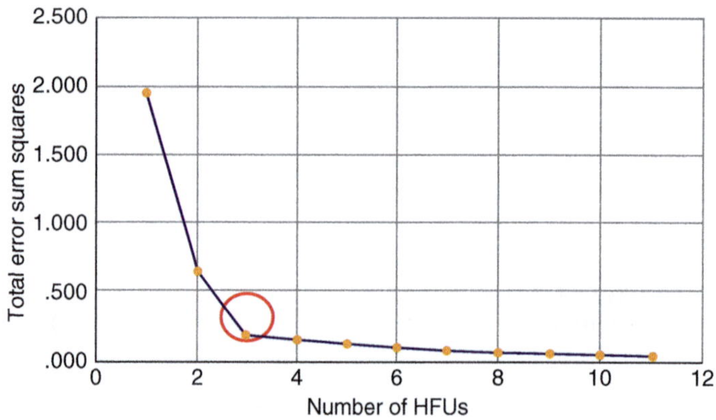

Figure 16.4 Determination of the optimal number of HFUs (HT-2).

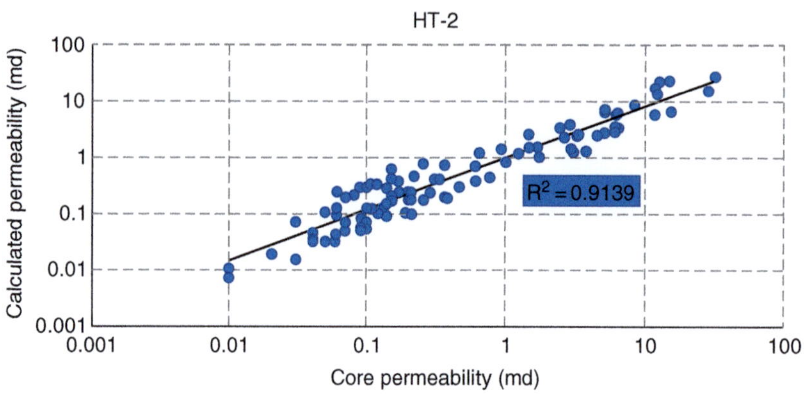

Figure 16.5 Calculated permeability considering 3 HFUs versus core permeability, Well HT-2.

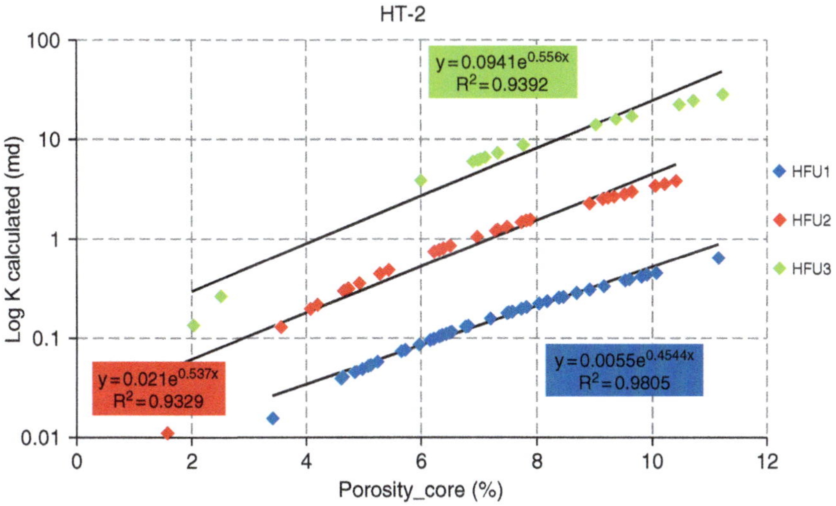

Figure 16.6 Permeability vs. porosity, Well HT-2.

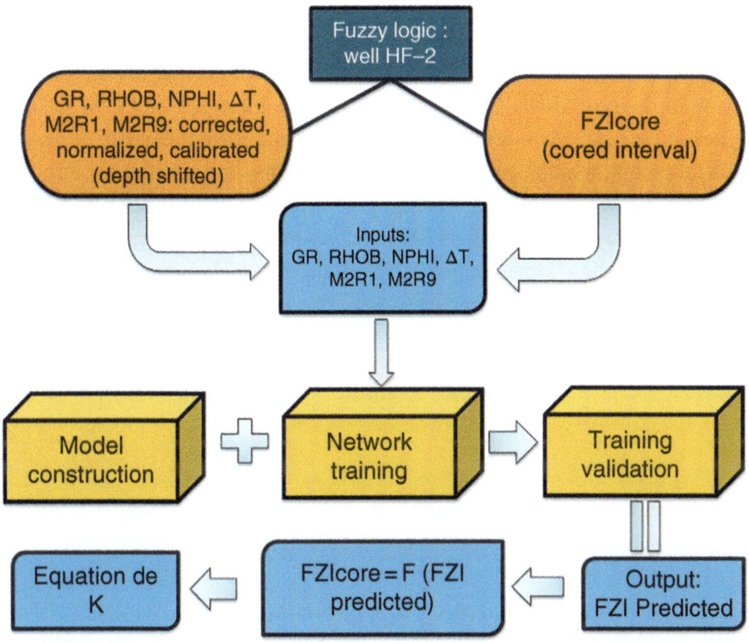

Figure 16.7 Different steps to predict FZI using FL in uncored intervals in well HF-2.

by the standard deviation, the data were each normalized. The system output was the log-scale FZI. As illustrated in Figure 16.8, the calculated FZI from FL was compared to the FZI obtained from effective porosity and core permeability, with a correlation coefficient of $R_2 = 0.97$.

The evaluation is consistent with equation (16.5) since the values of FZI estimated from core data were connected with FZI computed from well log data at its corresponding depth of core data:

$$FZI_{Core} = 0.9503 \times FZI\_log^{0.9507} \tag{16.5}$$

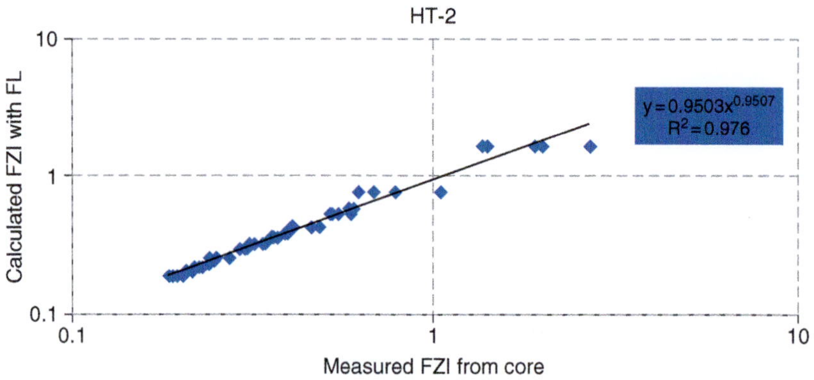

Figure 16.8 Measured FZI (core) vs. calculated FZI (FL), Well HT-2.

16.5 Conclusions

Three HFUs were identified and can be used to partition the reservoir after the HFU method was initially applied to the characterization of the reservoir in the Hassi Tarfa field. This work indicates how building consistent and reliable permeability models, which are required for geology and reservoir simulation modeling, can be done extremely well utilizing rock typing using the FZI approach. Due to its foundation in the pore throat radius and porous medium geometry, which are connected to the different types of petrophysical rock, the FZI is an appropriate measure for calculating HFUs. Additionally, visual errors, which are frequent in histograms, and normal probability errors have been reduced by applying Ward's method for figuring out the ideal number of HFUs. Using the FL algorithm in an uncored well, correlations between well log readings and FZI values from core data were in fact developed; they were successful in predicting FZI and, as a result, the values of permeability. The findings collected demonstrate that the FL approach is an effective method for predicting HFUs at wells without core data.

References

Amaefule, J.O., Mehmet, A., Djebbar, T. et al. (1993). Enhanced reservoir description: using core and log data to identify hydraulic (flow) units and predict permeability in uncored intervals/wells, SPE 26436.

Ebanks W. J (1987). Flow unit concept – integrated approach for engineering projects. Abstract presented June 8, during the roundtable sessions at the 1987 American Association of Petroleum Geologists.

Kracha, N. (2011) (Doctorat, UNIV DE LILLE France, IFP, SONATRACH 2011). Relation entre sédimentologie, fracturation naturelle et diagenèse d'un réservoir a faible perméabilité application aux réservoirs de l'ordovicien bassin de l'Ahnet, Sahara Central, Algerie. Doctorat 2011

Soto, B.R., Torres, F., Arango, S., and Cobaleda, G. (2001). Improved reservoir permeability models from flow units and soft computing techniques: a case study, Suria and Reforma- Libertad fields, Colombia, SPE 69625.

Tawadros, E.E. (2011). *xGeology of North Africa*. Boca Raton, NW: CRC Press, Taylor and Francis Group, A Balkema Book. ISBN-13:978-0-203-13061-2.

17

Integration of Rock Types and Hydraulic Flow Units for Reservoir Characterization. Application to Three Forks Formation, Williston Basin, North Dakota, USA

Aldjia Boualam and Sofiane Djezzar

University of North Dakota, Grand Forks, ND, USA

KEYPOINTS

- Three different techniques were used to classify rock types in the Three Forks Formation.
- The SOM technique showed high accuracy in classifying rock types and their depth distribution.
- Six rock types were defined, with PRT1 and PRT2 being the most abundant in the upper and middle Three Forks Formation and characterized by dolostones with relatively high permeability.
- Porosity and permeability analysis using Winland's R35 and RQI/FZI techniques revealed dominant pore throat diameters between 0.025 and 1.5 μm.
- Rock types with pore throat less than 0.5 μm had a significant impact on fluid flow.
- The identified hydraulic flow units exhibited similar normalized porosities but different permeabilities.
- PRT1 and PRT2 were identified as the best reservoir zones in Three Forks Formation.

17.1 Introduction

Understanding the architecture of the reservoir is critical for reservoir characterization and simulation purposes. In addition, the identification of the rock typing is critical for the spatial distribution of reservoir properties in the 3D static models. The porosity and permeability have long been used to describe the reservoir quality and to define the rock types. The reservoir quality is described by its hydrocarbon storage and flow capacity which are related to porosity and permeability, respectively. Saneifar et al. (2015) stated that a consistent characterization of carbonate reservoirs can be improved by rock classification based on both geological and petrophysical properties. However, the comparison and integration between geological and petrophysical properties in reservoir characterization are very complex. The concept of the hydraulic flow units has been developed to integrate geological heterogeneities and petrophysical properties into reservoir characterization.

The flow units have various definitions that are resultant of the depositional environment and diagenetic processes. A flow unit is a stratigraphic continuous interval of similar geological and petrophysical properties which does not always coincide with lithofacies. Based on that, many empirical equations and techniques have been developed to define the hydraulic flow units and rock types.

Unconventional Hydrocarbon Resources: Prediction and Modeling Using Artificial Intelligence Approaches, First Edition. Edited by Sid-Ali Ouadfeul.

As mentioned in the previous chapter, the reservoir characteristics measurements (permeability and porosity) at reservoir conditions are critical parameters in reservoir simulation and long-term production forecasts, in addition to the rock-type identification and their distribution to improve and predict the optimum reservoir production conditions. In this study, 21 wells were selected from the collected database. The permeabilities and porosities of the wells were measured at a net stress range between 2500 and 3000 psi. At this range of net stress, it was observed that the permeability reduction rate can reach up to 98% of the initial permeability. In this case, the porosity and permeability that have been measured at different confining stresses were not used for the identification of the petrophysical rock types (PRTs). This is to ensure that the differences in the data are due to rock types' assignment, not due to variations in confining stress between different studies. Porosity and permeability at in situ reservoir conditions are controlled not only by mechanical compaction but also by cementation, mineral composition, and diagenesis.

In this chapter, an integrated workflow was used to define the rock types and hydraulic flow units' distribution within the Three Forks reservoir based on well logs and conventional core data. In the first step, we assessed the use of the neural network technique based on indexation and probabilistic self-organizing map (IPSOM) for the rock-type classification and prediction in the Three Forks reservoir in order to integrate the carbonate reservoir heterogeneity and geological properties in rock-type characterization. The principal advantage of this technique is its capability to integrate core and log data with a neural network algorithm and captures the log responses to characterize the reservoir. In the second step, the deterministic approaches were applied to define the rock types including the rock-type classification based on pore throat distribution from the Winland technique (R_{35}). This technique uses porosity and permeability combined with pore throat lines to establish the PRTs. Finally, the reservoir quality index (RQI) and flow zone indicator (FZI) techniques were applied to define the flow units. This workflow ties between petrophysical and geological properties in the Three Forks Formation in order to evaluate the reservoir quality and predict the rock types which can be used in the distribution of the reservoir properties in building 3D static models.

17.2 Petrophysical Rock-Type Prediction

The PRTs are predicted at a log scale and are generally performed using deterministic methods combined with clustering methods, neural networks, or multivariate statistical tools. In this study, the dataset that included gamma-ray, bulk density, neutron porosity, compressional slowness, elemental capture spectroscopy, nuclear magnetic resonance, and deep induction resistivity was used to characterize the different rock types. Principal component analysis (PCA) was conducted in Chapter 4 to find the best curves in partitioning the dataset into the obtained rock types. Many studies have been performed to predict the rock types by using different approaches (Isleyen et al. 2019). In addition, the application of deep learning algorithms on rock-type prediction has been proven by many authors (Wang et al. 2018a, b; Yu et al. 2012). The accuracy of the predicted model depends mainly on the input data and on the selected machine learning algorithm. In this chapter, we evaluated the performance of the self-organizing map (SOM) method for rock-type prediction in a carbonate reservoir. Milijkovic (2017) demonstrated the powerfulness of the SOM technique for classifications where clusters of similar patterns can be detected without supervision.

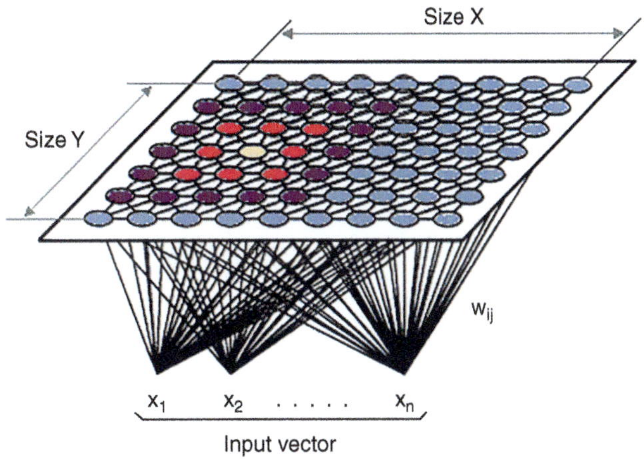

Figure 17.1 Self-organizing map (SOM). It defines a mapping from input space onto a two-dimensional map.

A clustering method, such as SOM, is an unsupervised data analysis technique. It uses a neural network algorithm. SOM is used to map high-dimension data into a 1- or 2-dimensional map. The two-layered structure of the SOM neural network is illustrated in Figure 17.1. It consists of an input layer fully connected to a competitive output layer and there are no hidden neurons.

The SOM is associated with a regular grid that is divided into a number of cells and each grid cell represents a particular group of points (cluster or nodes). SOM uses the input variables to identify patterns in the dataset by which the data points are classified into clusters based on the similarities of their neighbors.

The unsupervised hierarchical clustering technique is used to classify the rock types in conjunction with the dendrogram chart. This algorithm carries out a step-by-step grouping of the most similar log responses. In addition, in cluster analysis, the indexation technique Ward method was used. In this method, the attempt is to minimize intraclass variance. It uses both the number of depth points in each node and the distance of each point from the node value for indexation. Though, the dendrogram chart was used to define the optimum number of rock types to be achieved. Figure 17.2 shows that the height of the variables grouped together is quite low, which means that the similarities between the points are high. The outputs (rock types) were represented in different colors. Hence, a total of six rock types can be considered the optimum number to be used in hierarchical clustering analysis.

As mentioned earlier, SOM uses an unsupervised neural network learning algorithm. Initially, the nodes in SOM contain random values displayed in an unorganized SOM map, and also the weight for each variable was initiated randomly. When the model starts learning, the nodes' values keep changing based on the detected patterns from the dataset and the influence of each node on neighboring nodes. Also, the weights are updated for each node in every iteration. These weights show how influential an input variable is on the distribution of the data to each node. The SOM starts organizing the dataset spatially. Once the learning is complete, the nodes are updated and automatically indexed. The indexation is made based on a Ward technique to assign a classification to each node on the SOM map.

We repeated SOM analysis for several class numbers to examine the consistency of the performance of the SOM technique and see if it has any effect on the rock-type classification accuracy. The effect of the number of rock types is evaluated by comparing the dendrogram charts and the output statistics. It was observed that the 2nd to 5th clusters remain almost the same where the

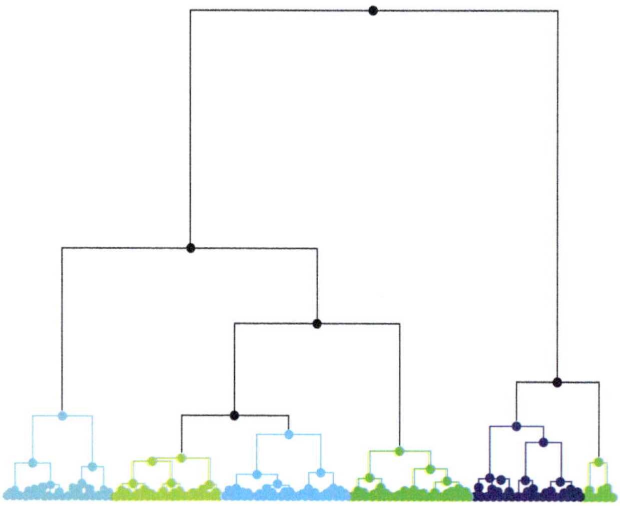

Figure 17.2 2D Dendrogram identifying the 6 optimum rock types.

content of samples for each cluster did not change significantly when changing the class number. However, the 1st cluster was the one that changed. Tables 17.1 and 17.2 illustrate the output statistics of the two models with 5 and 6 classes, respectively. It can be seen from Table 17.2 that the 1st cluster was divided into two subgroups with 18 samples for each when the class number increases from 5 to 6.

Figure 17.3 represents, based on the consistency of log patterns, the SOM map subdivided into six clusters. The rose diagrams show the normalized log responses related to each node. The result of indexation was represented by the colors which refer to the rock-type definition assigned to the nodes. The distribution of rock types versus depth is illustrated in Figure 17.4. This figure represents a type of log for the Three Forks Formation, which contains (a) the stratigraphic interval including upper, middle, and lower Three Forks; (b) NMR and GR; (c) petrophysical analysis model; (d) identified PRTs in the log domain; (e) the log facies model, and (f) core images of the predicted PRTs in well Charlie Sorenson 17-8-3THF. It is evident (track 8) that the rock types' classification of Three Fork Formation based on well logs with the SOM algorithm produced satisfactory results. Although the distribution of anhydrite in LTF was classified with high accuracy compared to the multimineral model. A good agreement was observed between the petrophysical model, cores, and the predicted rock types using SOM.

17.3 Rock Types' Classification Based on R_{35} Pore Throat Radius

A PRT classification using the Winland method was undertaken based on core measurements, porosity, and permeability, of the 21 wells. According to Gunter et al. (1997), the PRT are units of the reservoir that can be characterized by similar diagenetic and depositional conditions resulting in a typical porosity-permeability relationship and pore throat size distribution.

Winland developed an empirical relationship for estimating pore aperture-size parameters from the mercury injection-capillary pressure curve (Pittman 1992). He found out, by testing low

Table 17.1 Output statistics table with five rock types.

Cluster	Number of nodes	R_t mean	R_t variance	DT mean	DT variance	NPHI mean	NPHI variance	RHOB mean	RHOB variance	GR mean	GR variance
1	31	9.72	7.96	67.62	5.71	0.16	0.00033	2.67	0.00058	86.12	92.36
2	22	17.95	13.63	66.05	17.088	0.11	0.00046	2.71	0.00198	65.90	59.97
3	17	9.24	4.63	57.48	10.39	0.17	0.00021	2.66	0.00034	90.17	32.92
4	25	5.84	2.25	65.48	4.83	0.20	0.00047	2.63	0.00058	104.25	135.55
5	5	22.57	13.7	62.32	19.97	0.04	0.00019	2.85	0.00054	40.84	45.51

Table 17.2 Output statistics table with six rock types.

Cluster	Number of nodes	R_t mean	R_t variance	DT mean	DT variance	NPHI mean	NPHI variance	RHOB mean	RHOB variance	GR mean	GR variance
1	18	18.23	17.54	67.22	11.92	0.10	0.00047	2.72	0.00211	64.59	61.79
2	21	6.81	3.16	67.03	8.91	0.18	0.00023	2.66	0.00037	93.30	52.93
3	18	11.97	20.36	57.80	9.29	0.16	0.00062	2.66	0.00019	86.13	97.59
4	20	5.47	2.70	65.45	7.45	0.21	0.00017	2.62	0.00029	109.17	102.26
5	5	22.48	13.40	62.53	21.24	0.04	0.00015	2.85	0.00056	39.96	37.33
6	18	10.90	4.15	67.55	6.20	0.15	0.00012	2.68	0.00066	79.72	57.08

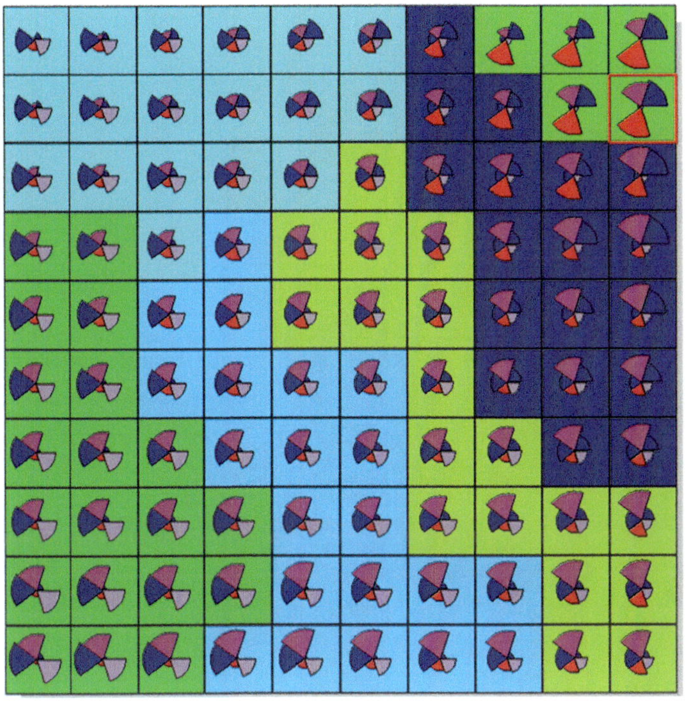

Figure 17.3 Sorted and organized self-organized map subdivided into six rock types indexed with different colors.

permeability sandstone and carbonate samples, that the effective pore network that dominates fluid flow corresponds to mercury saturation of 35%. R_{35} represents the size of the pore throats radius at 35% nonwetting phase saturation as determined from mercury injection capillary pressure (MICP) tests (Figure 17.5). In other words, R_{35} has been interpreted to be the reference where the pore network becomes interconnected. Winland used an R_{35} value of 0.5 μm as the definition of net pay for the Spindle Field (Colorado) based on the producing wells having $R_{35} > 0.5\,\mu\text{m}$. R_{35} is estimated from Winland's equation 17.1:

$$\text{Log}\,R_{35} = 0.732 + 0.588\,\text{Log}\,K - 0.864\,\text{Log}\,\phi \tag{17.1}$$

The Winland plot can be complemented with MICP measurements to determine pore throat entry pressures, the pore throat diameter controlling fluid flow, and the pressure at which hydrocarbons will not enter existing pores giving a measure of connate or immovable water saturation. This approach provides accurate rock types' classification based on pore throat distribution in the reservoir.

17.3.1 Upper Three Forks

Figure 17.6 illustrates the core porosity and permeability distribution based on 1100 samples combined with Winland R_{35} isopore throat lines. The distribution of the points along each line

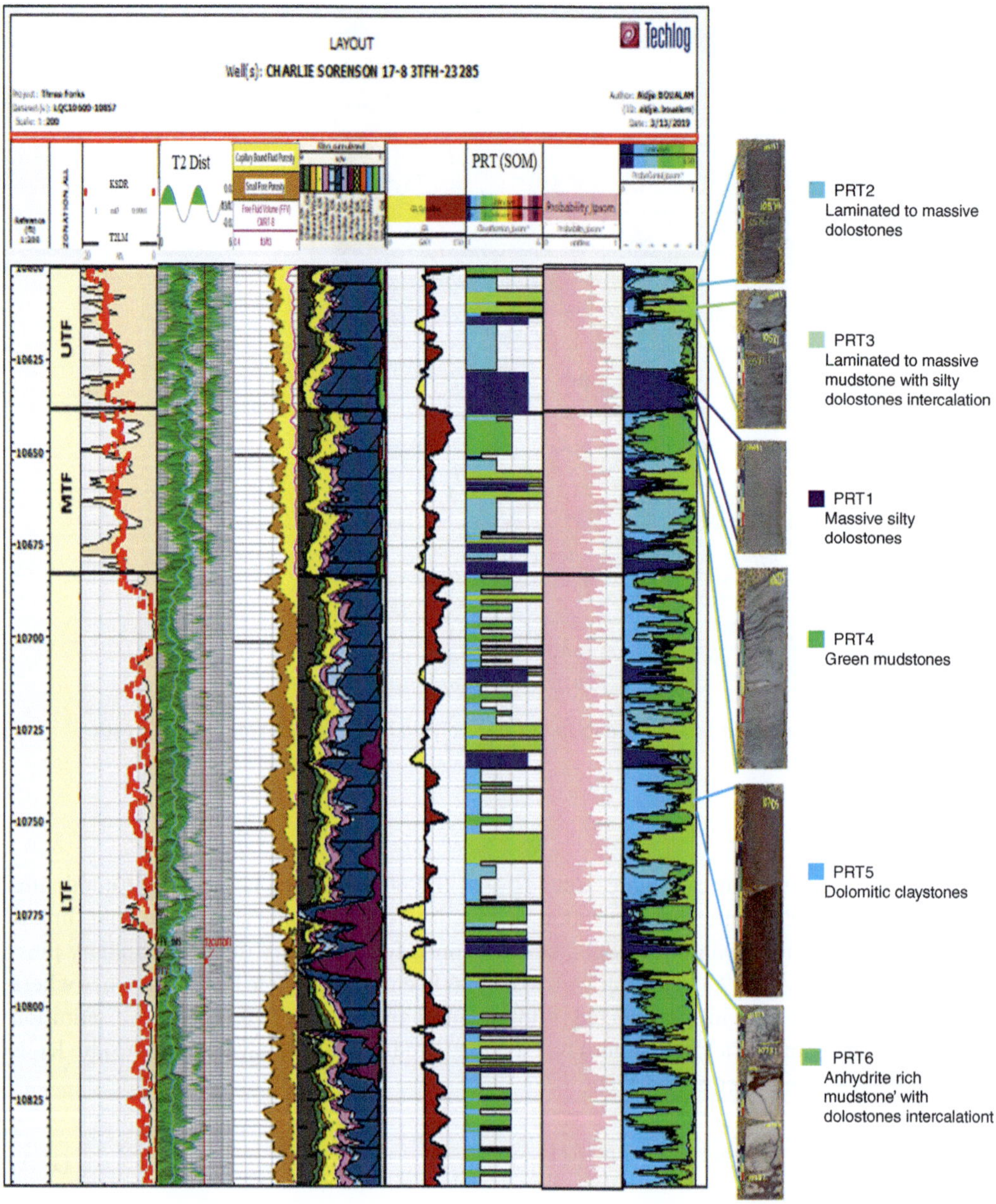

Figure 17.4 Well Charlie Sorenson 17-8-3TFH. Results of multimineral petrophysical model and the rock-type classification. From left to right, track 1, induction deep resistivity (AT90). Track 2, compressional slowness (DTCO). Track 3, neutron porosity (NPHI). Track 4, lithological model from the petrophysical analysis. Track 5, total gamma-ray (GR). Track 6, six rock types are correctly identified by the SOM algorithm. Track 8, distribution of rock types versus depth. Cores photographs for different lithofacies observed is studied well.

represents the same characteristics. Six PRTs were identified from the cross-plot of logarithmic permeability versus the porosity. PRT6 forms the poor reservoir interval, with porosity ranging from 1 to 10% and permeability less than 0.006 mD. In contrast, PRT1 has much better permeability (1–10 mD) and porosity ranges from 2.5 to 11%. Calculated R_{35} is 0.02–0.05 μm for PRT6; 0.05–0.1 μm for PRT5; 0.1–0.2 for PRT4; 0.2–0.5 μm for PRT3; 0.5–1.5 for PRT2; and 1.5–4.5 μm for PRT1.

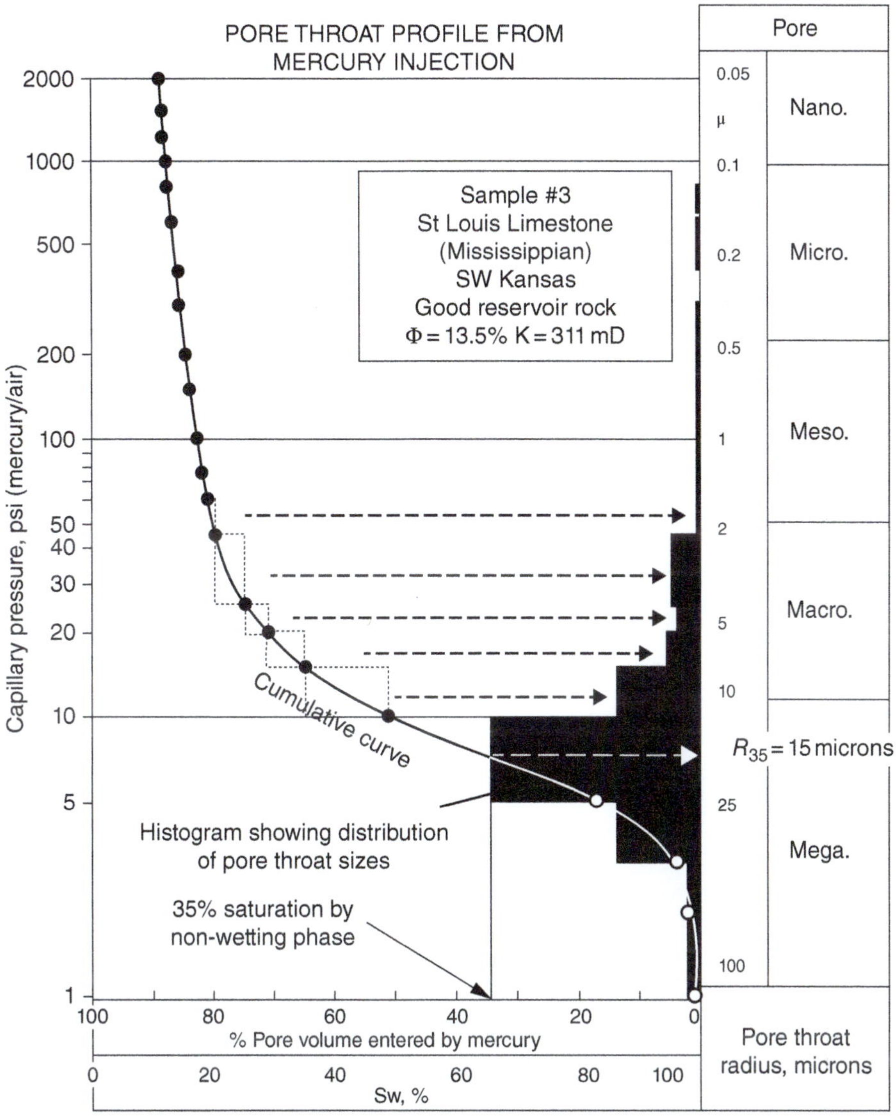

Figure 17.5 Pore throat size profile and capillary pressure curve from mercury injection.

17.3.2 Middle Three Forks

A total of six PRTs were identified from the cross-plot of logarithmic permeability versus porosity (Figure 17.7). PRT6 is characterized by a porosity ranging from 1.5 to 10% and permeability of less than 0.006 mD. PRT5 has a porosity ranging between 2.5 and 10.7% and permeability ranging from 0.02 to 0.02 mD. For PRT4, the porosity and permeability vary from 2.7 to 11% and from 0.03 to 0.09 mD, respectively. The porosities are consistent for PRTs 3, 2, and 1 (3–11%), while PRT1 exhibits much higher permeability (2–10 mD). The high permeability is generally related to the existence of natural fractures in tight formations. Calculated R_{35} are 0.02–0.05 μm for PRT6 and 1.5–4.5 μm for PRT1.

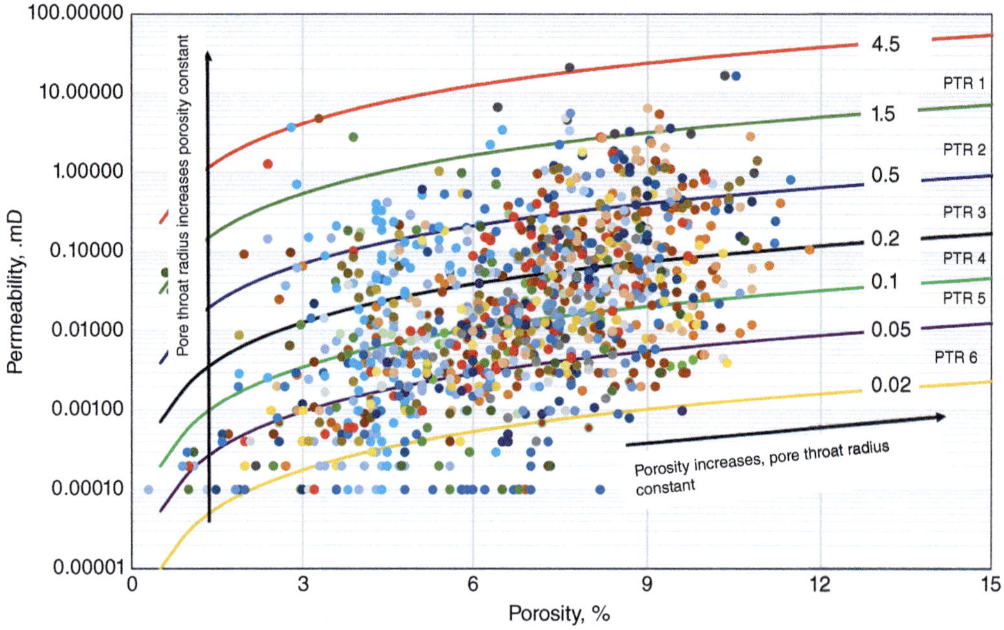

Figure 17.6 Permeability versus porosity core cross-plot combined with Winland R_{35} isopore throat lines, UTF. Winland R_{35} corresponds to the pore throat radius at 35% mercury saturation from the mercury injection test. Six rock types were defined with a pore throat radius ranging between 0.02 and 4.5 μm. The porosity is between 1 and 11.5% and the permeability is generally less than 10 mD.

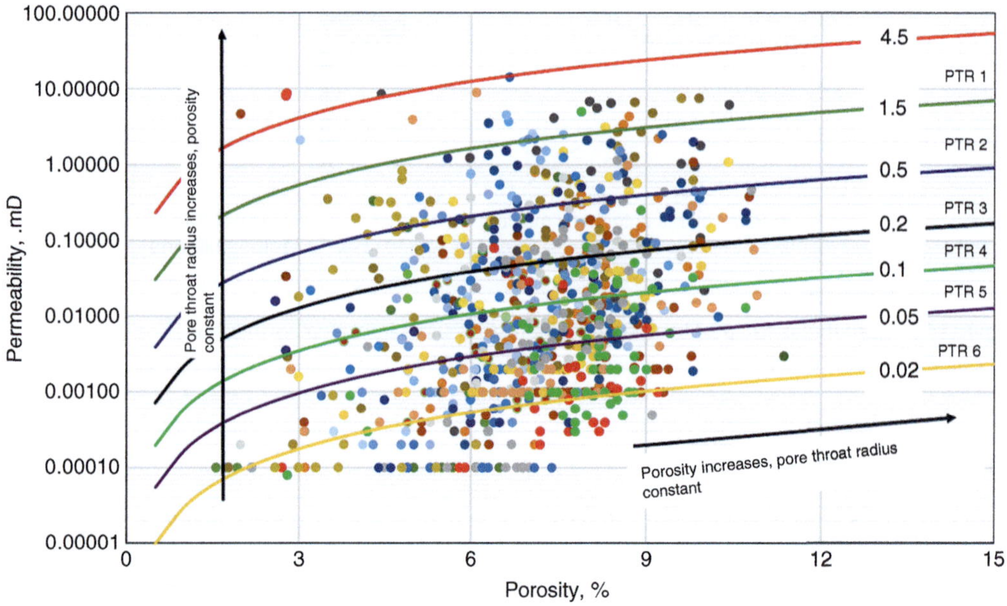

Figure 17.7 Logarithmic of permeability versus porosity core cross-plot combined with Winland R_{35} isopore throat lines, MTF. Six rock types were defined with a pore throat radius ranging between 0.02 and 4.5 μm. The porosity is between 2 and 11% and the permeability is generally less than 10 mD.

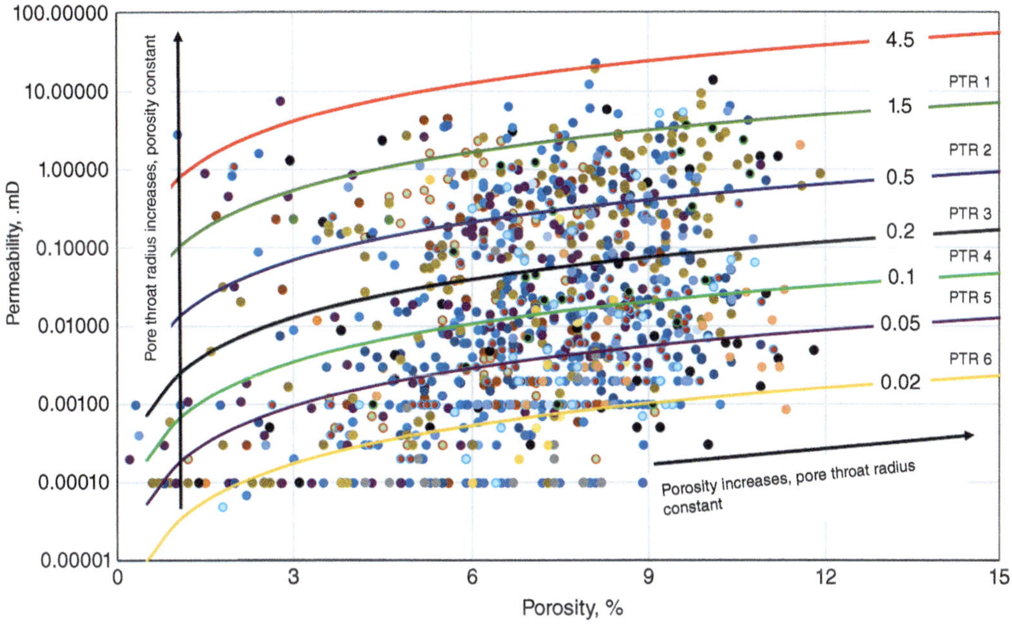

Figure 17.8 Logarithmic of permeability versus porosity core cross-plot combined with Winland R_{35} isopore throat lines, LTF. Six rock types were defined with a pore throat radius ranging between 0.02 and 4.5 μm. The porosity can reach 11% and the permeability is generally less than 10 mD.

17.3.3 Lower Three Forks

A total of six PRTs are illustrated in Figure 17.8 with a consistent porosity varying from 0.2 to 11%. However, the permeability is less than 0.004 mD for PRT6 and for the same range of porosity, PRT1 is characterized by higher values of permeability (0.2–10 mD).

17.4 Determination of Hydraulic Flow Units

The complexity of the unconventional reservoirs requires providing a better description of the reservoir and distinguishing the essential features of geological and petrophysical parameters affecting the fluid flow. The challenge is to group the reservoir volume into units that basically have consistent petrophysical properties which exclusively characterize its static behavior. Hence, the delineation of reservoir compartments is one of the critical factors to accomplish a vigorous development plan based on the static model and reservoir simulation model.

RQI and FZI have been introduced on the basis of stressed porosity and permeability for the identification of hydraulic flow units. The technique is based on the Kozeny-Carmen equation and the concept of the mean hydraulic radius. This technique has been successfully used in clastic and carbonate reservoirs. The generalized form of Kozeny-Carmen equation is given by equation 17.2:

$$K = \frac{\phi_e^3}{(1-\phi_e)^2}\left(\frac{1}{F_s\tau^2 S_{gv}^2}\right) \tag{17.2}$$

where F_s is the pore throat shape factor, τ is the tortuosity, S_{gv} is the specific surface area of the grain (μm^{-1}), $F_s\tau^2$ is Kozeny constant, K is the permeability, and ϕ_e is the effective porosity in fraction. It has been found that the term $F_s\tau^2$ can vary from 5 to 100 in the reservoirs. However, this term is constant within the same flow unit. Dividing equation 17.2 by ϕ_e (effective porosity):

$$\sqrt{\frac{K}{\phi_e}} = \left(\frac{\phi_e}{1-\phi_e}\right)\left(\frac{1}{\sqrt{F_s}\tau S_{gv}}\right) \tag{17.3}$$

If the permeability is in millidarcies, the RQI is defined as follows:

$$\mathrm{RQI}(\mu m) = 0.0314\sqrt{K/\phi_e} \tag{17.4}$$

And the Normalized Porosity Index, ϕ_z, is represented as:

$$\phi_z = \left(\frac{\phi_e}{1-\phi_e}\right) \tag{17.5}$$

It has been mentioned that RQI versus Normalized Porosity Index (ϕ_z) should yield a straight line with a unit slope and the intercept of unit slope line with $\phi_z = 1$, labeled as (μm), FZI, which is given by:

$$\mathrm{FZI} = \left(\frac{1}{\sqrt{F_s}\tau S_{gv}}\right) = \frac{\mathrm{RQI}}{\phi_z} \tag{17.6}$$

Substituting 17.6 to equation 17.3 and taking the logarithmic of both sides:

$$\log \mathrm{RQI} = \log \phi_z + \log \mathrm{FZI} \tag{17.7}$$

The method used to define the FZIs required calculating RQI and ϕ_z from equations 17.4 and 17.5 and then plot them on the log-log space of RQI versus ϕ_z. The data will lie along parallel lines with unit slope. The intercept of the unit slope line at $\phi_z = 1$ determines the FZI constant. Consequently, the sample values that lie on the same unit slope line constitute a hydraulic flow unit with similar pore throat characteristics. The RQI and FZI are related to the effective pore throat radius which gives a better understanding of the reservoir in terms of fluid flow.

17.4.1 Upper Three Forks

Plotting the RQI of the UTF core samples as a function of normalized porosity with the addition of the FZIs into the plots, as shown in Figure 17.9, leads to classifying the reservoir quality into HFUs. A significant degree of overlap between HFUs is observed on the RQI-ϕ_z cross plot which is related to the large variance of the permeability with a consistent porosity through the interval. In addition, the scattering of the data samples within a single HFU is relatively important. Six distinct HFUs can be determined over the UTF reservoir, and they are characterized by average FZI values ranging from 0.05 to 2.

The calculated Winland pore throat radius (R_{35}) and RQI are plotted in Figure 17.10. A strong correlation can be seen between RQI and Winland R_{35} in the UTF reservoir. One can see from the figure that an overall consistency is achieved from both techniques.

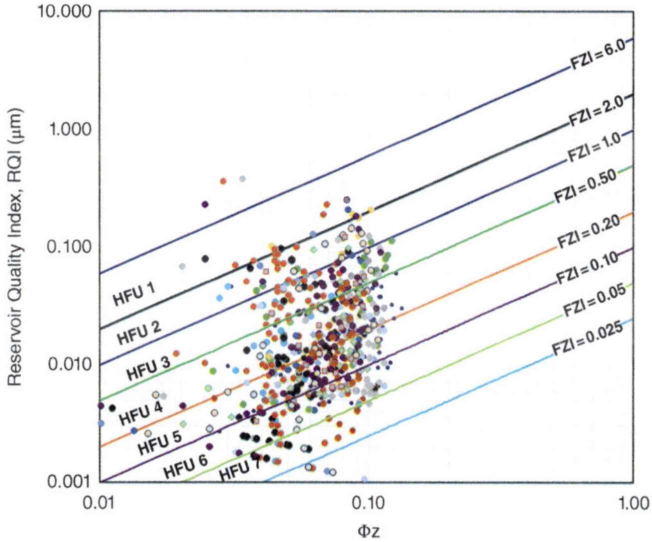

Figure 17.9 Reservoir quality index versus normalized porosity cross-plot (Log-Log scale) combined with the flow zone indicator (FZI), UTF. Six hydraulic flow units were defined with an FZI range between 0.05 and 2.

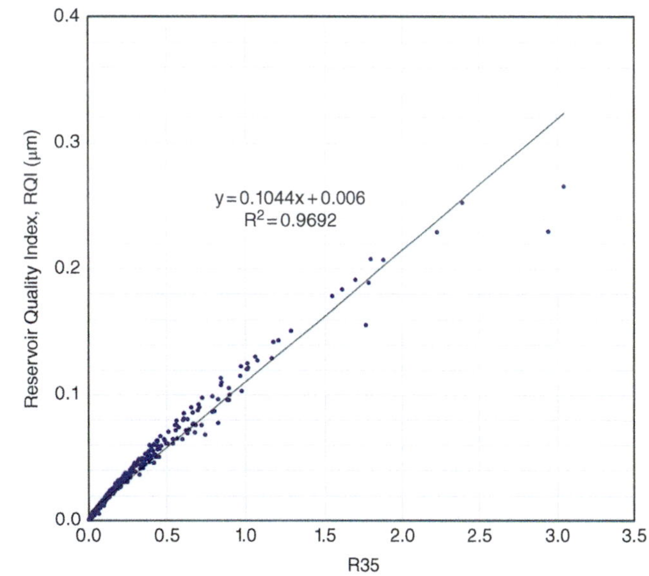

Figure 17.10 Reservoir quality index versus Winland R_{35}, UTF.

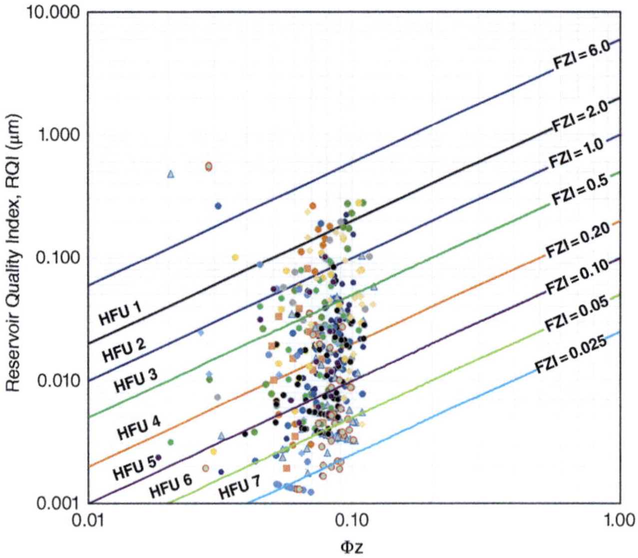

Figure 17.11 Reservoir quality index (RQI) versus normalized porosity cross-plot (Log-Log scale) combined with the flow zone indicator (FZI), MTF. Six major hydraulic flow units were defined with an FZI range between 0.05 and 2.

17.4.2 Middle Three Forks

Figure 17.11 shows the results of RQI/FZI technique of the conventional core samples from MTF. There appear to be six major HFUs within the MTF reservoir. The HFU 7 is characterized by poor properties with an average FZI of 0.025 because of its distinct mineralogy with high clay content. It can be a barrier which may have a significant impact on fluid flow. A high number of core samples are classified in HTU 3 to 6. On the other hand, HTU 1 and HTU 2 exhibit better FZIs.

17.4.3 Lower Three Forks

The RQI/FZI technique has been extended to the LTF reservoir. As shown in Figure 17.12, the LTF core samples can be grouped into 7 HFUs. Figure 17.12 shows a large number of data samples falling in the nanopores region with an average FZI ranging between 0.025 and 0.25 in this sequence. The lower values of FZI in the LTF appear to be mineralogically controlled. It is evident that FZI decreases with increasing the volume of clay minerals and anhydrite content in this unit.

17.5 Conclusion

Rock-type classification in Three Forks Formation was carried out based on three different techniques. We first explored the performance of SOM using an unsupervised clustering technique. This initial investigation on a few wells showed that the constrained SOM map illustrates high classification accuracy of the rock types and their distribution with depth. The distribution of the predicted rock types in the log domain is illustrated in Figure 17.4.

An overall verification of the consistency of the rock types' prediction was conducted using the output statistics, the petrophysical models, and cores. Six rock types were defined, PRT1 corresponds to massive silty dolostones characterized by low GR, high dry-weight calcium (DWCA), low dry-weight aluminum (DWAL), and high dolomite content. PRT2 corresponds to laminated to

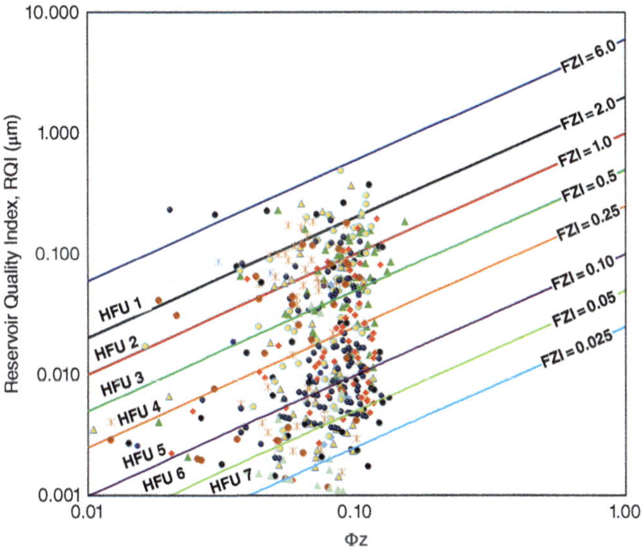

Figure 17.12 Reservoir quality index versus normalized porosity cross-plot (Log-Log scale) combined with the flow zone indicator (FZI), LTF. Seven hydraulic flow units were defined with an FZI range between 0.025 and 2.

massive dolostones and it is characterized by higher GR compared to PRT1. The two rock types (PRT1 and PRT2) are much more abundant in the upper and middle Three Forks Formation.

PRT3 is defined as laminated to massive mudstone with intercalation of silty dolostones. However, the nonreservoir PRT4 to PTR6 have very low T_2 distribution which indicates the nonpresence of vugs and high micropore porosity in this unit (Figure 17.4, tracks 4 and 5). They correspond to green mudstones, dolomitic claystones, and anhydrite-rich mudstone with dolostone intercalation, respectively. The latter is characterized by low GR with high resistivity and bulk density.

To investigate the impact of pore throat sizes on the fluid flow, the porosity and permeability from 21 wells were analyzed using both Winland's R_{35} and RQI/FZI techniques. Nevertheless, the reservoir MICP data was not available, which prevented to include the dynamic rock types in this study.

Winland R_{35} equation was used for the calculation of pore throat radius which is used as the main factor to classify the reservoir into six rock types. The dominant pore throat diameters in Three Forks Formation are between 0.025 and 1.5 μm .

The results from the plots of RQI versus normalized porosity show that the identified hydraulic flow units exhibit an almost similar range of normalized porosities but a different range of permeabilities. Consequently, some discrepancies are observed between upper, middle, and lower Three Forks where the estimated FZI ranges between 0.05 and 2 for UTF and between 0.025 and 2 for MTF and LTF which endorse the abundant of the nanopores and micropores in these two units (MTF and LTF). Furthermore, it was found that there is a good correlation between Winland's R_{35} and RQI/FZI results. The maximum matrix R_{35} appears to be 0.5–2 μm. The results indicate that a large number of core samples fall into the nanoporous (< 0.1 μm) and microporous ($0.1 < R_{35} < 0.5$ μm) rock types as classified in Figure 17.5. Only some points of macropore at R_{35} higher than 2 μm can be observed. The rock types with pore throats less than the threshold 0.5 μm will have a significant impact on the fluid flow.

Additionally, the PRT1 and PRT2 appear to be the best reservoir zones in Three Forks Formation because of the abundant of the dolostones with relatively high permeability and are characterized by higher R_{35}, having a higher percentage of mesoporous and macroporous rock types (>0.5 μm). Both rock types are predominantly distributed in the upper and middle Three Forks.

References

Gunter, G.W., Finneran, J.M., Hartmann, D.J., and Miller, J.D. (1997). Early determination of reservoir flow units using an integrated petrophysical method: Society of Petroleum Engineers. *SPE-38679-MS presented at the SPE Annual Technical Conference and Exhibition*, San Antonio, TX (5–8 October 1997), pp. 373–380. https://doi.org/10.2118/38679-MS.

Isleyen, E., Demirkan, D. C., Duzgun, H.S., Rostami, J., 2019. Lithological classification of limestone with self-organizing maps. *53rd U.S. Rock Mechanics/Geomechanics Symposium*, New York, NY, USA (23–26 June 2019). Paper Number: ARMA-2019-1791.

Milijkovic, D. (2017). Brief review of self-organizing maps. *40th International Convention on Information and Communication Technology, Electronics and Microelectronics (MIPRO)*, Opatija, pp. 1061–1066.

Pittman, E. D. (1992). Relationship of porosity and permeability to various parameters derived from mercury injection-capillary pressure curves for sandstone. *AAPG Bulletin*, 76: 191–198.

Saneifar, M., Skalinski, M., Theologou, P. et al. (2015). Integrated petrophysical rock classification in the McElroy field, West Texas, USA. *Petrophysics* 56 (5): 493–510.

Wang, W., Ren, X., Zhang, Y., and Li, M. (2018a). Deep learning based lithology classification using dual-frequency Pol-SAR data. *Applied Sciences* 8: 1513.

Wang, X., Yang, S., Zhao, Y., and Wang, Y. (2018b). Lithology identification using an optimized KNN clustering method based on entropy-weighed cosine distance in Mesozoic strata of gaoqing field, jiyang depression. *Journal of Petroleum Science and Engineering* 166: 157–174.

Yu, L., Porwal, A., Holden, E., and Dentith, M.C. (2012). Towards automatic lithological classification from remote sensing data using support vector machines. *Computers & Geosciences* 45: 229–239.

18

Stress-Dependent Permeability and Porosity and Hysteresis. Application to the Three Forks Formation, Williston Basin, North Dakota, USA

Aldjia Boualam and Sofiane Djezzar

University of North Dakota, Grand Forks, ND, USA

KEYPOINTS

- The study investigated the stress-dependent permeability and porosity of samples from upper and middle Three Forks Formation.
- The relationship between permeability, porosity, and net confining pressure was analyzed, and the results showed that the permeability and porosity decrease with an increase in net confining stress.
- The study observed irreversible loss in permeability during the unloading phase and an increase in permeability during stress unloading in some samples.
- The study highlights the importance of considering stress effects in permeability estimation and its impact on production forecast via reservoir simulation.
- Further experimental analysis on permeability hysteresis should be added for upper and middle Three Forks members to obtain more repeatable results.

18.1 Introduction

During the production stage, the reservoir losses its original pore pressure gradually, which means an increase in its net effective stress, or the stress that is applied to the rock's grain-to-grain contact. The decline in production rate is also manifested in a rapid decline of bottom-hole flowing pressure (Kurtogglu et al. 2013; Anderson et al. 2010). The reservoir properties can change in response to these changes in net stress within the pores due to the withdrawal of the fluids from the reservoir. Yildirım and Ers (2007) documented that the cyclic loads can cause damage and deformation in the rock samples, leading to a variation in rock permeability and porosity. The combination between the increasing stresses and multiple factors such as the lithology, rock fabric, content and location of clay particles, grain-contact type, the compressibility of the pores, and the existence of the microstructures in the rock can lead to variable decline rates of the permeability and porosity and their resulting paths. Under cyclic unloading, the deformation rock response also varies and these factors may, to some degree, cause irreversible loss of permeability and porosity. The amount of permanent deformation of the rock due to the cyclic unloading is known as hysteresis (Teklu and al. 2017). This phenomenon is observed when the unloading porosity and permeability paths

Unconventional Hydrocarbon Resources: Prediction and Modeling Using Artificial Intelligence Approaches, First Edition. Edited by Sid-Ali Ouadfeul.

are lower than the loading paths. According to Dong et al. (2010), the main reason for this difference is that the compaction of the geomaterial is not fully reversible. The stress-dependent permeability is significant in a tight formation where the rock can lose up to 99% of the original permeability and can affect the fluid flow mechanisms (Teklu and Li 2017).

Terzaghi (1943) defined the net effective stress as part of the total stresses that cause the deformation of the rock. He related the net effective stress to pore pressure and the external total stress as:

$$\sigma_{\text{eff}} = \sigma - \alpha P \tag{18.1}$$

Here, α is a poroelastic coefficient or Biot's coefficient, P is the pore pressure, σ_{eff} is the net effective stress, and σ is the total stress.

Warpinskin and Teufel (1992) defined a general effective stress law that can affect rock properties in the reservoir such as permeability and porosity. This law can be expressed mathematically as:

$$p = Q(\sigma - \alpha\, p) \tag{18.2}$$

where P is the studied property of the rock and the term $Q(\sigma - \alpha p)$ describes the effect of the effective stress on the property. This general equation is used by many authors with several studies being carried out on stress-dependent permeability and porosity on sandstone formations (Wang et al. 2014a; Dong et al. 2010; Ghabezloo et al. 2009; Davy et al. 2007; Klein and Reuschle 2003; Dana and Skoczylas 1999; Davies and Davies 1999; Davies and Holditch 1998; David et al. 1994; Luffel 1991; Morrow et al. 1986; Zoback and Byerlee 1975; and Brace et al. 1968).

Jones and Owens (1980), based on the experiment results of permeability measurements of tight gas sands, found that the cubic root of the normalized permeability can be correlated with a logarithm of net stress in the form of:

$$\left(\frac{k}{k_o}\right)^{1/3} = Q\left(\sigma - \alpha\, p\right) \tag{18.3}$$

In this equation, K is the permeability at a given net effective stress, and K_o is the permeability at a minimum or initial applied net effective stress.

Dong et al. (2010) documented that the results of the permeability and porosity on sandstone samples obtained from an integrated permeability and porosity measurement system showed that the gas permeability is related to confining pressure.

Klinkenberg (1941) measured the gas permeability of porous media where his research revealed that the permeability of a gas depends on factors that influence the mean free path, such as pressure, temperature, and the nature of the gas.

A pulse-decay method was used by Dana and Skoczylas (1999) to measure the relative gas permeability of three types of sandstones. Their test results showed that the gas permeability is closely dependent on pore structure parameters such as pore size, pore-throat radius, and pore shape.

Similarly, by using a liquid flow medium, Morrow et al. (1986) measured the permeability of Westerly granite under cyclic loading conditions, and they found that cyclic stress loading had a large effect on the permeability path of the samples where permeability decreased with each loading, regardless of the confining pressure.

David et al. (1994) directed laboratory experiments on five sandstones with high porosities, ranging from 14 to 35%. The results of their studies showed that the permeability of low-porosity samples presents relatively high-pressure sensitivity.

Permeability is a physical quantity that depends on the net effective stress state and pore structure parameters, such as porosity and pore-throat radius (Pape et al. 2000; David et al. 2001). It is clear that there is no unique permeability-porosity relationship that can be applied to describe the signature of porous media (Bernabe et al. 2003). Based on laboratory test results of different types of rocks (Wang et al. 2014a; David et al. 1994; Hoholick et al. 1984; Schmoker and Halley 1982), an exponential function was proposed to describe the relationship between gas permeability and net confining pressure. On the other hand, the experimental results of studies on sandstones presented by Dong et al. (2010) indicated that a power-law model is a better fit for the permeability versus confining pressure data than an exponential function.

It should be noted that the number of studies regarding stress dependency of permeability and porosity on tight carbonate formations is limited, particularly in Three Forks Formation, Williston Basin. The discussions presented in this chapter investigate the hysteresis response of permeability and porosity as a function of net effective stress and the parameters that influence this behavior. The study is focused on the upper Three Forks carbonate reservoir, where, to the best knowledge of the author, there are no similar studies conducted for this reservoir. The studies conducted here include a lab-based data analysis and both exponential and power-law models were used to define the stress-dependent permeability and porosity relationship of the upper Three Forks Formation. This relationship was updated using a new dataset (from well Charlie Sorenson 17-8-3TFH) under cyclic loading and unloading for a wide range of net confining stress. The results presented in this chapter will help in understanding the impact of stresses on reservoir characteristics and the hysteresis effect on the production strategies of the upper Three Forks tight reservoir.

18.2 Database

The database for this particular work includes the experimental data from well Charlie Sorenson 17-8-3TFH and a large volume of data collected from the NDIC Website. Permeability and porosity data under cyclic loading of effective stress for 48 wells were used in this study. All permeability values were measured using the unsteady-state approach and were corrected for the Klinkenberg effect. Measurements of porosity and permeability have been conducted by different laboratories using different measurement instruments and at different confining stresses (see Figures 18.1 and 18.2). At ambient conditions, the measurements were generally performed at net effective stress of either 500 psi or 800 psi. The highest frequency data in the dataset was obtained at initial net stress of 500 psi and a final net confining pressure of 4000 psi. The key was to have at least one consistent stress condition for stress-dependence analysis of permeability in order to compare the data. Therefore, 6 wells were selected in this net stress range, each with a different number of samples available (UTF: 120 samples, MTF: 165 samples, and LTF: 308 samples). The measurements of permeability and porosity under cyclic unloading were performed on four selected samples from upper Three Forks for well Charlie Sorenson 17-8-3TFH.

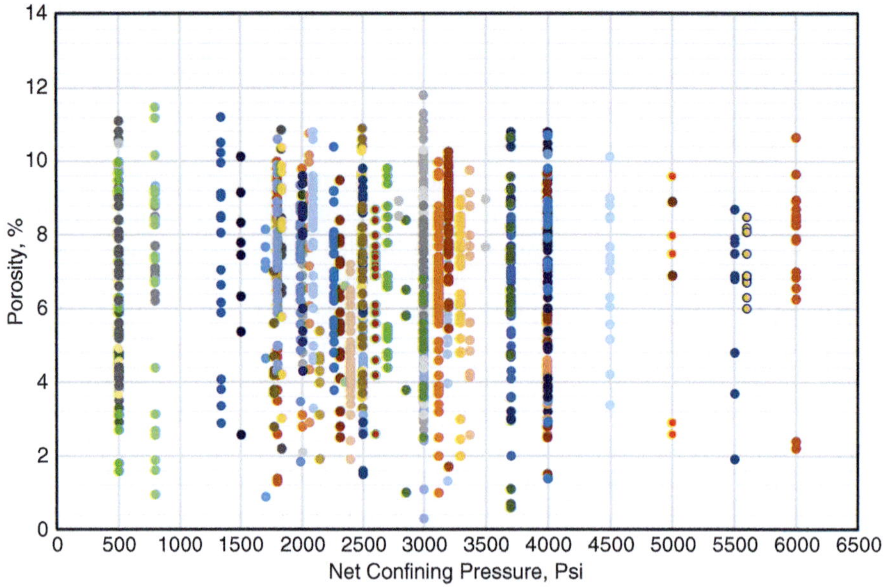

Figure 18.1 Variation of porosity at different net confining pressures for the studied wells, upper Three Forks.

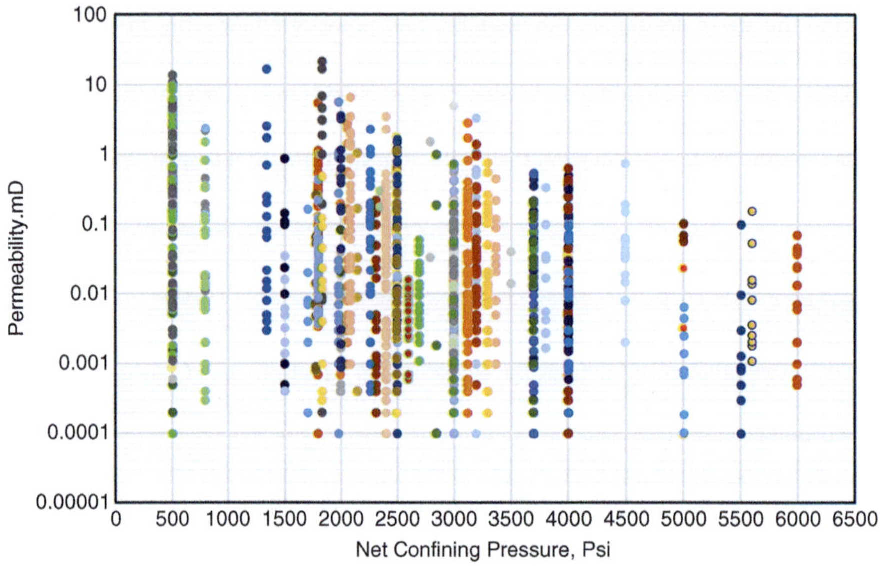

Figure 18.2 Variation of permeability at different net confining pressures for the studied wells, upper Three Forks.

18.3 Testing Procedure

Cores from five wells (Charlie Sorenson 17-8-3TFH, Jane Federal 11X-20, Mariana Trust 12X-20G2, Fort Berthold 150-94-3B-10-2H, and Anderson Federal 152-96-9-4-11H) were selected for a preliminary investigation of the best facies for sampling. Usually, the shaly dolomites will experience a greater proportional reduction in permeability due to the effective stress. One of the objectives is

Figure 18.3 (a) Core samples from upper Three Forks used for the experiments (well Charlie Sorenson 17-8-3TFH), (b) core sample 4A which shows natural fractures. The cores were drilled parallel to the lamination (horizontal plugs).

to determine if shaly dolomite facies act as permeability barriers under normal stress or become barriers due to reservoir depletion and an increase in the effective stress. However, it was not practical to drill any sample related to these facies due to the need for a special drilling facility to drill into these unconsolidated formations. Therefore, we limited the investigation just to the clean facies where we focused on the modeling of the changes in the clean reservoir zones.

Figure 18.3 shows the five horizontal plugs, 1.5 inches (3.9 cm) in diameter by approximately 2–2.75 inches (4.9–6.69 cm) in length, that were drilled from the upper Three Forks core (well Charlie Sorenson 17-8-3TFH) in a direction parallel to the slabbed face (i.e. parallel to the lamination). These samples were selected based on the petrographic and petrophysical models, and sample sizes. After the plugs were drilled from the core, they were placed in a trim saw machine, and the ends were cut flat and both ends were smoothened, as illustrated in Figure 18.4. The thin slices cut from the plug ends were used for X-ray diffraction. The API RP40 sample handling and preparation procedures were carefully followed. The samples contain significant amounts of liquid hydrocarbon. As per the API guidelines, the samples' liquid hydrocarbon, free water in the pores, and salt were removed by using the method of solvent extraction and then the samples were dried in the oven.

18.3.1 Core Samples Cleaning and Drying

Dean-Stark and Soxhlet apparatus was used for cleaning the samples. A mixture of toluene (2/3) and ethanol (1/3) was used as solvent fluids. The solvents were put in a volumetric flask and placed on a heating mantle. Once the ends of the core plugs were smoothened, they were placed into extraction thimbles (43 mm by 123 mm), and then placed in reflux core chambers where the samples were exposed to the boiling solvent. The cold water circulating above the inner chamber condenses the solvent's vapors to immiscible liquids which fall on the core samples in the thimble, and then the toluene soaks the core samples and dissolves any oil in the pores. The extraction process was terminated after four weeks when the solvent in the volumetric flask appeared clear and then compared to the CoreLab solvent color standards chart (see Figure 18.5a). Looking at Figure 18.5b, it is observed that the samples' solvents are between standards 1 and 2, so they are well cleaned.

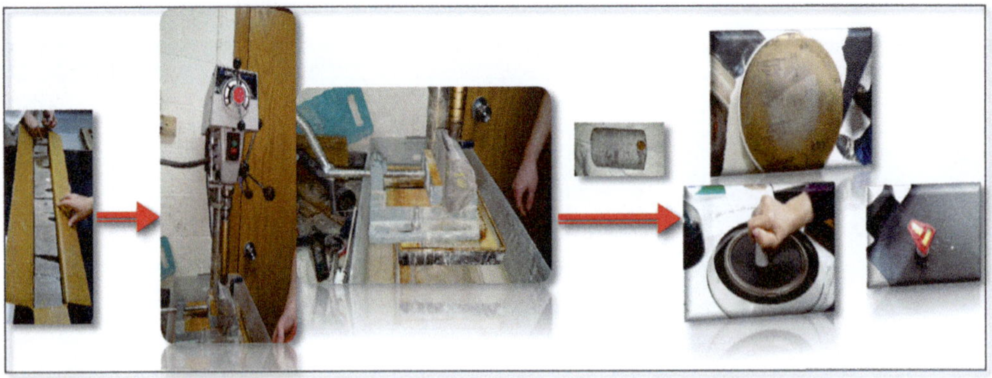

Figure 18.4 The preparation process of the cores samples is according to the API procedures and guidelines.

(a)

(b)

Figure 18.5 (a) Solvent Color Standards, CoreLab chart, (b) well Charlie Sorenson 17-8-3TFH, upper Three Forks samples. After cleaning, the solvents appear clean compared to CoreLab chart (between standard 1 and 2).

The objective of drying is to remove the water within the pore space without affecting the integrity of the core samples. After the cleaning process was complete, the upper Three Forks samples were placed in the oven at 140 °F (60 °C) temperature for 10 days. This temperature was recommended to retain the adsorbed water on clay surfaces. The drying process is completed when the samples reach a constant weight, which indicates that the free water in the pores has been removed. Also, the state of the samples was further checked at the CoreLab prior to any measurements of stress-dependent porosity and permeability. According to Bush and Jenkins (1970), clay minerals contain two types of bound water: Clay lattice water which is part of the crystal structure of the mineral, and adsorbed water which takes place between the clay plates under reservoir conditions. They point out that drying cores samples in an oven above 175 °F (80 °C) will remove not only the free pore water but also the adsorbed water from the clay minerals. They documented that the removal of the adsorbed water during the laboratory drying process will increase the sample pore space which leads to erroneously high porosity measurements.

18.3.2 Permeability and Porosity Measurements

The unsteady-state method (Core Measurement System, CMS-300) and steady-state method (Core Measurement System, CPMS) are the two methods used to measure the stress-dependency of the permeability and porosity and their hysteresis. The CPMS is a manually operated instrument, while the CMS-300 is an automated unit. The main difference though is that for the CPMS, a constant flow rate of gas is injected until stabilized pressure drawdown is established, and then the permeability is directly determined. The CMS-300, however, has three tanks of known volumes at known pressures of gas (helium and nitrogen). The gas is selected based on sample permeability. This gas is then bled through the core sample. The decaying tank pressure is monitored versus time, allowing the determination of gas flow rate and pressure drop across the core sample at any given time (Teklu and al. 2017). The permeabilities are then derived from the curve of pressure versus time. This makes the CMS-300 an unsteady-state measurement, compared to the CPMS, which is a steady-state measurement method. In addition, the CMS-300 system incorporates a porosimeter and permeameter and during the measurement, the confining stresses are applied equally in both radial and axial directions. The final output file includes the pore volume, the Klinkenberg permeability, the equivalent air permeability, Klinkenberg slip factor, Forchheimer inertial coefficient (alpha and beta), and the porosity are made automatically for each net confining pressure and for each sample. However, the CPMS, being manual, does not have the ability to calculate Klinkenberg permeability. It can measure the pore volume, porosity, and equivalent air permeability.

Also, the porosity measurements are obtained by the use of the expansion method, helium gas is allowed to expand from a known chamber into a core sample that was initially filled with helium at atmospheric pressure (Teklu et al. 2018). Then Boyle's law technique is used for both instruments, by measuring grain volume at ambient conditions and pore volume at indicated net confining stresses, to determine the porosity.

All the samples (1A, 2A, 3A, and 5A) were analyzed using the CPMS instrument (steady-state method). The size of the samples is 1.5 inches (3.9 cm) in diameter and 2.75 inches (6.69 cm) in length except for sample 1A where the length is 2 inches (4.9 cm). However, the unsteady-state method was conducted only on three samples, posterior to the CPMS measurements. Sample 3A was excluded because of the presence of fractures (see Figure 18.6). Additionally, it has the lowest porosity (less than 2% at 1000 psi) among all samples, which means that the storage capacity of this rock type is insignificant.

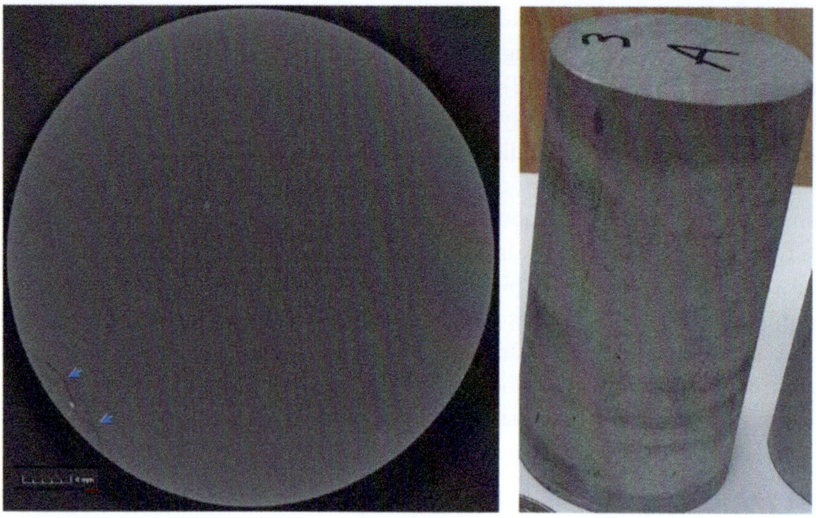

Figure 18.6 Sample 3A. The blue arrows show the fracture on the CT scan image.

Prior to the experiments using CMS-300, the original samples were under-drilled to approximately 1 inch in diameter and 1.8 inches in length to fit into the CMS-300 instrument and checked again for dry weight stability. According to Ghanizadeh et al. (2014), the unsteady-state method gives a better estimate of permeability in tight formation and also it takes into account the effect of slippage (Klinkenberg permeability). The experiments were conducted on CMS-300 by first gradually increasing the net confining pressure from 1, 2.5, 4, 5 to 6K psi and was followed by subsequently reducing it back gradually to 5, 4, 2.5, and 1K without relieving the stress between the two measurements, where the porosity and permeability were measured at each step. On the other hand, the CPMS experiments were conducted by gradually increasing the confining pressure from 1, 2.5, 4 to 5K psi but reducing it back subsequently to 1K psi. A Biot's coefficient of unity was assumed in the net effective stress calculation of all experiments. In addition, the selected core samples were used in CT scan and X-ray diffraction analyses.

Figure 18.7 illustrates the porosity and permeability values for the 4 core samples measured at initial effective stress of 1000 psi combined with Winland R_{35}. It can be seen from this figure that the samples fall between R_{35} equal to 0.1 and 1.5 μm (different flow units) and the storage capacity is decreasing from sample 1A to 3A, respectively. Sample 3A is characterized by a very low porosity (<2%) and a permeability of less than 0.1 mD. The permeability value of this sample may be affected by the fracture as depicted in Figure 18.6.

18.3.3 Mineral Composition Analysis

The ends of the plugs were used for analyzing the mineral composition and evaluating the overall mineralogy and volume of clay in the bulk sample. The samples were crushed to a fine powder using the McCrone micronizing mill and SPEX Sample Prep. mill machine (see Figure 18.8). The powdered samples were loaded into aluminum sample holders and analyzed with an X-ray diffractometer. The peak profile fitting method was used to interpret the raw data and identify the mineralogy of each sample. Table 1 (Appendix) illustrates the XRD results for the five samples. The results showed that for all samples, dolomite is the dominant mineral, ranging between 45 and

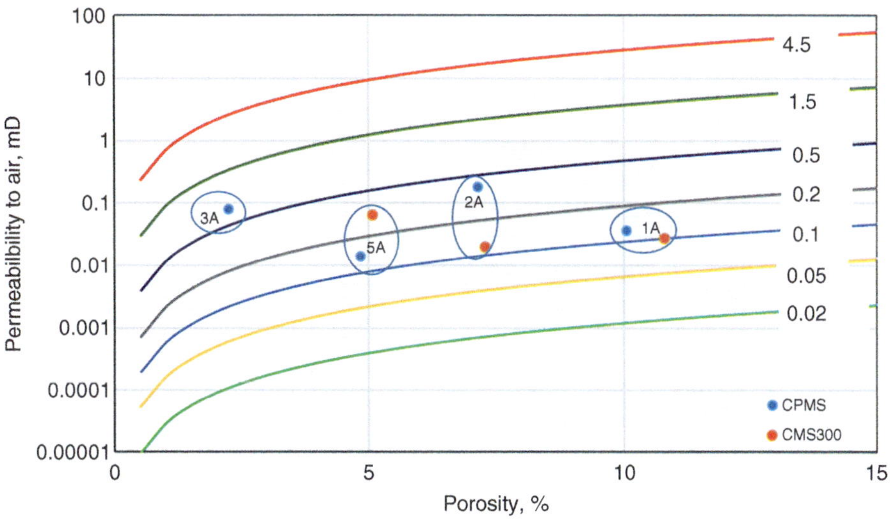

Figure 18.7 Core samples permeability versus the porosity at 1000 psi combined with Winland R_{35} isopore throat lines. Winland R_{35} corresponds to the pore throat radius at 35% mercury saturation from the mercury injection test. The permeability and porosity vary from 0.014 to 0.18 mD and from 2.25 to 10.05%, respectively.

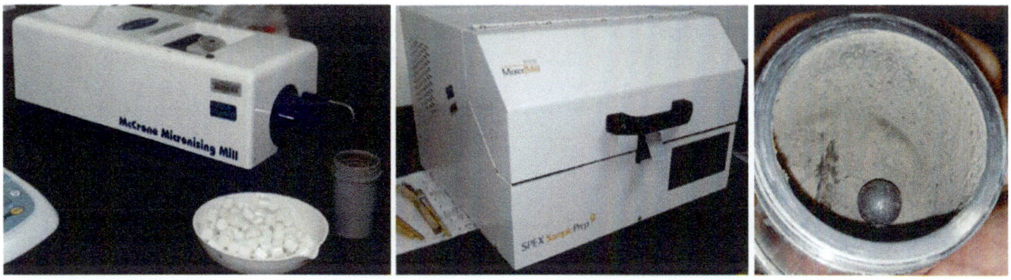

Figure 18.8 McCrone micronizing mill (left photo) and SPEX Sampleprep mill (middle photo) were used to ground the samples to very fine mesh.

68%. The volume of quartz is approximately constant (10–12.5%). The illite content is the most present as clay mineral in all samples and ranges from 9 to 14% (Figure 18.9).

18.3.4 Scanning Electron Microscope

The aim of the scanning electron microscope (SEM) is to characterize the geometry and the size of the pore systems in the upper Three Forks. According to Johnson et al. 2017, the largest pores in Three Forks Formation are less than 20 microns in size between dolomite crystals. To reveal these types of pores in SEM, the samples were polished by Ar-ion milling using the technique described by Loucks et al. (2009) (Johnson et al. 2017). The pores of the four samples can be resolved clearly at a magnification greater than 1000× with SEM. Irregularly distributed pores can be seen; dominantly small pores occur between dolomite crystals or micro intercrystalline porosity that are present in the clay that fills the intercrystalline pores between the dolomite crystals. Figure 18.10 displays Ar-ion-milled surface photos taken at 5000×, 4000×, and 2000×, respectively. It shows the

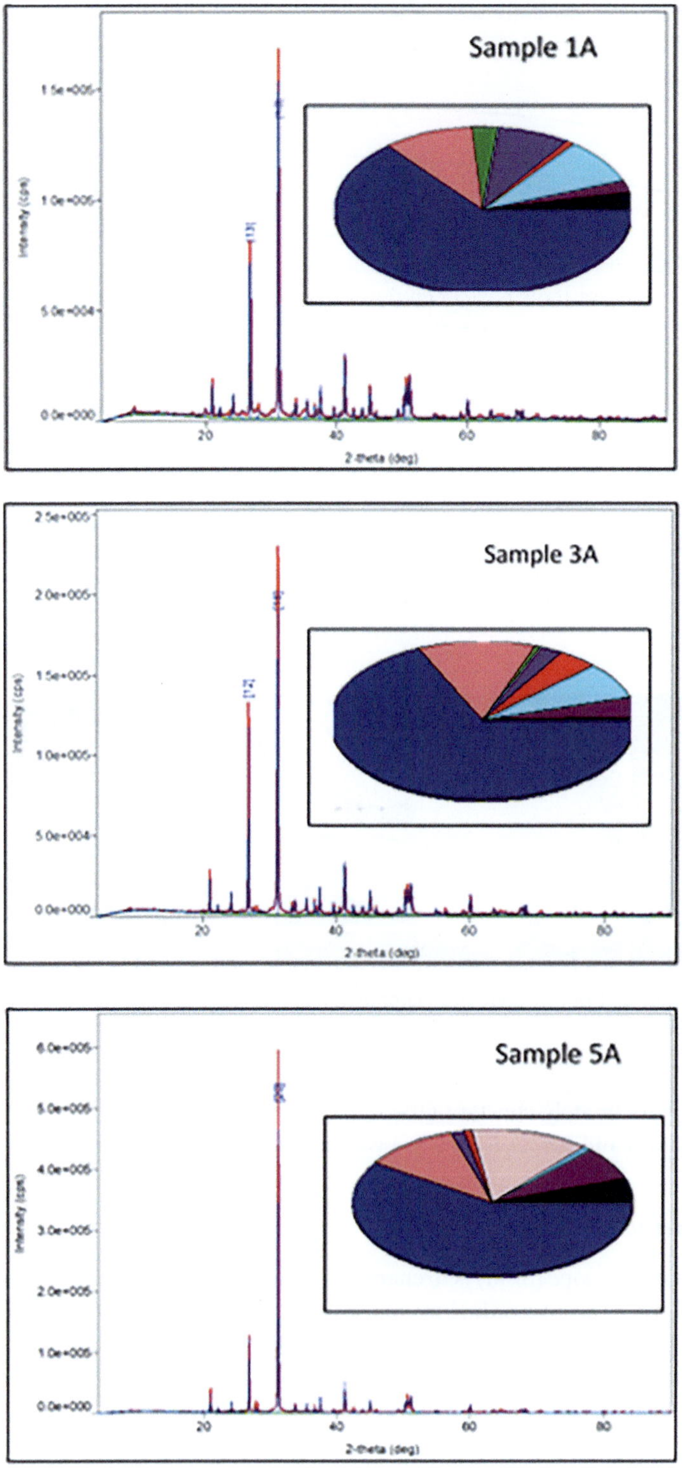

Figure 18.9 XRD signals of the mineralogy of the samples taken from well Charlie Sorensen 17-8-3TFH.

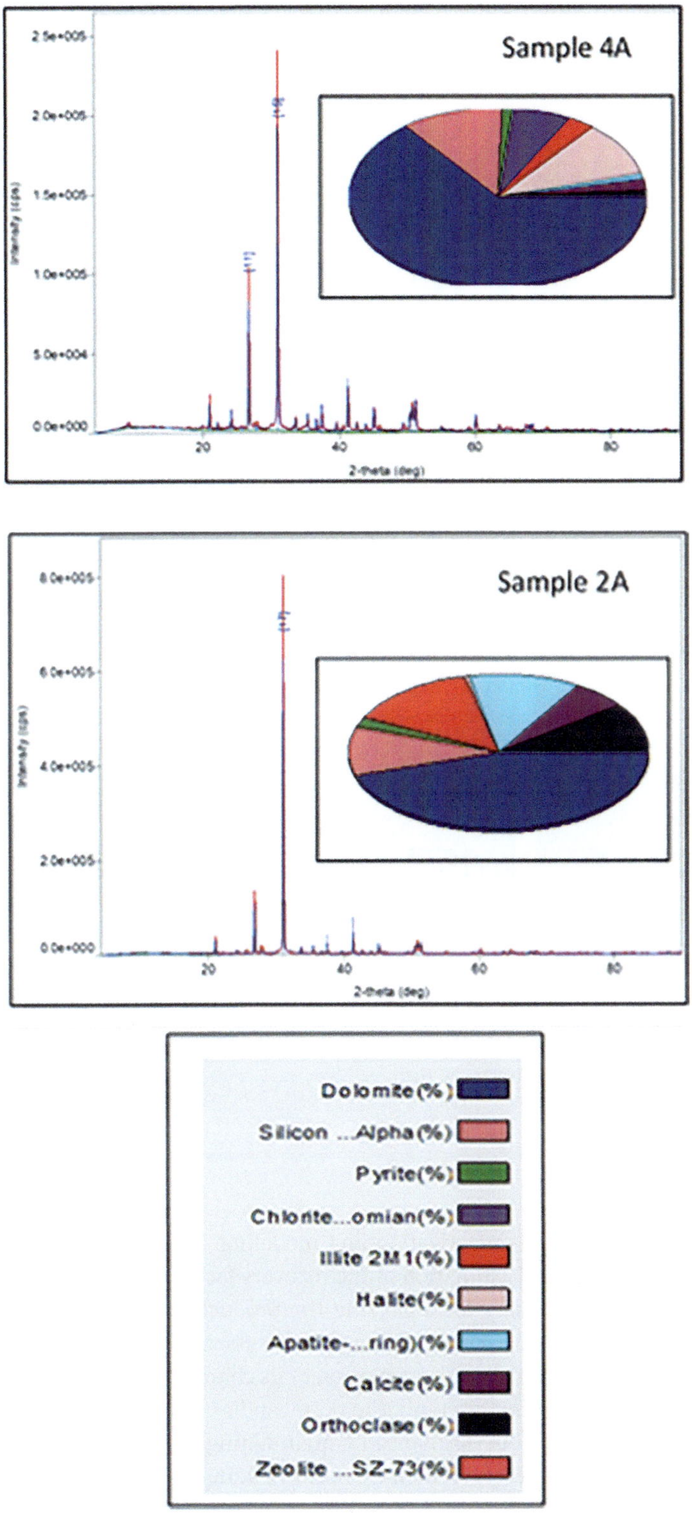

Figure 18.9 (Continued)

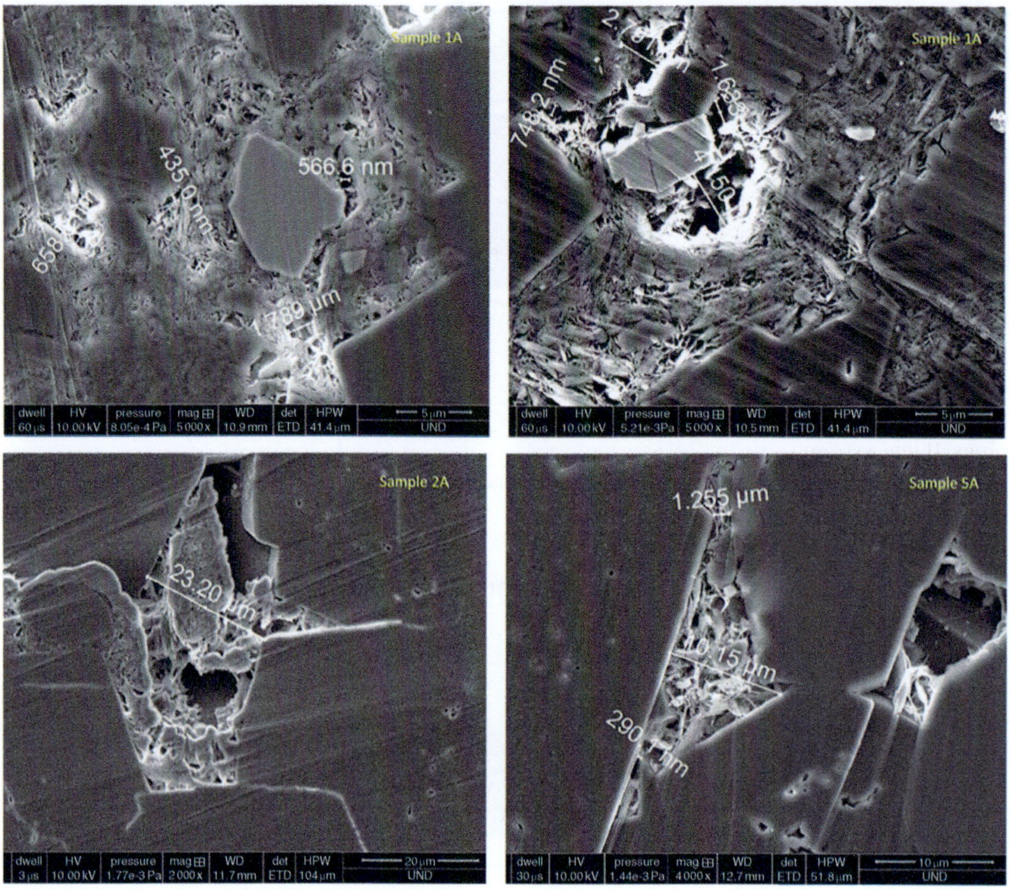

Figure 18.10 SEM photos of UTF (well Charlie Sorenson 17-8-3TFH).

interlaminated nature of the UTF and that dolomitic beds are interbedded with shale. The pores are characterized by a triangular shape. The dolomite has relatively large intercrystalline pores which range in size from less than one micrometer to 23 μm.

18.4 Results and Discussions

Permeability is a key parameter for reservoir characterization and modeling. Permeability variation with net stress can affect the production and estimation of the recovery factor. Understanding the stress-dependency of permeability is necessary for analyzing the production history (Teklu et al. 2016b). On the other hand, permeability and porosity hysteresis are essential parameters for understanding the behavior of the reservoir in terms of the evolution of its characteristics throughout its production life. During the depletion of the reservoir, the permeability hysteresis can help us enhance the production and forecast the well performance by maintaining the reservoir pressure above the threshold pressure under which reservoir damage occurs leading to irreversible loss of permeability and porosity. Additionally, the permeability hysteresis in repeated cyclic loading and unloading can give us an idea about the effective permeability value that can be used in the fracturing design and highlight the different rock types in the future refracturing operations.

According to Cui et al. (2013), permeability is one of the most challenging parameters to be precisely measured in the laboratory at in situ stresses because of the reservoir heterogeneity and characteristics of unconventional reservoirs. Furthermore, it is more difficult to measure the permeability by well logging due to its nature and relation with the flow in porous media (Dennis 2008; Roland and Olivier 2007).

As mentioned previously, the porosity and permeability at initial and final net confining pressures of 500 psi and 4000 psi are respectively the most measured values in the data set, and this range of stress was used for further analysis. In the first step, prior to the establishment of the permeability versus effective stress plots, the permeability values were normalized for each well and for each core sample by dividing the corresponding permeability at given effective stress by the permeability at the initial effective stress (500 psi) and by 1000 psi for Charlie Sorenson 17-8-3TFH's plugs. It lets a straightforward evaluation of the net stress effect on porosity and permeability independent of their initial values. In this chapter, the stress-dependent permeability and porosity of the Three Forks Formation are investigated, and also the permeability and porosity hysteresis for the four samples are discussed. Additionally, a stress law is determined for the upper and middle Three Forks.

18.4.1 Stress-Dependent Permeability and Hysteresis

Six wells with 120 and 165 samples from the upper and middle Three Forks members were used, respectively, to analyze the relationship between permeability, porosity, and net confining stress at selected net stress ranges. The lower Three Forks is not included in this study because it is considered a water-bearing interval in the basin. So, the results are presented here for the upper and middle Three Forks.

18.4.1.1 Upper Three Forks

Figure 1 (Appendix) shows the permeability as a function of effective stress for the 6 wells for the upper Three Forks. It can be seen that the permeability decreases as the net confining pressure increases for all the samples. However, the permeability's decline rate is highly variable. The samples exhibit a maximum variation in permeability at 2500 psi of net confining pressure, and a sharp drop in permeability was seen in almost all samples. The loss of permeability is between 50% and 98% of their initial permeability except for the well Charlotte 1-22H, which shows the highest reduction in permeability, varying from 70 to 99%, possibly related to microcracks closure. Overall, the loss of permeability at the final applied net stress ranges from 65 to 98% of the initial permeability for the studied wells and between 85 and 98% for Charlotte 1-22H.

Some discrepancies were also observed in this trend in the wells Debrecen 1-3H and Hawkinson 14-22H2 where two samples in each well exhibited a linear reduction in permeability during cyclic loading. The decline rate is from 28 to 45% at 2500 psi and from 42 to 72% at 4000 psi of effective stress, respectively.

Beyond 2500 psi effective stress, the plots illustrate an insignificant loss in permeability showing that the maximum reduction in permeability has already been reached, consequently, a relative plateau was observed. On the other hand, the samples that exhibit less loss at 2500 psi (<50%) are characterized by a reduction in permeability at final net stress (4000 psi), varying between 44 and 73%.

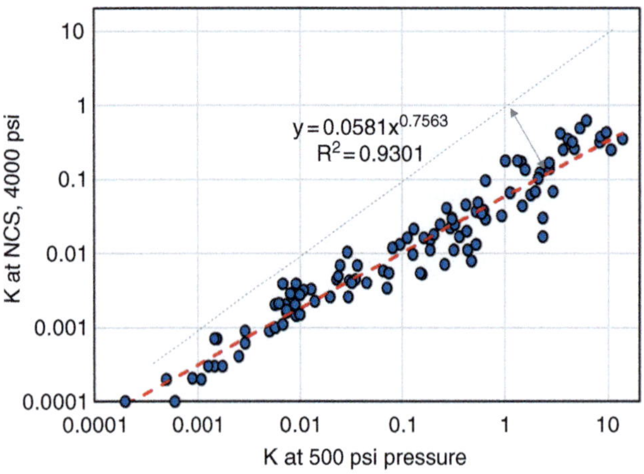

Figure 18.11 Comparison of core permeabilities for upper Three Forks (120 samples): Final applied net confining pressure (4000 psi) vs. initial net confining pressure (500 psi), showing a high stress-dependent permeability.

The stress-permeability sensitivity is shown in Figure 18.11, the Klinkenberg permeability at initial net stress (500 psi) was plotted against the Klinkenberg permeability at the final net stress for 100 samples from 6 wells. The comparison results show a much higher scatter in the permeability range above 0.05 mD and the best fit to the data is a power-law function as:

$$\text{Log}\, K_{\text{NCS}} = 0.0581 \times \left(\text{Log}\, K_{500\text{psi}}\right)^{0.7563} \tag{18.4}$$

Tables 2 and 3 (Appendix) present the results of the steady-state and unsteady-state permeability and porosity experiments during cyclic loading and unloading for the well Charlie Sorenson 17-8-3TFH, upper Three Forks member, respectively. The permeability was measured at different net confining pressures (1000–6000 psi). As one expects, the permeability decreases as the effective stress increases. The samples show different initial permeability values at initial confining pressure (1000 psi). However, the permeability is observed to be slightly higher using the steady-state method because it does not take into account the effect of slippage.

The samples were analyzed using both methods and the permeabilities during stress loading and unloading were normalized and plotted. As can be seen from Figure 18.12b (unsteady state), at net stress of 5000 psi, samples 1A and 2A retain a small amount of their original permeability, 21, and 28%, respectively. However, sample 5A retains more than 50% of its initial permeability.

A comparative result between steady-state and unsteady-state methods for cyclic loading indicates that sample 2A exhibits approximately the same permeability path with increasing in effective stress from 1000 to 5000 psi for both methods (Figure 18.12a and b). The highest loss (>70%) of permeability was observed at 5000 psi. This may be explained by the presence of microfractures as depicted in Figure 18.13. At high stresses, these microfractures will be closed, hence reducing permeability. Additionally, XRD analysis (Table 1, Appendix) shows that the sample contains 14% illite. The abundance of the clay minerals in the reservoir is critical. With the increase of stress, they tend to compact and occlude pore space, resulting in a reduction of permeability.

At the final applied stress of 6000 psi for the unsteady-state method, the sample (2A) exhibits an increase in permeability by 20%. The possible explanation may be that at a high level of stress, microcracks can be created or reopened, which enhances the permeability (Figure 18.12 b). In addition, the content of feldspars as shown by XRD results in Appendix (10%) may contribute to increasing the brittleness of the rock and its ability to be fractured more favorably (Davies et al. 2001).

Nevertheless, sample 1A retains 40% of the initial permeability more in the steady-state method (Figure 18.12 a,b). Presumably, this loss of permeability in the unsteady-state method could be

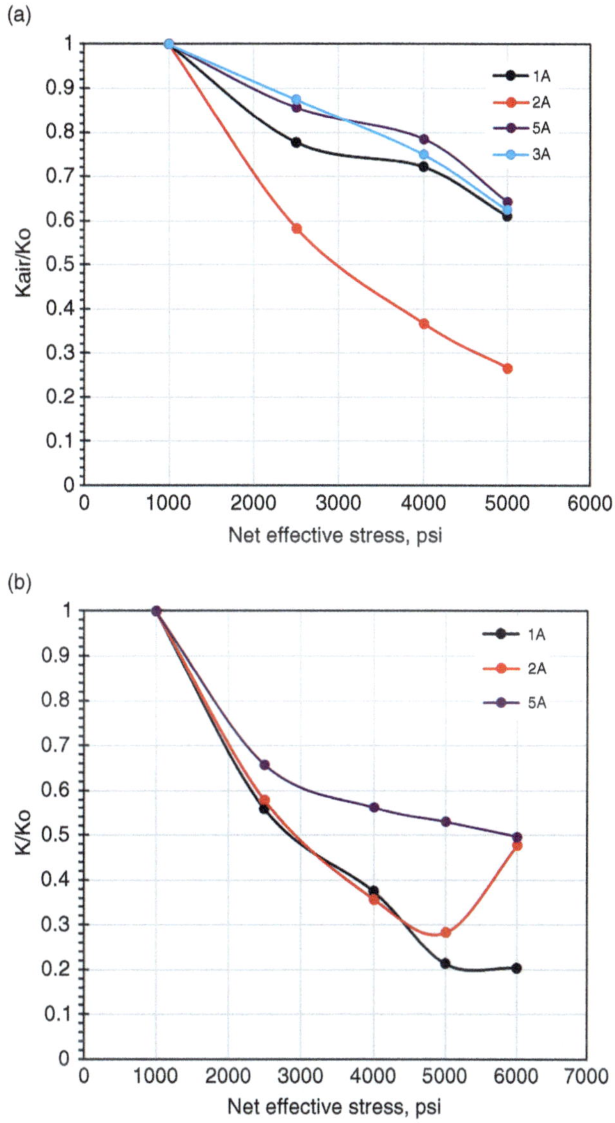

Figure 18.12 (a) Normalized permeability (CPMS, steady state) versus net stress showed approximately the same loss of permeability for the 3 samples except sample 2A (red color). (b) Normalized permeability (CMS-300, unsteady-state) versus net stress for three samples. As shown in this figure, samples 1A and 2A retain approximately 21–28% of their initial permeability at 5000 psi. Conversely, at 6000 psi sample 2A exhibits an increase in permeability by 20%.

related to the repeated stress loading/unloading process and nonelastic behavior of the connected pore network and microcracks closure. Furthermore, the analyzed sample shows that the UTF is characterized by angular-shaped porosities (Figure 18.14). During stress loading and unloading, this type of pore geometry in addition to pore-throat can collapse and lead to the loss of permeability. This may explain the strong stress-dependency of permeability characteristics for this sample.

The cubic root of the normalized permeability versus net confining stress (in semilogarithmic scale) shows a less linear behavior for all samples. Also, sample 1A exhibits two straight-line portions as shown in Figure 18.15. The intersection of the two segments occurs at the yield compressive strength of the sample (critical pressure) at 4000 psi. The second segment has a sharper slope than the first straight-line segment which shows a higher influence of net confining stress on the permeability of this sample.

During cyclic unloading (steady state), all samples show a clear hysteresis (Figure 18.16 (a) and (b), Table 18.1). Approximately, 7–32% of the original permeability could not be recovered. The highest

Figure 18.13 Sample 2A's CT Scan image shows an abundance of microfractures (the image was taken before the stress loading/unloading experiments).

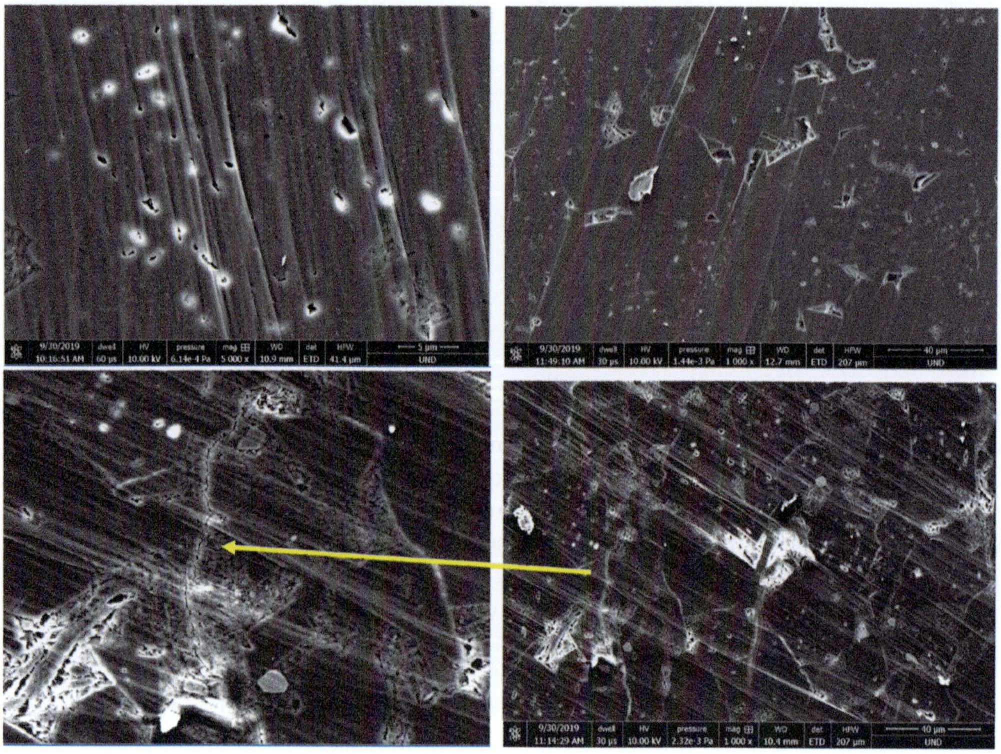

Figure 18.14 SEM images of upper Three Forks, showing abundant angular shaped porosities and microcracks (well Charlie Sorenson 17-8-3TFH, sample 1A).

Figure 18.15 The cubic root of normalized permeability (K/K_o) versus net effective stress (in semilogarithmic scale) showing nonlinear behavior for all samples. However, sample 1A shows two straight lines.

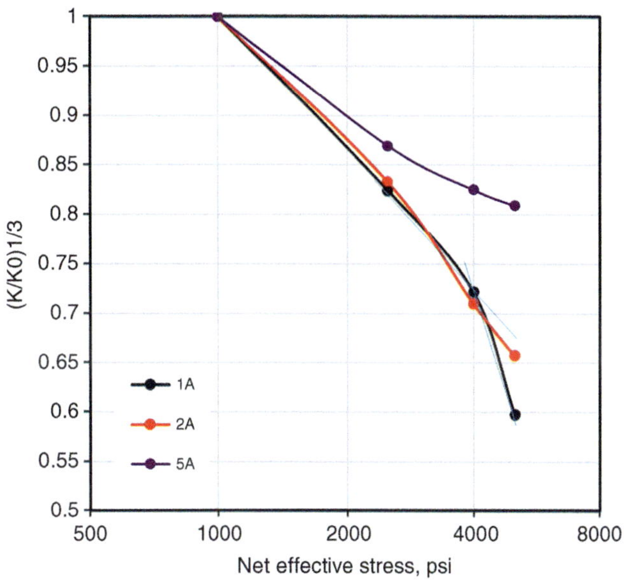

Figure 18.16 CPMS results, steady-state method: (a) permeability versus net effective stress, (b) normalized permeability versus net stress showing stress dependency of permeability of each sample during cyclic loading (solid line) and hysteresis during unloading process (dashed line) for upper Three Forks.

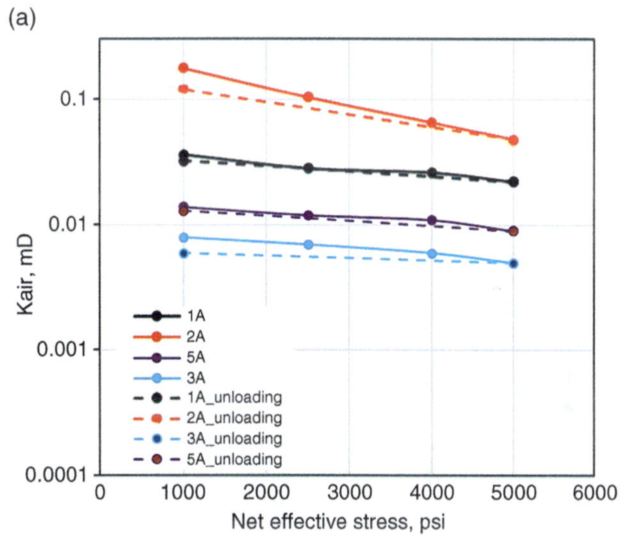

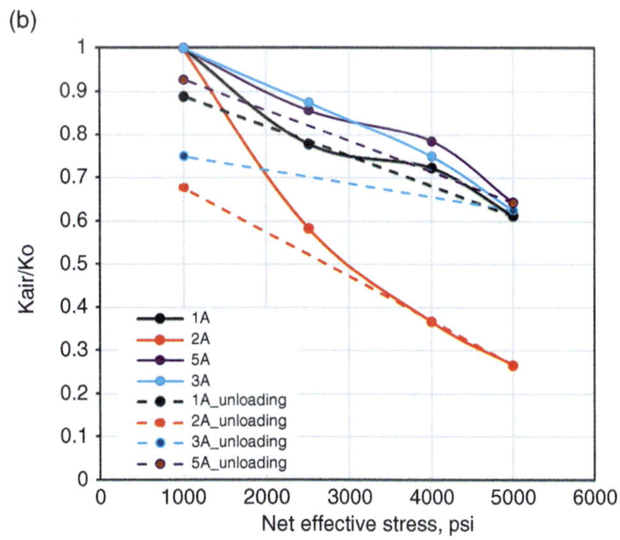

Table 18.1 Permeability hysteresis during stress unloading.

	Steady state (%)				Unsteady state (%)			
Samples N°	**5000 psi**	**4000**	**2500 psi**	**1000 psi**	**5000 psi**	**4000 psi**	**2500 psi**	**1000 psi**
1A	—	—	—	11.11	0.09	14.43	30.0	47.21
2A	—	—	—	32.22	2.52	7.77	8.03	Increase in K
3A	—	—	—	25	—	—	—	—
5A	—	—	—	7.14	2.92	Increase in K (no hysteresis)		

irreversible loss is observed in sample 2A followed by sample 3A at 25%. This means that during the stress loading process, the high stress can cause the closure of the microcracks/microfractures and during the unloading process, these microcracks could not be reopened or may be reopened partially. In addition, sample 3A has the lowest porosity value (<2%) but experiences a high permeability hysteresis during stress loading and unloading.

The studied samples show clearly that the permeability is strongly stress-dependent but the degree of stress dependency, the amount of the reduction, and hysteresis with stress loading and unloading can vary between samples.

Different stress paths were observed during the unloading process in the unsteady-state method as observed in Figure 18.17 (a) and (b). Only sample 1A shows a clear permanent hysteresis (47% at 1000 psi) with the initial Klinkenberg permeability decreasing from 0.0126 to 0.0067 mD (Figure 18.17b) despite the existence of microfractures as depicted in Figure 18.18.

In studies by Nelson and Handin (1977) on the cyclic matrix and fracture permeability hysteresis in tight rocks, they found that during cyclic unloading, when the matrix permeability increases, the fracture permeability can sometimes decrease with a decrease in effective stress. This may be a possible reason for the behavior observed for sample 1A.

Nevertheless, both samples 2A and 5A show different unloading paths (Figure 18.17c and d). The expectation is that the stress-dependent permeability changes during the unloading path are lower than those of the loading path, suggesting that there is permanent deformation (hysteresis) of the sample due to cyclic loading. However, at the beginning of the unloading process (high stresses), an insignificant hysteresis of permeability was observed (see Table 18.2). The permeability of sample 5A is slightly higher than the permeability during cyclic loading when the net confining pressure was decreased from 5000 to 4000 psi and then to 2500 psi. This means that microcracks may have been developed due to the cyclic loading which can result in higher-than-expected permeability. At the end of the unloading process (corresponding to 1000 psi confining pressure), the permeability increased dramatically for both samples (2A and 5A). The increase is more than 2.5 times the initial permeability (Figure 18.17 c and d). The possible reason for this is that at low stresses, the fractures were reopened and connected with the microcracks that were created at high stresses and led to the enhancement of permeability (Figures 18.19 and 18.20). Teklu and al. (2016) reported that repeated stress loading and unloading can improve the effective permeability and the main fracture could be connected to microcracks leading to the potential creation of a new flow path. In addition, the cyclic loading will shear the existing fractures and that may increase the aperture at low stresses when dilation can happen. This conclusion may explain the changes in the production of the upper Three Forks after shutting and reopening the wells or during the depletion and injection operations.

18.4.1.2 Middle Three Forks

Similar procedures presented for the upper Three Forks were followed to analyze the effect of load cycling on measured permeability values in the middle Three Forks. The results are presented in Figures 3 and 4, Appendix. The samples were subjected to a cyclic increase in effective stress during permeability measurements (500–4000 psi). It can be seen that the measured permeabilities reduced up to 99% of their original permeability at the largest confining pressure. At a low net effective stress range of 500–2500 psi, the core samples can lose between 30 and 99% of their initial permeability. The results show a nonlinear stress-dependent permeability relationship. In all samples, the permeability is strongly stress dependent. Additionally, some discrepancies were observed, for example, 2% of the core samples show a linear reduction in permeability with increasing net stress, with a decline rate from 48% to 80% at final effective stress.

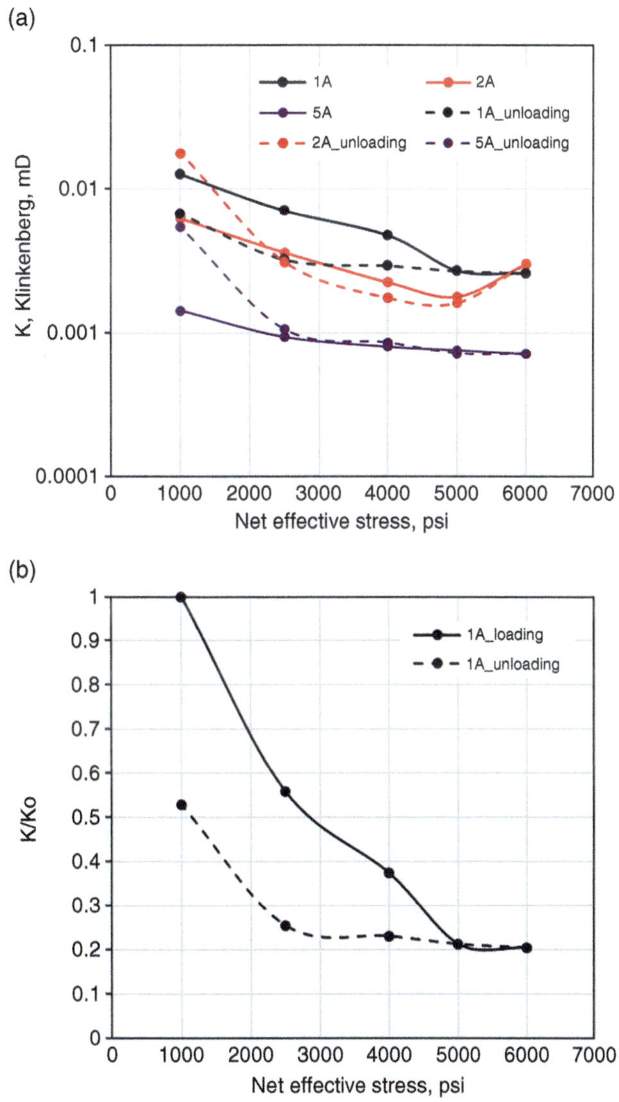

Figure 18.17 CMS-300 results, unsteady-state method: Normalized permeability versus net stress showing loss of permeability and hysteresis during cyclic loading (solid lines) and unloading (dashed lines) for upper Three Forks.

(c)

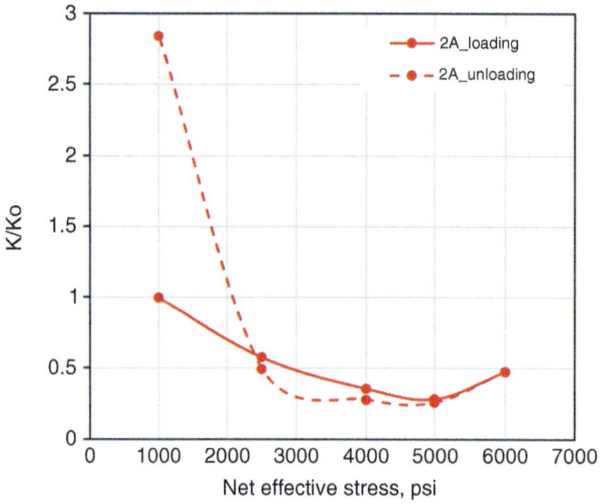

Figure 18.17 (Continued)

(d)

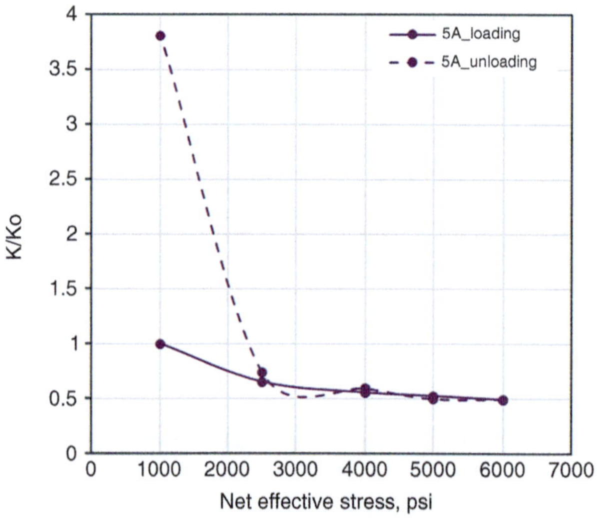

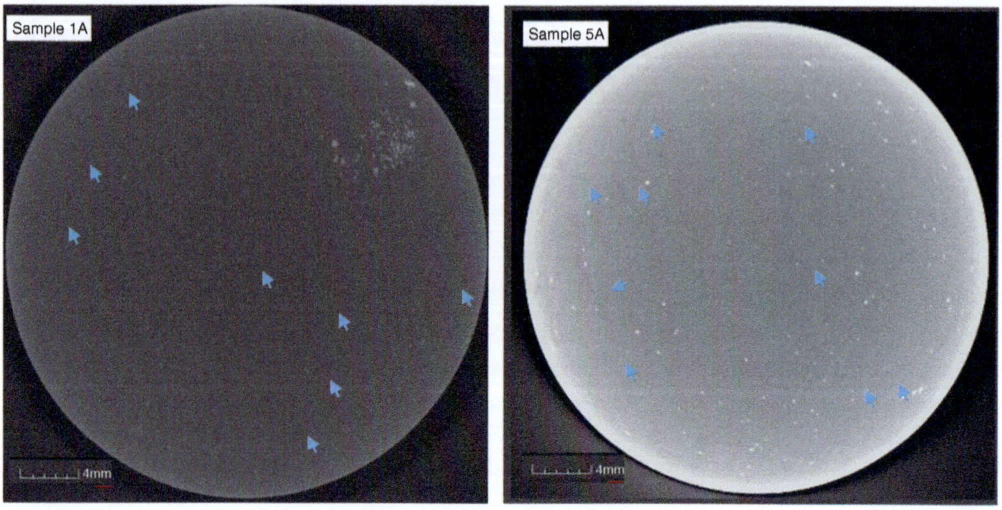

Figure 18.18 CT scan images (samples 1A and 5A) show a less abundance of microfractures compared to sample 2A, the image was taken before the experiment of stress loading/unloading.

Table 18.2 Reduction in permeability compared to permeability at initial net stress (1000 psi) during stress loading.

	Steady state (%)				Unsteady state (%)			
Samples N°	**2500 psi**	**4000 psi**	**5000 psi**	**6000 psi**	**2500 psi**	**4000 psi**	**5000 psi**	**6000 psi**
1A	22.2	27.8	38.9	—	44.1	62.5	78.7	79.5
2A	41.7	63.3	73.3	—	42.5	64.2	71.5	52.0
3A	12.5	25.0	37.5	—	—	—	—	—
5A	14.3	21.4	35.7	—	34.2	43.7	47.0	50.3

The Klinkenberg permeability at initial net stress versus Klinkenberg permeability at final net stress for 165 samples from 6 wells is plotted in Figure 18.21. The results show a much higher scatter in the permeability values than the upper Three Forks, but the stress-permeability sensitivity is within the same range. The best fit to the data is a power-law function in the form of:

$$\text{Log}\, K_{\text{NCS}} = 0.0833 \times \left(\text{Log}\, K_{\text{initial}}\right)^{0.825} \tag{18.5}$$

18.4.2 Permeability Evolution with Net Stress

The results presented in the earlier sections demonstrate that permeability is not a simple function of net stress where accurate values of permeability can be predicted solely based on values of net stresses. It was shown, based on Figures 2 and 4 (Appendix), that the cubic root of normalized permeability versus logarithmic of net stress for upper and middle Three Forks that the permeability is nonlinearly related to net stress over the examined net stress range of 500–4000 psi. In addition, the core samples analyzed in this study (Well Charlie Sorenson 17-8-3TFH) show a nonlinear reduction in permeability with increasing net stress (Figure 18.22).

Hoholick et al. (1984) and David et al. (1994) suggested an exponential relationship to correlate the net stress with the permeability:

$$K = K_o \times \exp\left[-\gamma\left(P_{\text{eff}} - P_o\right)\right] \tag{18.6}$$

Here, k denotes the permeability under net confining pressure (P_{eff}), k_0 represents the permeability under atmospheric pressure which is assumed to be 0.1 MPa (14.5 psi), γ is the pressure sensitivity coefficient (material constant). The high values of γ correspond to a higher decrease in permeability with increasing effective stress (Davis et al. 1994). On the other hand, the stress dependence of permeability can be described by a power law (Dong et al. 2010):

$$K = K_o \times \left(\frac{P_{\text{eff}}}{P_o}\right)^{-P} \tag{18.7}$$

where P is a material constant.

In general, for the studied wells (see Appendix), the changes in permeability with increasing net stresses were observed to follow an exponential function in the selected stress range (500–4000 psi). The permeability results of well Charlie Sorenson's samples, upper Three Forks, show the same

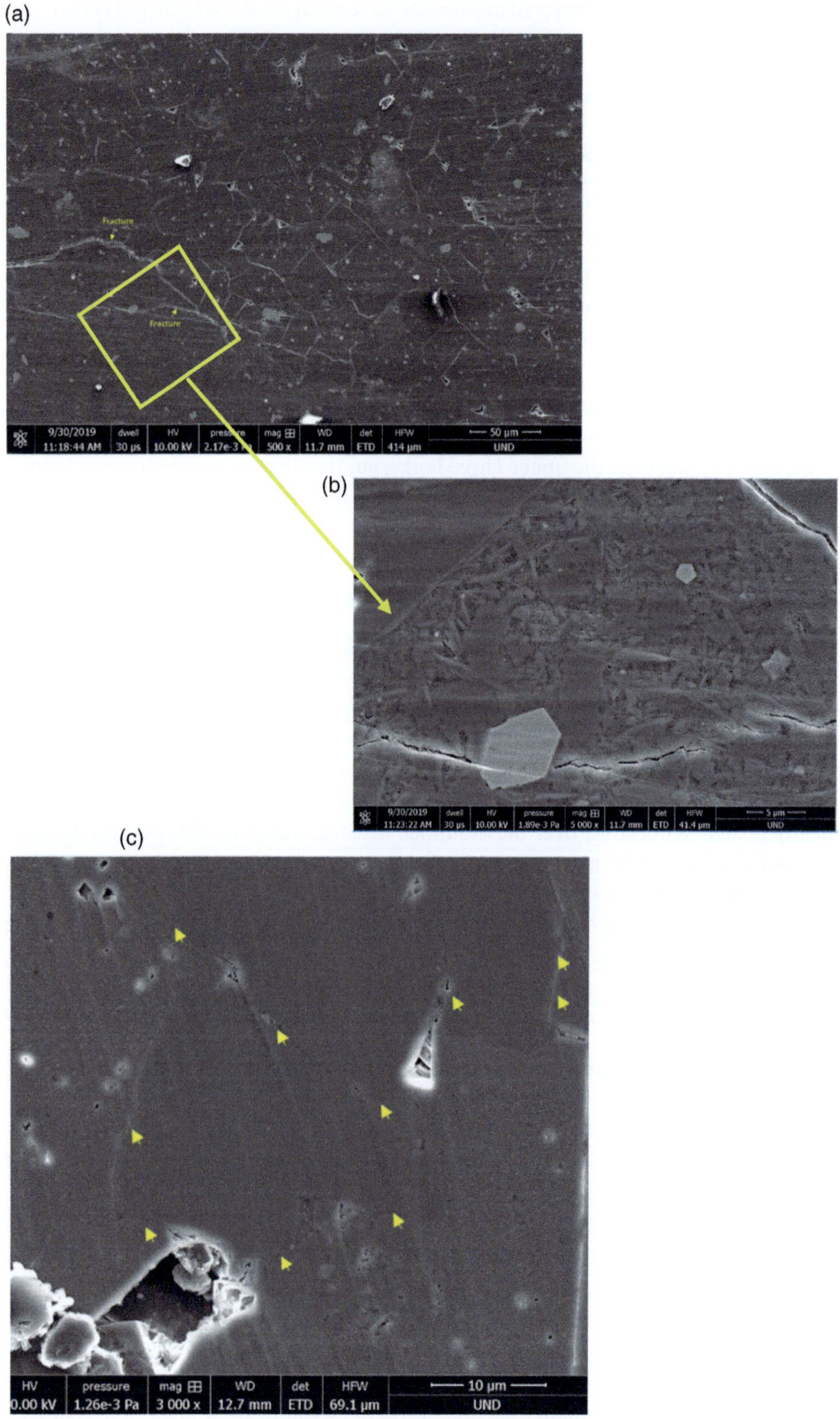

Figure 18.19 (a) Sample 2A and 5A SEM photos. (b) 2A, a magnification of 5000× of Ar-ion milled surface showing fractures and microcracks. (c) 5A a magnification of 3000x of Ar-ion-milled surface showing microfractures (well Charlie Sorenson 17-8-3TFH).

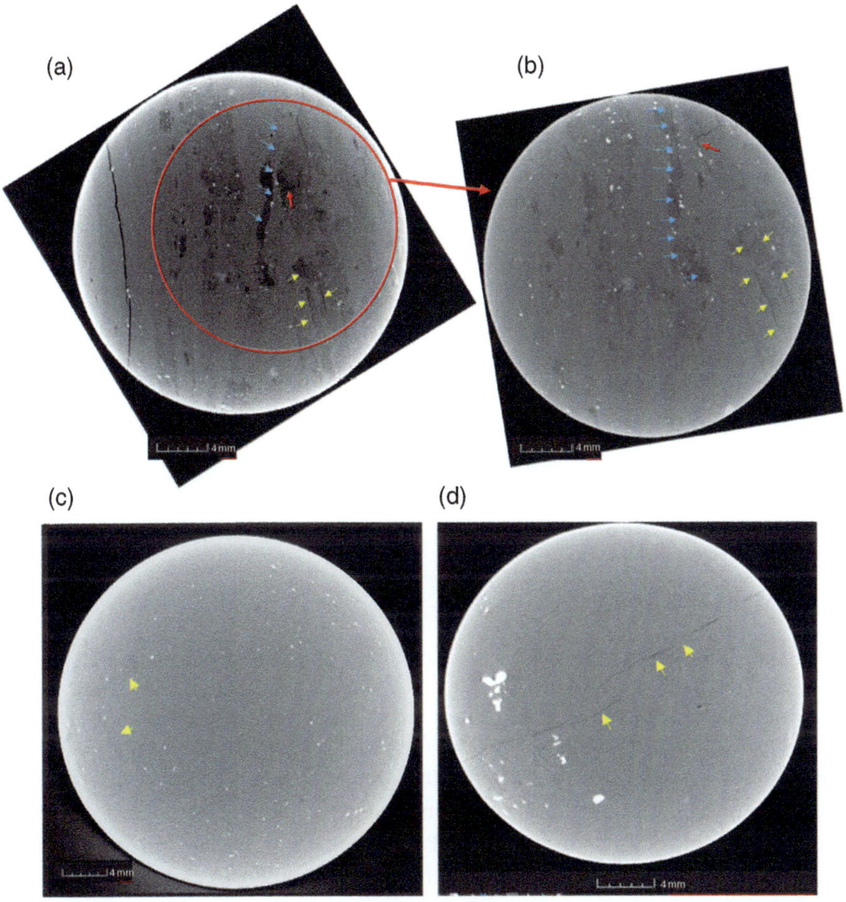

Figure 18.20 Sample 2A, CT scan images: (a) before the experiment of stress loading/unloading, (b) after the experiment of stress loading/unloading. Sample 5A, CT scan images: (c) before the experiment of stress loading/unloading, (d) after the experiment of stress loading/unloading.

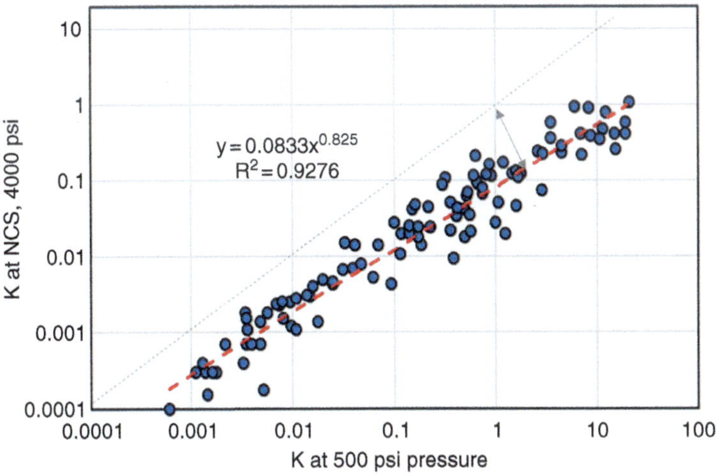

Figure 18.21 Core permeabilities measured for middle Three Forks (165 samples): Final applied net confining stress, NCS (4000 psi) vs. initial net confining pressure (500 psi), showing a high stress-dependent permeability. High permeabilities due to induced fractures were excluded.

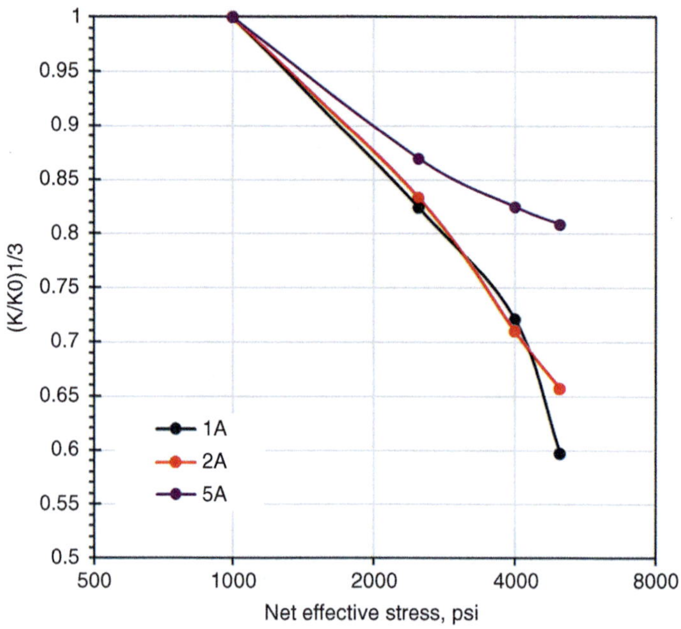

Figure 18.22 Cubic root of normalized permeability versus logarithmic of net stress (unsteady-state).

relationship between permeability and net stress. The data closely fit an exponential function under cyclic loading as observed in Figure 18.23 in the stress range between 1000 and 5000 psi. K_o and γ parameters were determined from the fitted equations and their values are listed in Table 18.3. Compared with the parameters for sandstone reported by David et al. (1994), the values of γ obtained in this study under cyclic loading are lower for the upper Three Forks carbonate samples. For the power-law model, the values of parameter P under cyclic loading for the three samples change from 0.4 to 0.879. It can be seen from Figure 18.24 that a power-law sample fits the data better than an exponential function for sample 5A.

Also, the curves fitting techniques were used for cyclic unloading curves dependency of permeability as shown in Figures 18.25 and 18.26. For the power-law model, much higher values of P were obtained for samples 2A and 5A, which indicates a high-stress sensitivity compared to sample 1A and also K_o has a wider range of values. The model describing better the permeability relationship with decreasing net stress (cyclic unloading) is a power-law function with a correlation coefficient greater than 0.93.

18.4.3 Stress-Dependent Porosity and Hysteresis

In this section, the results of stress dependency of porosity for cyclic loading for upper and middle Three Forks will be presented.

18.4.3.1 Upper Three Forks

The same wells, as in the previous section, with 120 samples were used to examine the stress dependency of porosity for cyclic loading for the upper Three Forks. The measured porosities were selected for the same net stress ranges (500–4000 psi) and the values were normalized. However, the new dataset, belonging to well Charlie Sorenson's core samples includes data from porosity

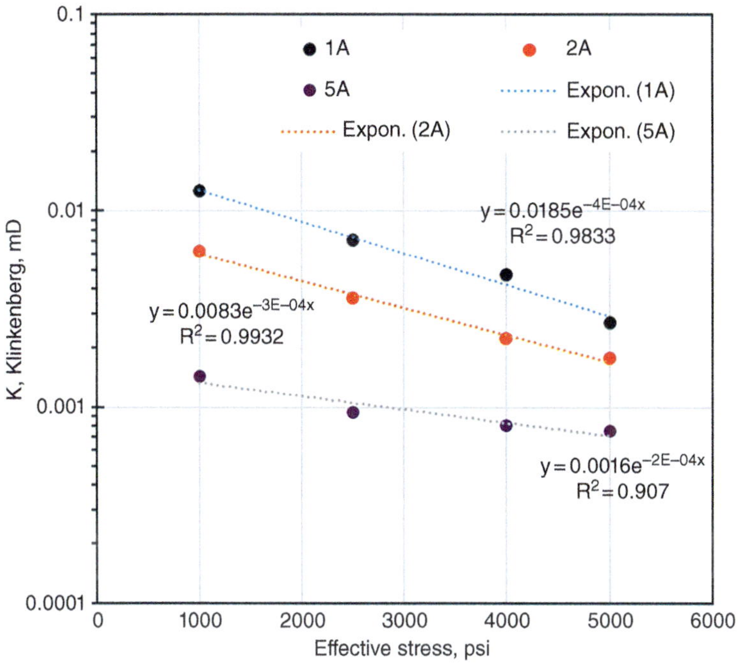

Figure 18.23 Cyclic loading curves of stress-dependent permeability, well Sorenson 17-8-3TFH. The permeability versus net stress relation is described using an exponential function.

Table 18.3 Stress-dependent permeability parameters determined using the curve fitting technique on measured permeability.

	Permeability							
	Cyclic loading				Cyclic unloading			
	Exponential relationship		Power-law relationship		Exponential relationship		Power-law relationship	
Sample No	K_o	γ	K_o	P	K_o	γ	K_o	P
1A	0.018	4×10^{-4}	5.9849	0.879	0.0071	2×10^{-4}	0.3232	0.57
2A	0.0083	3×10^{-4}	1.3746	0.774	0.0225	6×10^{-4}	30.658	1.542
5A	0.0016	2×10^{-4}	0.022	0.4	0.0061	5×10^{-4}	676.99	1.27

measurements during cyclic loading and unloading at net stress varying between 1000 and 6000 psi. The porosities measured tend to be the same magnitude as those previously reported in the collected data.

The results showed that the porosity decreases progressively with the increase of the net stress, while the deformation is minor compared to the stress dependency of permeability (Figure 5, Appendix). The porosity decreases by 2–6% at low stresses (500–2500 psi), and 1–3.5% when the

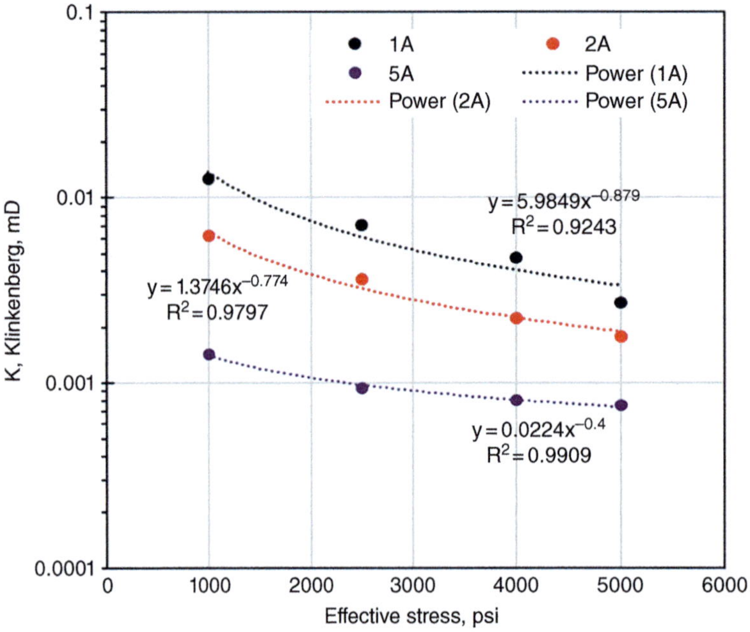

Figure 18.24 Cyclic loading curves of stress-dependent permeability, well Sorenson 17-8-3TFH. The permeability versus net stress relation is described using a power-law function.

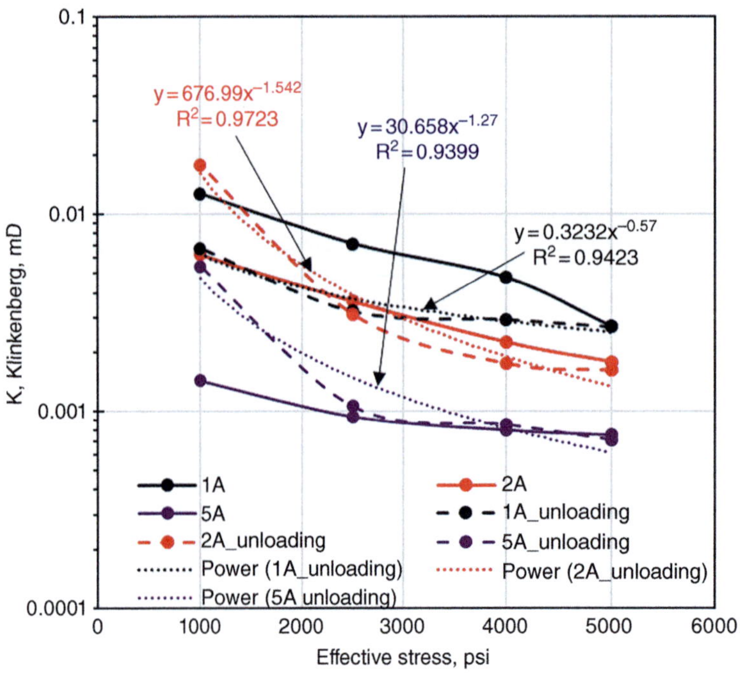

Figure 18.25 Cyclic-unloading curves of stress dependency of permeability, well Sorenson 17-8-3TFH. The relationship between permeability and net stress is described using a power-law function.

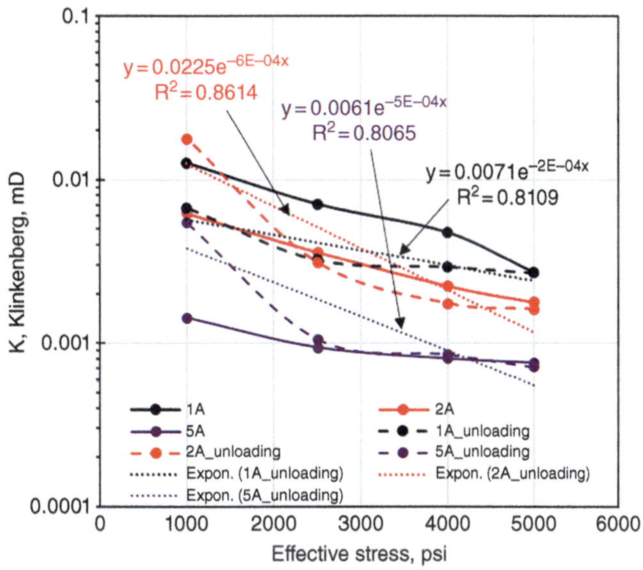

Figure 18.26 Cyclic unloading curves of stress dependent permeability, well Sorenson 17-8-3TFH. The relationship between permeability and net stress is described using an exponential function.

effective stress is changing from 2500 to 4000 psi except for the Pojorlie 21-2-1H well, where the reduction rate is from 2 to 12% at low stresses and from 1 to 5% at final stress.

Figure 18.27 presents the comparison results between the porosities at final net stress and the porosities at initial net stress for the same wells. It can be seen that the porosities are less stress-dependent than permeability with a linear function best fitting the data is:

$$\phi_{\mathrm{NCS}} = 1.0015 \times \phi_{500\mathrm{psi}} - 0.2962 \tag{18.8}$$

The analysis of the measured porosities for the well Charlie Sorenson 17-8-3TFH shows the same magnitude of reduction in porosity as the studied wells as is seen in Figure 18.27. The core samples experienced almost a nonlinear reduction of porosity for both methods (steady-state and unsteady-state) with increasing the loads from 1000 to 4000 psi. At the latter stress ranges, most of the samples retain more than 92% of their initial porosity, the amount of loss ranges between 1.7 and 7.9% for both methods except sample 3A, which experienced a higher reduction (refer to Table 18.4). This means porosity is less stress-sensitive within this range of net stress

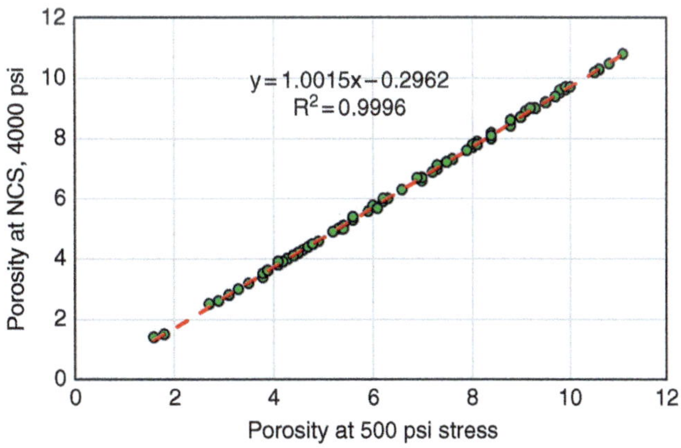

Figure 18.27 Comparison of core porosities for upper Three Forks (120 samples): final applied net confining pressure (4000 psi) versus initial net confining pressure (500 psi), showing a less stress-dependent porosity then the permeability.

Table 18.4 Percentage reduction in porosity at different net confining stresses compared to porosity at initial net stress (1000 psi) during stress loading.

	Steady state (%) Net confining stresses				Unsteady state (%) Net confining stresses			
Samples no	**2500 psi**	**4000 psi**	**5000 psi**	**6000 psi**	**2500 psi**	**4000 psi**	**5000 psi**	**6000 psi**
1A	5.1	6.1	19.8	—	1.3	1.7	1.9	11.2
2A	6.0	6.3	7.2	—	3.3	5.7	7.0	9.2
3A	8.7	12.8	22.6	—	—	—	—	—
5A	3.6	5.5	6.7	—	4.3	7.9	9.6	11.3

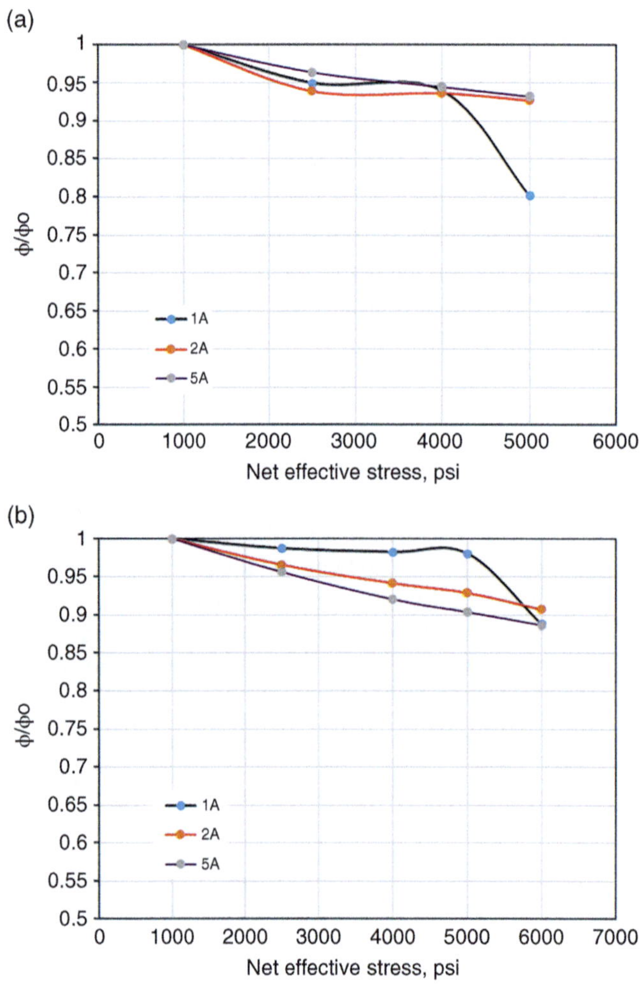

Figure 18.28 Normalized porosity versus net effective stress for well Charlie Sorenson 17-8-3TFH. (a) porosity from CPMS, (b) porosity from CMS-300.

values. However, at final net stress (cyclic loading), sample 1A exhibits an abrupt loss in porosity of 13.7 and 9.5% during the increase of net stress from 4000 to 5000 psi for steady state and from 5000 to 6000 psi for unsteady state, respectively, as shown in Figure 18.28 and Table 18.5. This abrupt loss confirms the nonelastic behavior of the pore network of the sample (1A). In addition, a porosity hysteresis was observed during the unloading path in both methods (Figure 18.29).

Table 18.5 Percentage porosity hysteresis at different net confining stresses during stress unloading.

	Steady state (%) Net confining stresses				Unsteady state (%) Net confining stresses			
Samples no	**5000 psi**	**4000 psi**	**2500 psi**	**1000 psi**	**5000 psi**	**4000 psi**	**2500 psi**	**1000 psi**
1A	—	—	—	3.2	8.9	9.0	8.7	8.14
2A	—	—	—	4.6	1.8	2.2	2.8	3.9
3A	—	—	—	11.9	—	—	—	—
5A	—	—	—	4.3	1.6	2.8	2.9	Slight increase

(a)

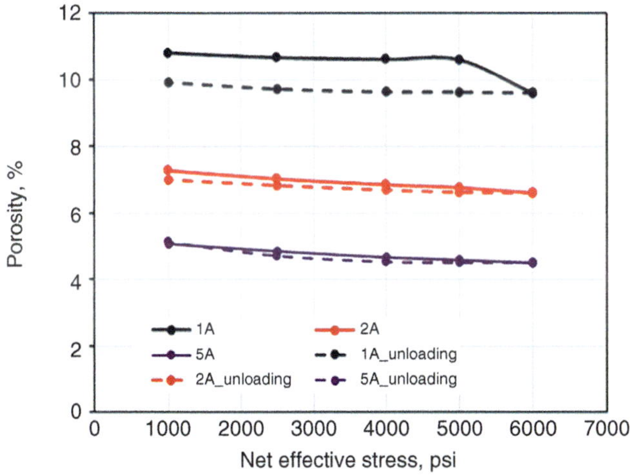

(b)

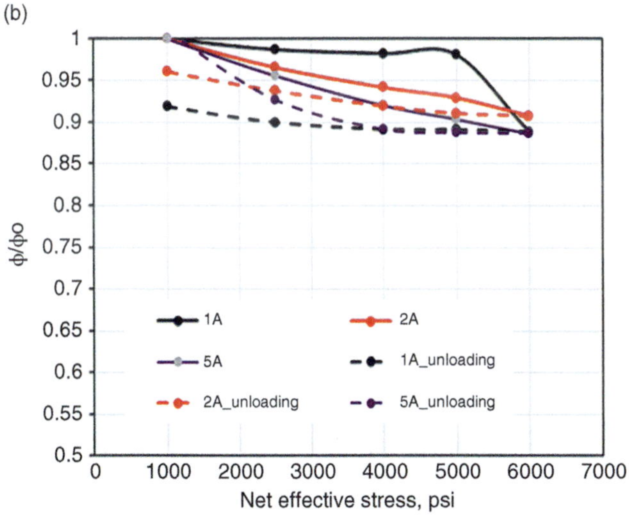

Figure 18.29 Unsteady-state method: (a) Porosity versus net effective stress, (b) Normalized porosity versus net stress. Steady-state method: (c) Porosity versus net effective stress, (d) Normalized porosity versus net stress showing stress dependency of porosity for each sample during cyclic loading (solid lines) and hysteresis during unloading process (dashed lines) for upper Three Forks, well Charlie Sorenson 17-8-3TFH.

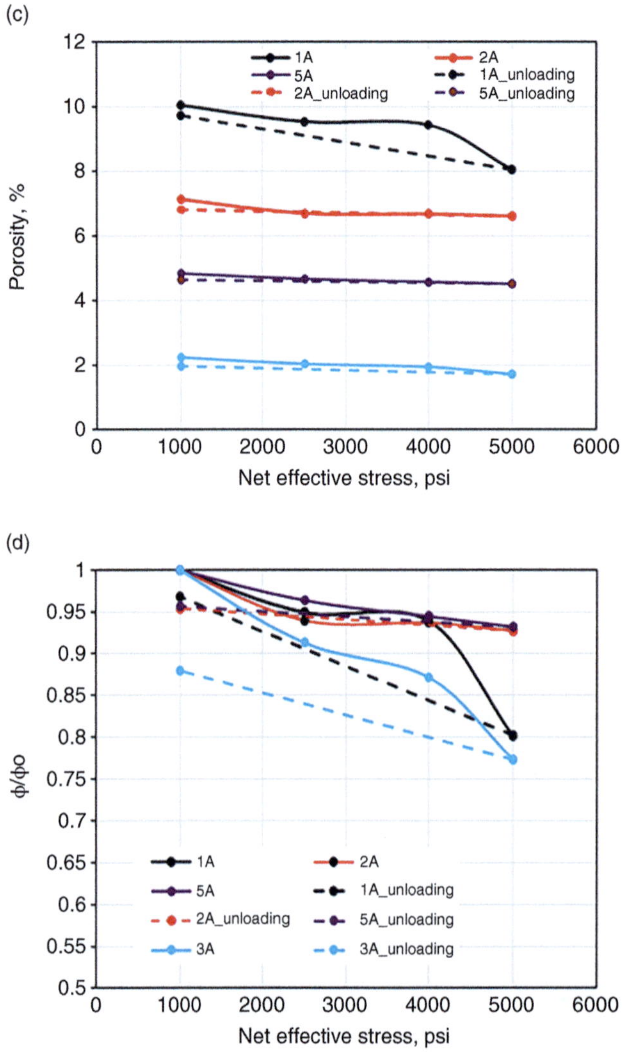

Figure 18.29 (Continued)

Sample 1A and 2A experienced a permanent loss in the porosity of 8% and 3.9%, respectively, in the unsteady-state method (Figure 18.30 a&b). This hysteresis in porosity gives a better understanding of the permeability hysteresis for these samples: apparently, a small loss in porosity (8%) has a large effect on its permeability (permanent loss of 47%). However, a slight increase in porosity is observed in sample 5A (Figure 18.30c).

18.4.3.2 Middle Three Forks

Figure 7 (Appendix) presents the rate of porosity reduction after the samples have been subjected to a range of stress from 500 psi to 4000 psi for the middle Three Forks. It can be seen that the measured porosity data for six wells (165 samples) reveals that 98% of all samples retain more than 93% of their initial porosity at the final applied stress value. Less than 3% of the samples retain less than 90% of their initial porosity at 4000 psi. At low net stress ranges (500–2500 psi), the core samples can lose from 1 to 10% of their initial porosity.

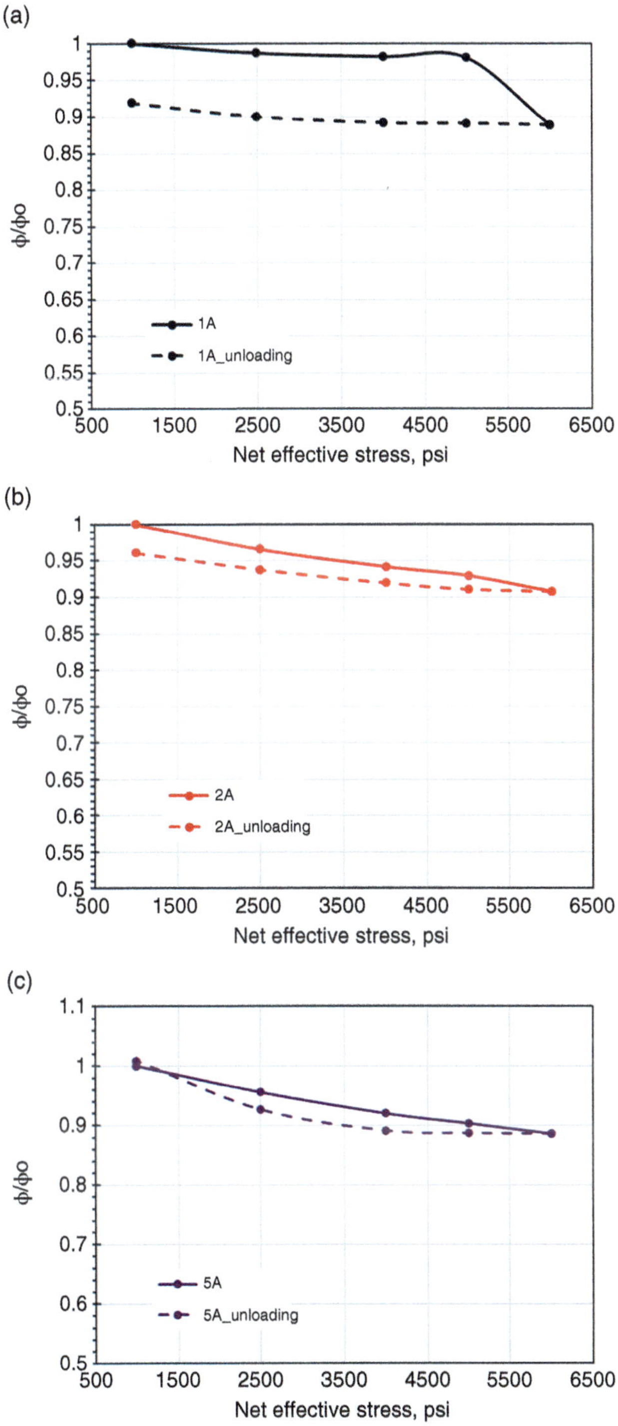

Figure 18.30 Unsteady-state method: Normalized porosities versus net effective stresses showing stress dependency of porosity during cyclic loading (solid lines) and hysteresis during unloading process (dashed lines) for each sample: (a) Sample 1A, permanent deformation of 9%. (b) sample 2A, permanent deformation of 4%. (c) slight increase in porosity is observed in sample 5A. Well Charlie Sorenson 17-8-3TFH.

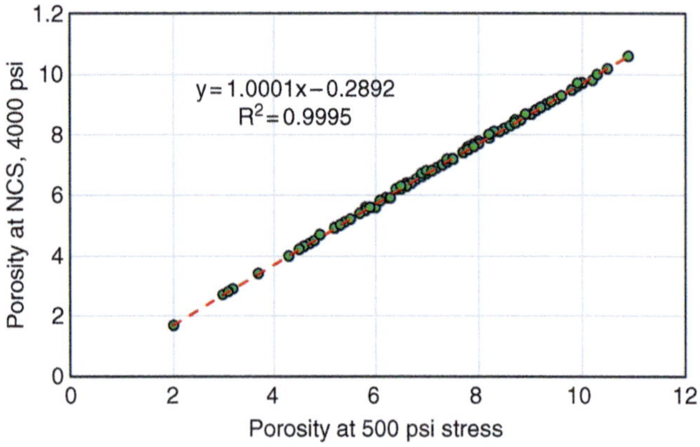

Figure 18.31 Comparison of core porosities for middle Three Forks (165 samples): final applied net confining pressure, NCS, (4000 psi) versus initial net confining pressure (500 psi), showing a less stress-dependent porosity. The fractured samples were excluded.

The results show that the porosities of all studied samples are less sensitive to changes in net effective stress than permeability and looking at Figure 18.31 the best linear fit to the data is:

$$\phi_{\text{NCS}} = 1.0001 \times \phi_{500\ \text{psi}} - 0.2992 \tag{18.9}$$

18.4.4 Porosity Evolution with Net Stress

Two empirical functions have been used by many authors to describe the relationship between effective porosity and net stress similar to that of permeability. The exponential function for carbonate reservoir was developed by Schmoker and Halley (1982) as:

$$\phi = \phi_o \times \exp\left[-\beta\left(P_e - P_o\right)\right] \tag{18.10}$$

where β is the material constant.

The power law function expressing the relationship between effective porosity and net stress (Dong et al. 2010) is written as:

$$\phi = \phi_o \times \left(\frac{P_e}{P_o}\right)^{-P} \tag{18.11}$$

where p is a material constant.

Curves-fitting techniques were used to determine the parameters in both of these empirical functions for the porosity data for 6 wells (Appendix) and three samples from well Charlie Sorenson 17-8-3TFH.

Figure 18.32 shows that the cubic root of normalized porosity versus net stress shows a nonlinear relation between porosity and effective stress. A power-law function shows a good fit to samples 1A and 2A, however, the data corresponding to sample 5A fits better an exponential function (Figures 18.33 and 18.34). Table 18.6 illustrates the parameter values of Φ_0, β and P. This table shows that for samples 1A, 2A, and 5A, P varies from 0.012 to 0.062 (cyclic loading) and from 0.019 to 0.076 (cyclic unloading), respectively. The values of Φ_0 obtained for the power-law model under cyclic loading and unloading are greater than those of the exponential function.

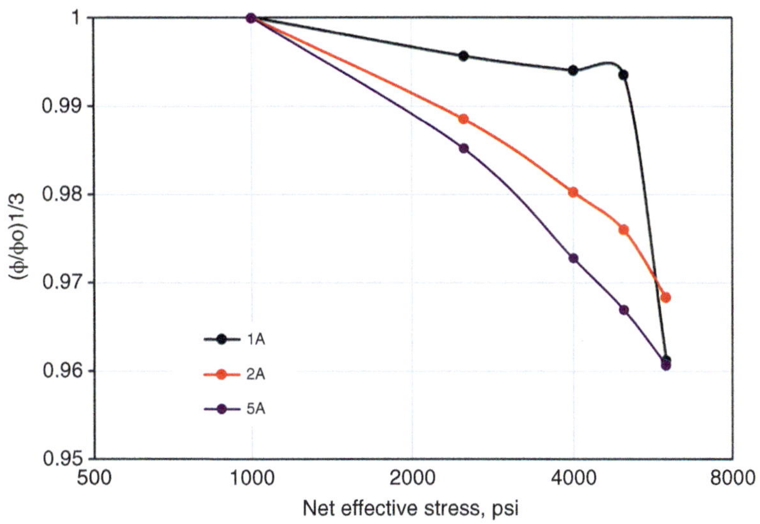

Figure 18.32 Cubic root of normalized porosity (ϕ/ϕ_o) versus logarithmic of net stress (unsteady-state), showing a nonlinear relationship between porosity and net stress (well Charlie Sorenson 17-8-3TFH).

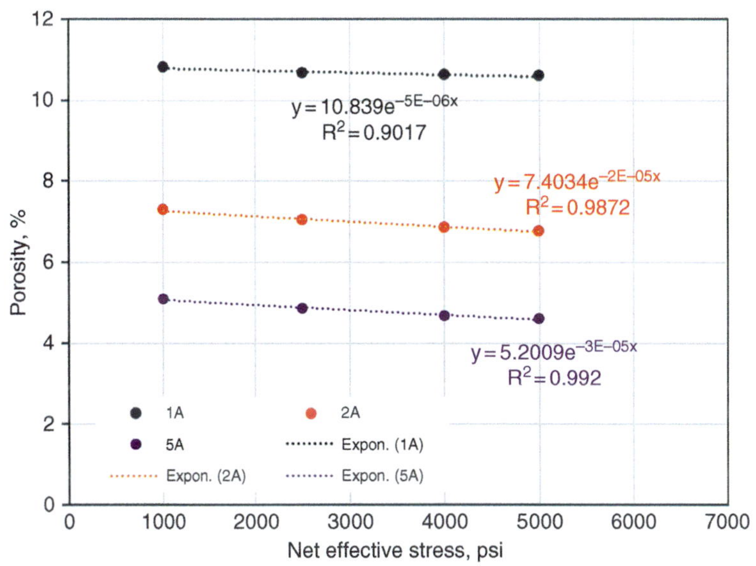

Figure 18.33 Cyclic loading curves of stress-dependent porosity, well Charlie Sorenson 17-8-3TFH. The relationship between porosity and net stress fits an exponential function.

18.4.5 Permeability Evolution with Porosity

The results of permeability evolution with porosity are shown in Figure 18.35. The same data were plotted to determine the correlation between porosity and permeability with increasing net confining pressure. An exponential correlation was found to fit the data corresponding to the three samples from the upper Three Forks.

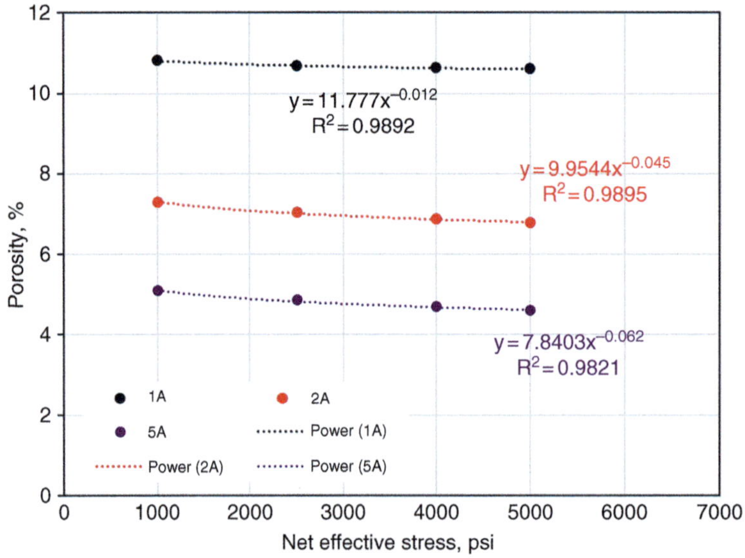

Figure 18.34 Cyclic loading curves of stress dependency of porosity, well Charlie Sorenson 17-8-3TFH. The relationship between porosity and net stress fits a power-law function.

Table 18.6 Constant parameters in two models fit the lab data of porosity versus confining pressure.

	Porosity							
	Cyclic loading				Cyclic unloading			
	Exponential relationship		Power-law relationship		Exponential relationship		Power-law relationship	
Sample no	Φ_o	B	Φ_o	P	Φ_o	β	Φ_o	P
1A	10.839	5×10^{-6}	11.777	0.012	9.937	6×10^{-6}	11.308	0.019
2A	7.403	2×10^{-5}	9.9544	0.045	7.059	1×10^{-5}	8.800	0.033
5A	5.200	3×10^{-5}	7.8403	0.062	5.133	3×10^{-5}	8.599	0.076

18.5 Conclusion

This chapter investigated the stress-dependent permeability and porosity of a set of samples from the upper and middle Three Forks Formation varying the net confining pressure from initial net stress to reservoir net effective stress. Experimental results during the loading and unloading phases were presented and discussed for four samples from well Charlie Sorenson 17-8-3TFH, upper Forks Formation. The relationship between permeability, porosity, and net confining pressure was presented. The results are summarized as follows:

- The permeability and porosity decrease with an increase in net confining stress. It was shown that 95% of the samples experience a loss of permeability between 50% to 98% of the initial

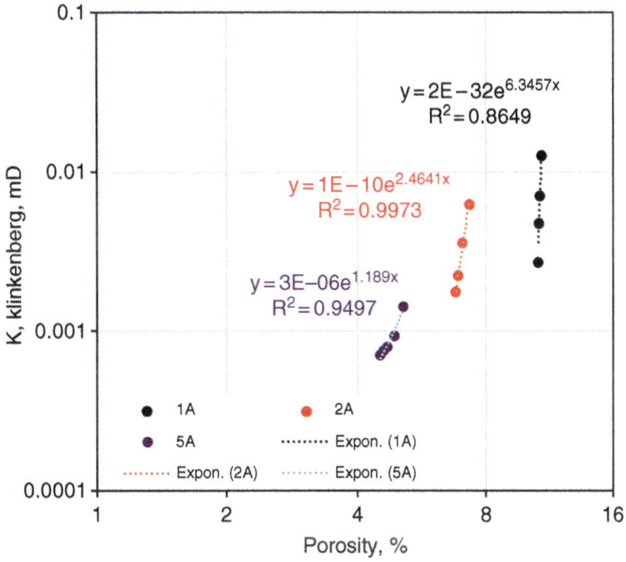

Figure 18.35 Permeability versus porosity (cyclic loading) in log-log scale with increasing net confining stress (well Charlie Sorenson 17-8-3TFH, upper Three Forks).

permeability during cyclic loading. However, the results suggest that the porosity in the upper and middle Three Forks is less sensitive to changes in net effective stress compared to permeability.

- The maximum reduction in permeability occurs at low net stresses (500–2500 psi) where some samples showed up to 98% reduction in their original permeability. In some samples, the microcracks play an important role in permeability changes.
- Only 5% of the samples exhibit different sensitivities to changes in net effective stress. They retain more than 50% of their initial permeability at the final applied net stress.
- The core sample 2A from the upper Three Forks, well Charlie Sorenson 17-8-3TFH showed an abundance of microfractures. The two cycles of stress loading using both methods (steady-state and unsteady-state) presented the loss of permeability at 5000 psi but an increase in permeability was observed at the final stress, 6000 psi, which may be due to the creation of microcracks at a high level of net stress.
- Irreversible loss in permeability was seen during the unloading phase, up to 47% of the initial permeability for sample 1A. This hysteresis is a function of the pore geometry (angular-shaped porosities), with the repeated loading and unloading potentially leading to the collapse of the pore throat and the existing microcracks, which has a large effect on the permeability.
- An increase in permeability during stress unloading was observed in samples 2A and 5A. The increase is more than 2.5 times the initial permeability. At low stresses, the existing fractures may have been reopened and connected with the microcracks, which may enhance the effective permeability, and consequently create new flow paths. In addition, the cyclic loading will shear the existing fractures and that may increase the aperture at low stresses when dilation can happen. This conclusion may explain the changes in the production of the upper Three Forks after shutting and reopening the wells or during the depletion and injection operations.
- Sample 3A has the lowest porosity value (< 2%) but experienced a high reduction in permeability during loading and unloading cycles.
- Also, the porosity hysteresis was observed to be much smaller than that of permeability. The unrecoverable deformation in porosity can reach 11.9% of the origin porosity (Sample 3A) and

8.14%, sample 1A. The results of the stress-dependent porosity showed to be almost identical in terms of stress sensitivity.

- The cubic root of the normalized permeability versus the logarithm of net effective stress showed a nonlinear relationship in the majority of the studied samples. Also, the cubic root of the normalized porosity was found to be nonlinearly related to the net stress over the net confining pressure ranges experienced by the samples.
- Despite some discrepancies (sample 5A), the changes in permeability with increasing net stresses are typically fitted an exponential law function but a power-law function described better the porosity relationship with net stress.
- Based on the obtained results from Charlie Sorenson 17-8-3TFH, in the upper Three Forks, the permeabilities experienced a rapid and significant loss. As cited earlier, just one sample shows a high hysteresis (47%) in both methods. Therefore, further experimental analysis on permeability hysteresis should be added for upper and middle Three Forks members to obtain more repeatable results.
- Permeability at reservoir conditions is a critical parameter in reservoir characterization. Ignoring stress effects can lead to erroneous estimation of permeabilities which will have an impact on long-term production forecast via reservoir simulation.
- As discussed earlier, the permeabilities decreased dramatically at 2500 psi. Thus, the permeabilities and porosities at net stress range between 2500 and 3000 psi were selected to be used in rock types characterization.

References

Adams, J.A. and Weaver, C.E. (1958). Thorium-uranium ratios as indicators of sedimentary processes: example of concepts of geochemical facies. *Bell. Am. Assoc. Petrol. Geol.* 42 (2): 387–430.

Adeniran, A., Elshafei, M., and Hamada, G. (2009). Functional network soft sensor for formation porosity and water saturation in oil wells. *2009 IEEE Instrumentation and Measurement Technology Conference*, Singapore, pp. 1138–1143. https://doi.org/10.1109/IMTC.2009.5168625.

Aguilera, R. (1990). A new approach for analysis of the nuclear magnetic log-resistivity log combination. *Journal of Canadian Petroleum Technology* 29 (1): 67–71.

Aguilera, R. (2002). Incorporating capillary pressure, pore aperture radii, height above free water table, and Winland r 35 values on Pickett plots. *AAPG Bulletin* 86 (4): 605–624.

Aguilera, R. (2003). Determination of matrix flow units in naturally fractured reservoirs. *Journal of Canadian Petroleum Technology* 42 (12): 54–61.

Aguilera, R. and Aguilera, M.S. (2002). The integration of capillary pressures and Pickett plots for determination of flow units and reservoir containers. *Society of Petroleum Engineers Reservoir Evaluation and Engineering* 5 (6): 465–471.

Al Duhailan, M. (2014). Petroleum-expulsion fracturing in organic-rich shales: genesis and impact on unconventional pervasive petroleum systems. Ph.D dissertation. Colorado School of Mines, 206p.

Allen, D., Crary, S., Freedman, S. et al. (1997). How to use borehole nuclear magnetic resonance. *Oilfield Review* 9 (3): 34–57.

Allen, D.C., Flaum, T.S., Ramakrishnan, J. et al. (2008). Trends in NMR logging. *Oilfield Review* 12 (3): 2–19.

Allen, D., Flaum, C., Ramakrishnan, T.S. et al. (2000). Trends in NMR logging. *Oilfield Review* 12 (3): 2–19.

Al-Bulushi, N., King, P.R., Blunt, M.J., and Kraaijveld, M. (2009). Development of artificial neural network models for predicting water saturation and fluid distribution. *J. Pet. Sci.Eng.* 68 (3–4): 197–208.

Alexeyev, A., Ostadhassan, M., Bubach, B. et al. (2017). Integrated reservoir characterization of the Middle Bakken in the Blue Buttes field, Williston Basin, North Dakota. *Society of Petroleum Engineers* https://doi.org/10.2118/185664-MS.

Allwein, E.L., Schapire, R.E., and Singer, Y. (2000). Reducing multiclass to binary: a unifying approach for margin classifiers. *J.Mach.Learn.Res.* 1: 113–141.

Al-Wardy, W. (2002). Measurement of the Poroelastic Parameters of Reservoir Sandstones. Ph.D. thesis. Imperial College, London.

Bateman, R.M. (1940). Open hole log analysis and formation evaluation.

Bell, J.S. and Gough, D.I. (1979). Northeast-southwest compressive stress in Alberta-evidence from oil wells. *Earth and planetary science letters* 45: 475–482.

Bernabé, Y. (1991). Pore geometry and pressure dependence of the transport properties in sandstone. *Geophysics* 56: 436–446.

Bernabé, Y., Mok, U., and Evans, B. (2003). Permeability-porosity relationships in rocks subjected to various evolution processes. *Pure Appl. Geophys.* 160: 937–960.

Bernabe, Y., Brace, W.F., and Evans, B. (1982). Permeability, porosity and pore geometry of hot-pressed calcite. *Mech. Maters.* 1982 (1): 173–183.

Berwick, B.R. (2008). Depositional environment, mineralogy, and sequence stratigraphy of the Late Devonian Sanish Member (upper Three Forks Formation), Williston Basin, North Dakota. Master's thesis. Colorado School of Mines, Golden, Colorado, 262p.

Bottjer, R., Sterling, R., Grau, A., and Dea, P. (2011). Stratigraphic relationships and reservoir quality at the Three Forks - Bakken unconformity, Williston basin, North Dakota. In: *The Bakken–Three Forks Petroleum System in the Williston Basin* (ed. J.W. Robinson, J.A. LeFever, and S.B. Gaswirth), 173–228. Rocky Mountain Association of Geologists.

Boualam, A. (2019). Impact of stress on the characterization of the flow units in the complex three forks reservoir, williston basin. Theses and Dissertations. 2839. https://commons.und.edu/theses/2839.

Boualam, A., Rasouli, V., Dalkhaa, C., and Djezzar, S. (2020a). Advanced petrophysical analysis and water saturation prediction in three forks, Williston basin. *SPWLA 61st Annual Logging Symposium.* Virtual Online Webinar (June 2020). https://doi.org/10.30632/SPWLA-5104.

Boualam, A., Rasouli, V., Dalkhaa, C., and Djezzar, S. (2020b). Stress-Dependent Permeability and Porosity in Three Forks Carbonate Reservoir, Williston Basin. *54th U.S. Rock Mechanics/ Geomechanics Symposium.*

Bush, D.C. and Jenkins, R.E. (1970). Proper hydration of clays for rock property determinations. *J. Pel. Tech.* 22: 800–804.

Chen, S., Ostroff, G., and Georgi, D.T. (1998). Improving estimation of NMR log T2 cutoff value with core NMR and capillary pressure measurements. *SCA-9822, International Symposium of the Society of Core Analysts*, The Hague (14 September 1998), 12p.

Coalson, E.B., Hartmann, D.J., and Thomas, J.B. (1985). Productive characteristics of common reservoir porosity types. *Bulletin of the South Texas Geological Society* 25 (6): 35–51.

Cristianini, C. and Shawe-Taylor, J. (2000). *An Introduction to Support Vector Machine and Other Kernel-based Learning Methods*, 1e, 189. New York, NY: Cambridge University Press.

Cui, A., Wust, R., Nassichuk, B. et al. (2013). A nearly complete characterization of permeability to hydrocarbon gas and liquid for unconventional reservoirs: a challenge to conventional thinking.

SPE Paper 168730 Presented at the SPE Unconventional Resources Technology Conference, Denver (12–14 August 2013).

Davies, J.P. and Davies, D.K. (1999). Stress-dependent permeability: Characterization and modeling. *SPE Annual Technical Conference and Exhibition*. Houston. SPE 56813.

Dennis, D. (2008). Measuring porosity and permeability from drill cuttings. *Journal of Petroleum Technology* 60 (8): https://doi.org/10.2118/0808-0051-JPT.

Dong, J.J., Hsu, J.Y., Wu, W.J. et al. (2010). Stress-dependence of the permeability and porosity of sandstone and shale from TCDP Hola-A. *Int. J. Rock Mech. Min.* 47: 1141–1156.

Dorfman, M.H. (1984). Discussion of reservoir description using well logs. *Journal of Petroleum Technology* 36: 2196–2197.

Doveton, J.H., Guy, W., Watney, L.W. et al. (1996). Log analysis of petrofacies and flow units with microcomputer spreadsheet software: Kansas Geological Survey, University of Kansas 66047, 10p.

Basak, D., Pal, S., and Patranabis, D.C. (2007). Support vector regression. *Neural Information Processing – Letters and Reviews.* 11 (10): 203–224.

Berryman, J.G. (1992). Effective stress for transport properties of inhomogeneous porous rock. *J. Geophys. Res.* 97 (17): 409–424.

David, C. and Darot, M. (1989). Permeability and conductivity of sandstones. In: *Rock at Great Depth* (ed. V. Maury and D. Fourmaintraux), 203–209. Balkema.

Djebbar, T. and Donaldson, E.C. (2012). *Petrophysics: Theory and Practice of Measuring Reservoir Rock and Fluid Transport Properties*, 3e. Elsevier.

Drucker, H., Burges, C.J., Kaufman, L. et al. (1997). Support Vector Regression Machines. ARPA contract number N00014-94-C-1 086.

Droege, L.A. (2014). Sedimentology, facies architecture and diagenesis of the middle Three Forks Formation – North Dakota, U.S.A. Master's thesis. Colorado State University, Fort Collins, CO, 14p.

Dumonceaux, G.M. (1984). Stratigraphy and depositional environments of the Three Forks formation (upper Devonian), Williston basin, North Dakota. Master thesis.

Dunham, R.J. (1962). Classification of carbonate rocks according to depositional texture. In: *Classification of Carbonate Rocks* (ed. W.E. Ham), 108–121. AAPG Memoir.

Dunn, K.J., Bergman, D.J., and Lattoraca, G.A. (2002). *Nuclear Magnetic Resonance Petrophysical and Logging Applications (Handbook of Geophysical Exploration, Seismic Exploration, v. 32)*. Amsterdam: Pergamon an Imprint of Elsevier Science. 293p.

Ebanks, W.J. Jr. (1987). Flow unit concept – Integrated approach to reservoir description for engineering projects. *AAPG Bulletin* 71 (5): 551–552.

Efron, B. and Tibshirani, R. (1994). *An Introduction to the Bootstrap*. New York: Chapman & Hall.

Egbe, S.J., Omole, O., Diedjomahor, J., and Crowe, J. (2007). Calibration of the elemental capture spectroscopy tool using the Niger Delta formation. *31st Nigeria Annual International Conference and Exhibition held in Abuja*. SPE 111910.

El-sebakhy, E.A., Asparouhov, O., Abdulraheem, A. et al. (2010). Data mining in identifying carbonate litho-facies from well logs based from extreme learning and support vector machines. *AAPG GEO, Middle East Geoscience Conference & Exhibition*.

Farber, D.L., Bonner, B.P., Balooch, M. et al. (2001). Observations of water induced transition from brittle to viscoelastic behavior in nano-crystalline swelling clay. *EOS* 82: 1189.

Focke, J.W. and Munn, D. (1987). *Cementation Exponents in Middle Eastern Carbonate Reservoirs*. SPE Formation Evaluation. SPE, Qatar General Petroleum Corp.

Franklin, A. (2017). Deposition, Stratigraphy, Provenance, and reservoir characterization of carbonate mudstone: The Three Forks Formation, Williston Basin. Ph.D thesis. Colorado School of Mines.

Galford, J., Quirein, J., Shannon, S. et al. (2009). Field test results of a new neutron induced gamma-ray spectroscopy geochemical logging tool. *SPE Annual Techinical and Exibition Held in New Orleans*. SPE 123992.

Ganpule, S.V., Srinivasan, K., Izykowski, T. et al. (2015). Impact of geomechanics on well completion and asset development in the Bakken Formation. *Society of Petroleum Engineers* https://doi.org/10.2118/173329-MS.

Gerhard, L.C., Anderson, S.B., and Fischer, D.W. (1990). Petroleum geology of the Williston Basin. *American Association of Petroleum Geologists Memoir* 51: 507–559.

Ghanizadeh, A., Amann-Hildenbrand, A., and Gasparik, M. (2014). Experimental study of fluid transport processes in the matrix system of the European organic rich shales: II Posidonia Shale (Lower Toarcian, Noerthern Germany). *International J. Coal Geology* 123: 20–33. https://doi.org/10.1016/j.coal.2013.06.009.

Gottlib Zeh, S., Briqueu, L., and Veillerette, A. (1999). Indexed Self-Organizing Map: a new calibration system for a geological interpretation of logs. *Proc IAMG*, pp. 183–188.

Gunn, S.R. (1998). Support vector machine for classification and regression. Tech. Rep., Univ. Southampton, Southampton, U.K.

Gunter, G.W., Finneran, J.M., Hartmann, D.J., and Miller, J.D. (1997). Early determination of reservoir flow units using an integrated petrophysical method: Society of Petroleum Engineers. *Paper 38679 Presented at the SPE Annual Technical Conference and Exhibition*, San Antonio, TX, pp. 373–380.

Hashem, S. (1997). Optimal linear combinations of neural networks. *Neural Net-works* 10 (4): 599–614.

He, J., Ling, K., Wu, X. et al. (2019). Static and dynamic elastic moduli of bakken formation. *International Petroleum Technology Conference* https://doi.org/10.2523/IPTC-19416-MS.

Heck, T.J., LeFever, R.D., Fischer, D.W., and LeFever, J.A. (2002). Overview of the petroleum geology of the North Dakota Williston Basin: North Dakota Geological Survey. https://www.dmr.nd.gov/ndgs/Resources/ (accessed June 2016).

Herron, S., Herron, M., Pirie, I. et al. (2014). Application and quality control of core data for the development and validation of elemental spectroscopy log interpretation. *Petrophysics* 55 (5): 392–414.

Hassan, M., Selo, M., and Combaz, A. (1975). Uranium distribution and geochemistry as criteria of diagenesis in carbonate rocks. *9th International Sedimentological Congress*, Nice, France, 7p.

Hassan, M., Hossin, A., and Combaz, A. (1976). Fundamentals of the differential gamma-ray log. In terpretation technique. SPWLA. *17th Ann. Log. Sym. Trans., Paper H.*

Havens, J. B. and Batzle, M. L., 2011. Minimum horizontal stress in the bakken formation. *45th U.S. Rock Mechanics/Geomechanics Symposium*, San Francisco, CA (June 2011). Paper Number: ARMA-11-322.

Hizem, M., Budan, H., Devillé, B. et al. (2008). Dielectric dispersion: a new wireline petrophysical measurement. *Paper SPE-116130 presented at the SPE Annual Technical Conference and Exhibition*, Denver, CO, USA (21–24 September).

Holubnyak, Y.I., Bremer, J.M., Mibeck, B.A.F. et al. (2011). *Understanding the Sourcing at Bakken Oil Reservoirs*. Woodlands, TX, USA: Society of Petroleum Engineers. http://doi.org/10.2118/141434-MS.

Isleyen, E., Demirkan, D.C., Duzgun, H.S., and Rostami, J. (2019). Lithological classification of limestone with self-organizing maps. *53rd U.S. Rock Mechanics/Geomechanics Symposium*, New York, NY, USA (23–26 June 2019). Paper Number: ARMA-2019-1791.

Jin, H., Sonnenberg, S.A., and Sarg, J.F. (2015). Source rock potential and sequence stratigraphy of bakken shales in the Williston Basin. *Unconventional Resources Technology Conference* https://doi.org/10.15530/URTEC-2015-2169797.

Johnson, R., Longman, M., and Ruskin, B. (2017). Petrographic and petrophysical characteristics of the upper Devonian Three Forks Formation, southern Nesson anticline, North Dakota. *The Mountain Geologist.* 54 (3): 181–201.

Jones, S.C. (1988). Two-point determinations of permeability and PV vs Net confining stress. SPEFE 235.

Jones, F.O. and Owens, W.W. (1980). A laboratory study of low-permeability gas sand. SPE

Kamalyar, K., Sheikhi, Y., and Jamialahmadi, M. (2012). Using an artificial neural network for predicting water saturation in an Iranian oil reservoir. *Pet. Sci. Technol.* 30 (1): 35–45.

Kausik, R., Fellah, K., Feng, L. et al. (2016). High and low field NMR relaxometry and diffusometry of the Bakken petroleum system: SPWLA-2016-SSS, Society of Petrophysicists and Well-Log Analysts. *SPWLA 57th Annual Logging Symposium* (25–29 June), 7p. https://www.onepetro.org/conference-paper/SPWLA-2016-SSS (accessed 12 April 2017).

Kerans, C., Lucia, F.J., and Senger, R.K. (1994). Integrated characterization of carbonate ramp reservoirs using Permian San Andres Formation outcrop analogs. *AAPG Bulletin* 78: 181–216.

Kenyon, W., Day, P., Straley, C., and Willemsen, J. (1988). A three-part study of NMR longitudinal relaxation properties of water-saturated sandstones. *SPE Formation Evaluation* 3 (3): 622–636.

Kingma, D.P. and Lei Ba, J. (2015). *Adam: A method for stochastic optimization.* ICLR.

Klenner, R., Braunberger, J., Sorensen, J. et al. (2014). A formation evaluation of the middle Bakken member using multimineral petrophysical analysis approach. *Unconventional Resources Technology Conference*, URTeC:1922735.

Kwon, O., Kronenberg, A.K., Gangi, A.F., and Johnson, B. (2001). Permeability of Wilcox shale and its effective pressure law. *J. Geophys. Res.* 106 (19): 339–353.

LeFever, J.A., Le Fever, R.D., and Nordeng, S.H. (2011). Revised nomenclature for the Bakken Formation (Mississippian-Devonian), North Dakota. The Bakken-Three Forks petroleum system in the Williston basin. Chapter 1. pp. 11–26.

LeFever, J.A. and Nordeng, S.H. (2009). The three forks formation–north dakota to Sinclair field. Manitoba. North Dakota Geological Survey Geological Investigations-GI-76. *One postor.*

LeFever, R.D., LeFever, J.A., and Nordeng, S.H. (2008). Correlation cross-sections for the Three Forks Formation, North Dakota. North Dakota Geological Survey Geologic Investigations No. 65. 1p.

LeFever, J.A., LeFever, R.D., and Nordeng, S.H. (2014). Reservoirs of the Bakken Petroleum System: A Core-based Perspective: North Dakota Geological Survey Geologic Investigations, no. 171. https://www.dmr.nd.gov/ndgs/Publication_List/gi.asp (accessed June 2016).

Leon, M., Lafournere, A.N., Bourge, J.P. et al. (2015). Rock typing mapping methodology based on indexed and probabilistic self-organized map in shushufindi field. *Society of Petroleum Engineers* https://doi.org/10.2118/177086-MS.

Loucks, R.G., Reed, R.M., Ruppel, S.M., and Jarvie, D.M. (2009). Morphology, genesis, and distribution of nanometer-scale pores in siliceous mudstones of the Mississippian Barnett Shale. *Journal of Sedimentary Research* 79: 848–861.

Loucks, S.G.R., Zhang, T., and Peng, S. (2017). Origin and characterization of eagle ford pore networks in the south Texas upper cretaceous shelf. *AAPG Bulletin* 101 (3): 387–418.

Loucks, R.G., Reed, R.M., Ruppel, S.C., and Hammes, U. (2012). Spectrum of pore types and networks in mud-rocks and a descriptive classification for matrix-related mud-rock pores. *AAPG Bulletin* 96: 1071–1098.

Lucia, F.J. (1995). Rock fabric and petrophysical classification of carbonate pore space for reservoir characterization. *AAPG Bulletin* 79 (9): 1275–1300.

Lucia, F.J., Kerans, C., and Senger, R.K. (1992). Defining flow units in dolomitized carbonate-ramp reservoirs. *Proceedings, Society of Petroleum Engineers, paper SPE* 24702: 399–406.

Lucia, F.J. (1983). Petrophysical parameters estimated from visual description of carbonate rocks: a field classification of carbonate pore space. *Journal of Petroleum Technology* 35 (3): 629–637.

Martin, A.J., Solomon, S.T., and Hartmann, D.T. (1997). Characterization of petrophysical flow units in carbonate reservoirs. *AAPG Bulletin* 81: 734–759.

Mardi, M., Nurozi, H., and Edalatkhah, S. (2012). A water saturation prediction using artificial neural networks and an investigation on cementation factors and saturation exponent variations in an Iranian oil well. *Pet. Sci. Technol.* 30 (4): 425–434.

Mayergoyz, I.D. (1986). Mathematical models of hysteresis and their application. *IEEE Trans. Magnet.* 22: https://doi.org/10.1109/TMAG.1986.1064347.

Merkel, R., Machesney, J., and Tompkins, K. (2018). Calculated determination of variable wettability in the middle Bakken and Three Forks, Williston basin, USA. *SPWLA 59th Annual Logging Symposium*, London, UK (June 2018).

McKenon, D., Cao Minh, C., Freedman, R. et al. (1999). An improved NMR tool design for faster logging, Paper CC, transactions. *SPWLA 40th Annual Logging Symposium*, Oslo, Norway (30 May–3 June 1999).

Millard, M. and Brinkerhoff, R. (2016). The integration of geochemical, stratigraphic, and production data to improve geological models in the bakken-three forks petroleum system, Williston Basin, North Dakota. In: *Hydrocarbon Source Rocks in Unconventional Plays* (ed. M.P. Dolan, D.K. Higley, and P.G. Lillis), 190–211. Denver: Rocky Mountain Region: RMAG.

Milijkovic, D. (2017). Brief review of self-organizing maps. *40th International Convention on Information and Communication Technology, Electronics and Microelectronics (MIPRO)*, Opatija, pp. 1061–1066.

Mitchell, R. (2013). Sedimentology and reservoir properties of the Three Forks dolomite, Bakken Petroleum System, Williston Basin, U.S.A: AAPG. Search and Discovery Articles, no. 120079.

Mitra, P.P., Sen, P.N., and Schwartz, L.M. (1993). Short-time behavior of the diffusion coefficient as a geometrical probe of porous media. *Phys. Rev. B* 47: 8565–8574.

Mohaghegh, S., Arefi, R., Ameri, S. et al. (1996). Petroleum reservoir characterization with the aid of artificial neural networks. *J. Petrol. Sci. Eng.* 16 (1996): 263–274.

Mollajan, A., Memarian, H., and Jalali, M.R. (2013). Prediction of reservoir water saturation using support vector regression in an Iranian carbonate reservoir. *ARMA* 13–311.

Moss, A.K. and Jing, X.D. (2001). An investigation into the effect of clay type, volume and distribution on NMR measurements in sandstones. *SCA 2001- 29, Society of Core Analysts Symposium*, Edinburgh. http://www.jgmaas.com/SCA/2001/ SCA2001-29.pdf (accessed 13 April 2017).

Newman, J., Edman, J., Howe, J., and LeFever, J. (2013). The bakken at parshall field: inferences from new data regarding hydrocarbon generation and migration. *Unconventional Resources Technology Conference* https://doi.org/10.15530/URTEC-1578764-MS.

Nordeng, S.H. and LeFever, J.A. (2009). *Three Forks Formation Log to Core Correlation*. North Dakota Geological Survey Geologic Investigations No. 75. 1p.

Passey, Q.R., Dahlberg, K.E., Sullivan, K.B. et al. (2004). A systematic approach to evaluate hydrocarbons in thinly bedded reservoirs.

Peterson, K.J. (2017). Pore-size distributions from nuclear magnetic resonance and corresponding hydrocarbon saturations in the Devonian Three Forks Formation, Williston Basin, North Dakota (abstract): RMS-AAPG Regional Meeting, Billings.

Peters, K.E., Walters, C., and Moldowan, J.M. (2005). *The Biomarker Guide*, vol. 2. Cambridge, UK: Cambridge University Press. 490p.

Petty, D. (2014). Mineralogy and petrology controls on hydrocarbon saturation in the Three Forks reservoir, North Dakota. AAPG Search and Discovery Article-10623.

Pittman, E.D. (1992). Relationship of porosity and permeability to various parameters derived from mercury injection-capillary pressure curves for sandstone. *AAPG Bulletin* 76: 191–198.

Prammer, M.G., Drack, E.D., Bouton, J.C. et al. (1996). Measurements of clay-bound water and total porosity by magnetic resonance logging: SPE- 36522-MS. *Presented at the SPE Annual Technical Conference and Exhibition*, Denver, pp. 311–320. http://dx.doi.org/10.2118/36522-MS (accessed 14 April 2017).

Price, L.C. (2000). Origins and characteristics of the basin-centered continuous reservoir unconventional oil-resource base of the Bakken source system, Williston Basin, paper. Available at: http://www.unddeerc.org/Price/.

Qi, L. and Carr, T.R. (2006). Neural network prediction of carbonate lithofacies from well logs, big bow and sand arroyo creek fields, South west Kansas. *Comput. Geosci.* 32 (7): 947–964.

Ramakrishna, S., Balliet, B., Miller, D, and Sarvotham, S., 2010. Formation evaluation in the Bakken Complex using laboratory core data and advanced logging technologies. *SPWLA 51st Annual Logging Symposium,* Perth, Australia (19–23 June 2010).

Rezaee, M., Ilkhchi, A., and Alizadeh, P. (2008). Intelligent approaches for the synthesis of petrophysical logs. *Journal of Geophysics and Engineering* 5: 12–26.

Rider, M. and Kennedy, M. (2011). *The Geological Interpretation of Well Logs*, 3e.

Roland, L. and Olivier, F. (2007). Advances in measuring porosity and permeability from drill cuttings. *SPE/EAGE Reservoir Characterization and Simulation Conference*, Abu Dhabi, UAE (3–28 October). https://doi.org/10.2118/111286-MS.

Rosepiler, M. J., 1981. Calculation and significance of water saturation in low porosity shaly gas sands. *SPWLA 22nd Annual Symposium*, Mexico City, Mexico (June 1981).

Saffarrzadeh, S. and Shadizadeh, S.R. (2012). Reservoir rock permeability prediction using support vector regression in an Iranian Oil Field. *Journal of Geophysics and Engineering* 9 (3): 336–344.

Sandberg, C.A. and Hammond, C.R. (1958). Devonian system in Williston basin and central Montana. *AAPG Bulletin* 42: 2293–2334.

Saneifar, M., Skalinski, M., Theologou, P. et al. (2015). Integrated petrophysical rock classification in the McElroy Field, West Texas, USA. *Petrophysics* 56 (5): 493–510.

Schmoker, J.W. and Halley, R.B. (1982). Carbonate porosity versus depth: a predictable relation for south Florida. *AAPG Bull* 1982 (66): 2561–2570.

Serra, O. (1984). *Fundamentals of well-Log Interpretation. Developments in Petroleum Science 15A*. Elsevier.

Simpson, G., Hohman, J., Pirie I., and Horkowitz, J. (2015). Using advanced logging measurements to develop a robust petrophysical model for the Bakken petroleum system. *SPWLA 56th Annual Logging Symposium*, Long Beach, CA, USA (18–22 July 2015).

Sloss, L.L. (1984). Comparative anatomy of cratonic unconformities. In: *Interregional Unconformities and Hydrocarbon Accumulation* (ed. J.S. Schlee), 7–36. American Association of Petroleum Geologist Memoir 36.

Smola, A.J. and Schölkopf, B. (1998). A Tutorial on Support Vector Regression, NeuroCOLT. Technical Report NC-TR-98-030, Royal Holloway College, University of London, UK.

Soeder, D.L. and Doherty, M.G. (1983). The effects of laboratory drying techniques on the permeability of tight sandstone core. *SPE 11622 presented at the SPEIDOE*, Denver (13–16 March 1983).

Soeder, D.L. (1986). Laboratory drying procedures and the permeability of tight sandstone core. *SPE, Inst. of Gas Technology*.

Sonnenberg, S.A. (2017). Sequence stratigraphy of the Bakken and Three Forks Formations, Williston basin, USA. *AAPG Rocky Mountain Section Annual Meeting*, Billings, Montana.

Sonnenberg, S.A., Gantyno, A., and Sarg, R. (2011). Petroleum potential of the upper Three Forks Formation, Williston Basin, USA. *AAPG Annual Convention and Exhibition*, Houston, TX.

Sonnenberg, S.A. (2015). Keys to production, Three Forks, Williston Basin. SPE-178510-MS/URTeC:2148989.

Sørland, G.H., Djurhuus, K., Widerøe, H.C. et al. (2007). Absolute pore size distributions from NMR. *Diffusion Fundamentals* 5: 4.1–4.15.

Gaswirth, S.B. and Marra, K.R. (2015). *U.S. Geological Survey 2013 Assessment of Undiscovered Resources in the Bakken and Three Forks Formations of the U.S.* Williston Basin Province.

Swanson, V.E. (1960). Oil yield and Uranium Content of black shales. Geol. Survey, Prof. Paper 356-A.

Teklu, T., Zhou, Z., Li, X., and Abass, H. (2016a). Cyclic Permeability and Porosity Hysteresis in Mudrocks-Experimental Study. *50th US Rock Mechanics/Geomechanics Symposium*, vol. 1, pp. 1–12.

Teklu, T., Zhou, Z., Li, X., and Abass, H. (2016b). Experimental investigation on permeability and porosity hysteresis in low-permeability formations. *The SPE Low Perm Symposium held in Denver, CO, USA*. https:/doi.org/10.2118/180226-MS.

Terzaghi, K. 1943. *Theoretical Soil Mechanics.*

Thrasher, L. (1987). Macrofossils and stratigraphic subdivisions of the Bakken Formation (Devonian-Mississippian). In: *Williston Basin, North Dakota, in D* (ed. W. Fischer), 53–67. Fifth International Williston Basin Symposium.

Vapnik, V.N. and Lerner, A. (1963). Pattern recognition using generalized portrait method. *Automation and Remote Control* 24: 774–780.

Vapnik, V.N. and Chervonenkis, A. (1964). A note on one class of perceptron. *Automation and Remote Control* 25.

Vapnik, V.N. (1995). *The Nature of Statistical Learning Theory*. Springer-Verlag New York, Inc. 99-39803.

Walstrom, J.E., Mueller, T.D., and McFarlane, R.C. (1967). Evaluating uncertainty in engineering calculations. *Journal of Petroleum Technology* 19: 1595–1599.

Walls, J. and Nur, A. (1979). Pore pressure and confining pressure dependence of permeability in sandstone. *7th Form. Eval. Symp. Can. Well Logging Soc.*, Calgary.

Wang, W., Ren, X., Zhang, Y., and Li, M. (2018a). Deep learning based lithology classification using dual-frequency Pol-SAR data. *Applied Sciences.* 8: 1513.

Wang, X., Yang, S., Zhao, Y., and Wang, Y. (2018b). Lithology identification using an optimized KNN clustering method based on entropy-weighed cosine distance in Mesozoic strata of gaoqing field, jiyang depression. *Journal of Petroleum Science and Engineering.* 166: 157–174.

White, F.M. (1974). *Viscous Fluid Flow*. New York: McGraw-Hill.

Xu, C. and Torres-Verdin, C. (2013). Quantifying fluid distribution and phase connectivity with a simple 3D cubic pore network model constrained by NMR and MICP data. *Computers and Geosciences* 61: 94–103.

Xu, J. and Sonnenberg, S. A. (2017). An SEM study of porosity in the organic-rich lower Bakken lower Member and Pronghorn member, Bakken Formation, Williston Basin. *Unconventional Resources Technology Conference*, Austin, TX, USA. https://doi.org/10.15530/URTEC-2017-2697215.

Yu, L., Porwal, A., Holden, E., and Dentith, M.C. (2012). Towards automatic lithological classification from remote sensing data using support vector machines. *Computers & Geosciences* 45: 229–239.

Zoback, M.D. and Byerlee, J.D. (1975). Permeability and effective stress. *Amer. Assoc. Pet. Geol. Bull.* 154–158.

Zoback, M. D. (1975). High pressure deformation and fluid flow in sandstone, granite, and granular materials. Ph.D. thesis. Stanford Univ., Stanford, Calif.

Zoback, M.D. and Byerlee, J.D. (1976). Effect of high-pressure deformation on permeability of Ottawa sand. *Amer. Assoc. Pet. Geol. Bull.* 1976 (60): 153–142.

19

Petrophysical Analysis of Three Forks Formation in Williston Basin, North Dakota, USA

Aldjia Boualam and Sofiane Djezzar

University of North Dakota, Grand Forks, ND, USA

KEYPOINTS

- This chapter discusses alternative approaches to thin beds petrophysical analysis for water saturation estimation.
- A complex petrophysical model was developed using advanced logging techniques to assess the output component and fluid volumes.
- The model was compared to routine and special core analysis for accuracy.
- The second approach used array dielectric measurements to estimate water saturation independently of resistivity.
- The petrophysical model was extrapolated to remote wells by rescaling input components and setting appropriate parameters.
- The Three Forks reservoir is composed of dolostones, quartz, feldspars, and clay minerals mainly illite, with dolomite being the most abundant mineral.
- Clean dolo-siltite facies are characterized by high values of T2 Log Mean and free fluid index, while dolo-mudstone facies have high water saturation.
- The upper Three Forks is the most prospective interval due to reservoir parameters, pore size distribution, and low clay content.

19.1 Introduction

The successful application of horizontal drilling and completion and multistage hydraulic fracturing technologies along with a better understanding of the critical reservoir and source-rock characteristics have contributed to the successful development and production of the Bakken and Three Forks Formations in the Williston Basin. Many studies have been carried out on the Bakken Formation including the formation evaluation (Simpson et al. 2015; Klenner et al. 2014; Ramakrishna et al. 2010), sedimentary (Sonnenberg 2017; Alexeyev et al. 2017; Bottjer et al. 2011; Dumonceaux 1984), geochemistry (Xu and Sonnenberg 2017; Jin et al. 2015; Newman et al. 2013; Holubnyak et al. 2011), and geomechanics (He et al. 2019; Ganpule et al. 2015;

Unconventional Hydrocarbon Resources: Prediction and Modeling Using Artificial Intelligence Approaches, First Edition. Edited by Sid-Ali Ouadfeul.

Havens 2011). However, a limited number of studies on the Three Forks Formation has been conducted so far and with the exploration of the Three Forks as a new target for drilling and production, therefore, effort is needed to characterize the Three Forks reservoir. A detailed petrophysical analysis is required in order to fully characterize the formation. It is important to note that the Three Forks reservoir is a much more complex unit than the middle Bakken. The presence of thinly dolo-mudstone and clay layers alternating with dolomitic-silt facies, in addition to its complex lithology, makes the petrophysical analysis more challenging. The use of conventional logs such as gamma-ray (GR), deep resistivity (R_t), shallow resistivity (R_{xo}), neutron porosity (NPHI), bulk density (RHOB), and compressional slowness (DTCO) result in an underdetermined model by using multimineral solver due to the high number of mineral components in the reservoir. Therefore, it is important to investigate alternative approaches for the water saturation (S_w) estimation to gain a better understanding of the reservoir characteristics. Two substitute approaches to thin beds analysis will be described in this chapter. The first approach includes the integration of the advanced logging in petrophysical evaluation such as elemental capture spectroscopy (ECS), nuclear magnetic resonance (NMR), multifrequency array dielectric measurements (DL), triaxial induction resistivity (R_V-R_H), and core analysis. The second approach to thin beds petrophysical analysis consists of the estimation of the S_w from dielectric dispersion measurements.

In addition, the NMR logs were used to characterize the pore size distribution through the T_2 Log Mean and its relationship with the water saturation for each unit. The software used for petrophysical analysis was Techlog, a SLB software. In the initial phase of the petrophysical analysis work, digital logs and core data were loaded into the Quanti.Elan program. Well header information was recovered from scanned logs.

19.2 Petrophysical Database

Approximately, 2000 wells were examined, and 120 wells with digital logs were selected based on the availability of core analysis and digital logs through the North Dakota Industrial Commission (NDIC) website. However, many of these digital files were incomplete or consisted of very limited log suites. The digital files were supplemented by additional digitizing of curve data from the scanned logs. The first step was the selection of data for specific evaluation associated with the complexity of the reservoir. 52 vertical wells with a complete set of data such as triple combo, ECS, and NMR were selected including two wells with multifrequency array dielectric measurements and triaxial induction resistivity. The most abundant data were the conventional logs (GR, NPHI, RHOB, R_t, and DTCO). 10 wells with porosity and permeability measurements at a net confining pressure and 11 wells with porosity and permeability measurements at multiple net confining pressures ranging from 500 psi to 5000 psi were available from different companies including Weatherford, TerraTek, and CoreLab. The core porosity, permeability, and water saturation data were extracted from well files, compiled, and imported into the Quanti.Elan program in Techlog, SLB software. Well header information was recovered from image logs, where available. In addition, X-ray diffraction (XRD) data was available for 16 wells including those two wells whose XRD data was provided by the Energy and Environmental Research Center (EERC). Unfortunately, capillary pressure analysis was not available for any of these wells. However, the thin section data was available for two wells (well Rasmussen 1-21-16H, well Muller 1-21-16H) and petrographic data was available for only one well (Trigger 1-31H). All these data described were used to assess the Three Forks reservoir quality.

19.2.1 Curve Editing and Environmental Correction

After importing the data, the well logs for each well were processed in preparation for the export and analysis phase. The goal of this phase of the study was to generate, as much as possible, a complete set of continuous log and core data curves. The curve editing and environmental correction were carried out where needed in order to generate a set of continuous curves where it is possible to do so for the Three Forks Formation after uploading the digital logs and core data into the Quanti.Elan program. Logging runs were spliced into one continuous curve for some wells. After continuous log curves were generated, all curves were displayed into the log composite and were depth-shifted to the reference GR curve as shown as an example in Figure 19.1. Core data was

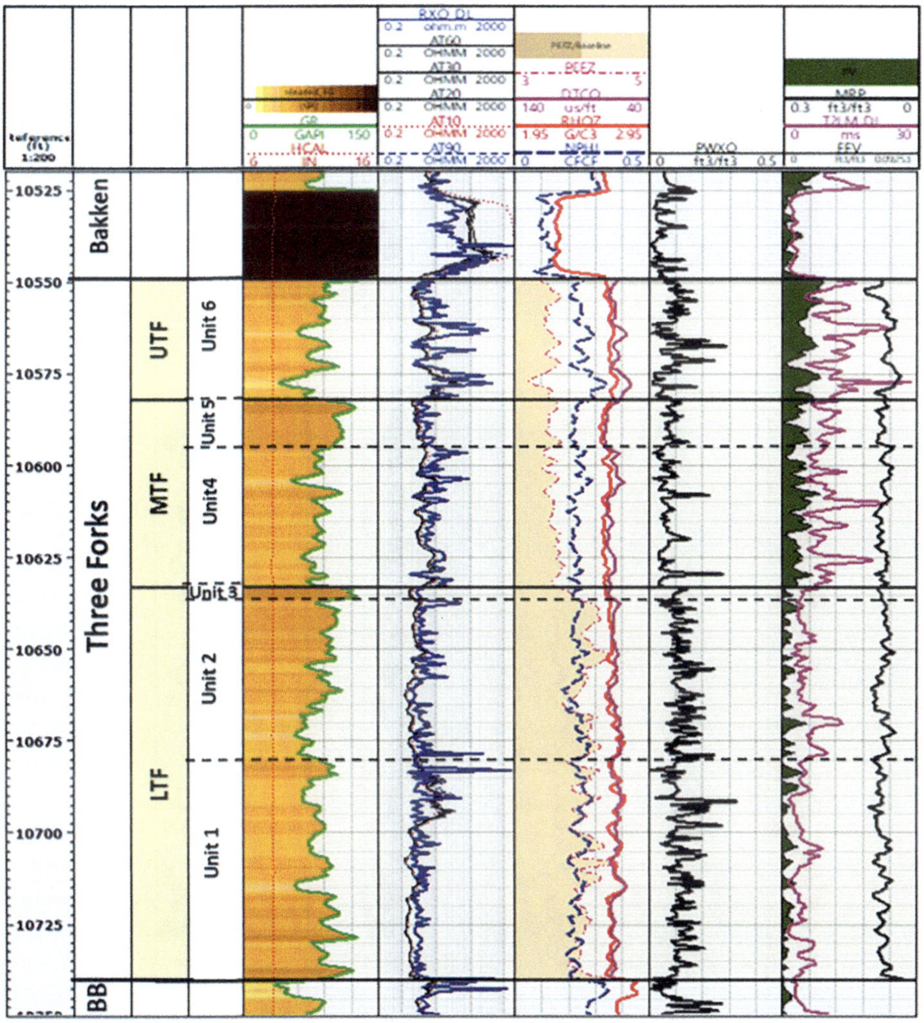

Figure 19.1 Anderson Federal 152-96-9-4-11H Well. Type log for the TF Formations. Tracks from left to right include Track 1, total GR, and HCAL. Track 2, induction resistivities and shallow resistivity from the dielectric. Track 3, NPHI, RHOB, DTCO, and PEF. Track 4, water-filled porosity from the dielectric. Track 5, T_2 Log Mean, free fluid index (FFI), and total porosity (MRP) from NMR. The uppermost sequence contains what is referred to as UTF (Unit 6). The second sequence contains what is referred to as MTF (Units 4 and 5). The third and fourth sequences correspond to the LTF (Units 1, 2, and 3).

block-shifted to prevent data stretching and to preserve the integrity of the data (see Figure 19.2). In some wells, the spectral GR log of the cores is available, this information is critical to a depth matching the core with the wireline logs. In addition, during this phase, the erroneous data such as cycle skips on the sonic travel time logs were eliminated. Well measurements such as GR, RHOB, and NPHI were then environmentally corrected where needed. Tool calibration checks, which are

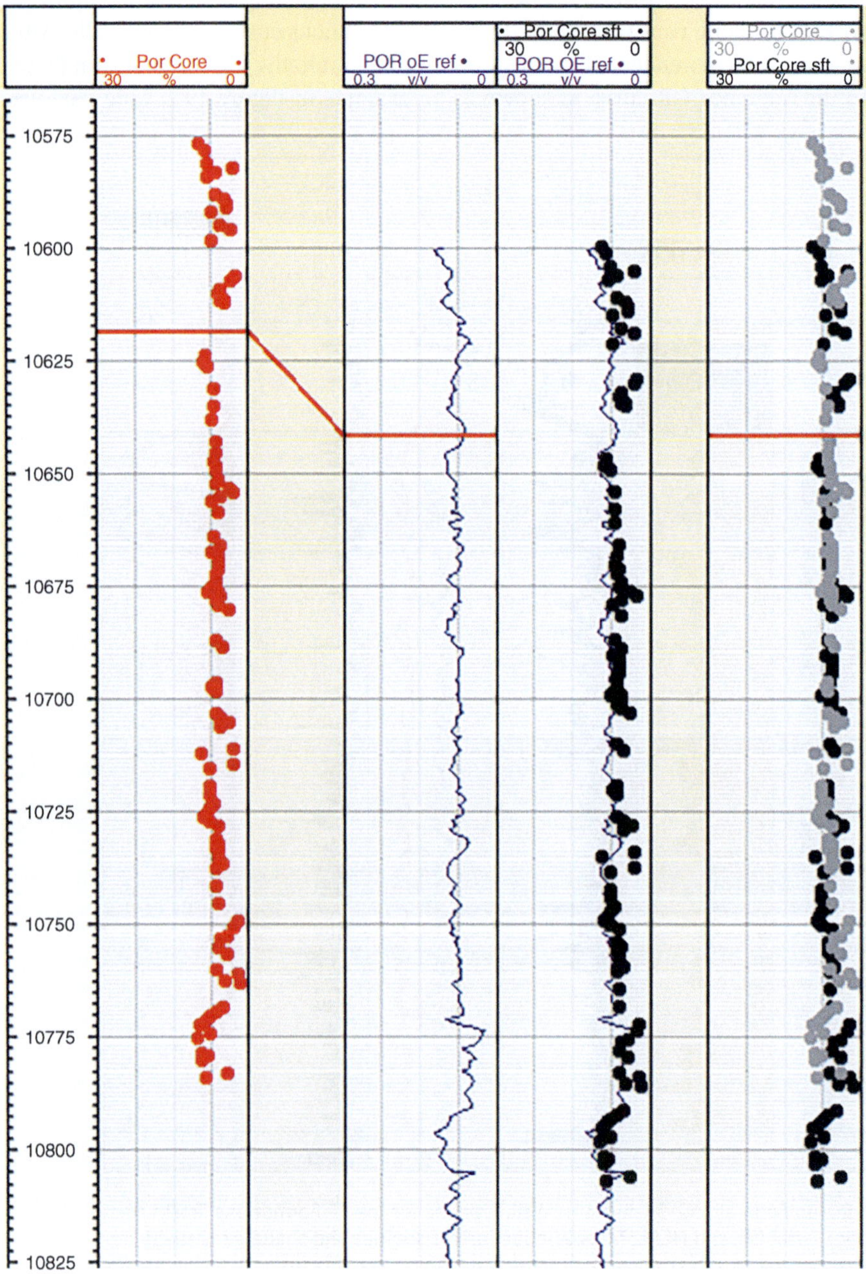

Figure 19.2 Well Charlie Sorenson 17-8-3TFH. Track 1, core porosity before the depth shifting. Track 2, effective porosity from the petrophysical analysis. Track 3, core porosity data blocked shifted to the effective porosity. Track 4 shifted interval.

part of the log quality control measures, were performed to ensure the validity of measurements of downhole tools. Environmental corrections were made in the field on all sensor measurements. The corrections applied on each sensor measurement are listed at the bottom of each log print. The following environmental corrections were done:

- The resistivity tool was corrected for the borehole environment (mud properties and Borehole profile) to obtain the true formation resistivity (R_t).
- The bulk density measurement, RHOB, was corrected for the borehole size and mud weight.
- The neutron log, NPHI, was corrected for temperature, pressure, borehole size, salinity, and mud weight and is presented in limestone units.

Log normalization is necessary for any petrophysical analysis because log measurements obtained from different logging service companies are not consistent from well to well even where the lithology is the same in the reservoir. Normalization must, therefore, be performed to ensure consistent log analysis results from well to well within the study area. For example, shale volumes calculated from GR logs that have not been normalized will differ significantly from shale volumes computed from normalized GR. Unfortunately, log normalization was not performed in this study due to the absence of a normalization module license required.

19.2.2 Preanalysis Processing

Preanalysis processing included the identification of the zones corresponding to the upper, middle, and lower Three Forks and the generation of a bad hole flag as a function of borehole rugosity and the density correction curve. The curve is generated in areas in which the density correction curve exceeds a ratio of $\pm 0.15\%$. The bad hole flag was generated using either the density correction curve or the caliper curve, or both for some wells. Almost, the entire wells had a good hole shape and no rugosity was observed on the caliper logs, for that reason, the bad hole flag was applied just for some wells.

19.3 Methods and Background

19.3.1 Wireline Logs

The logging tools measure the physical properties of formations in the subsurface which are plotted continuously against the depth. These tools have been developed to cope with the complexity of the reservoirs in addition to the type of mud used for drilling. The data recorded are used for reservoir characterization, reservoir management, and decision strategy. In this chapter, a brief overview of the logging tools principles and applications relevant to this study will be presented including the importance of each tool at each step of the formation analysis.

19.3.1.1 Caliper Tool

Wireline caliper tools measure borehole size and give a detailed borehole shape. Caliper log is a continuous measurement of a hole diameter with depth. The geometry of the tool can vary up to six arms and also some tools contain orientation equipment to define the azimuth. An extremely important use of the caliper in this study is in the quality control of the logs when bad hole conditions and caving are present. Also, it is used for recognition of the permeable zones which are indicated by the decrease in borehole diameter and a mud cake build-up as well as detection of the breakouts (shear failures), which are resulted from the drilling-induced stresses around the borehole (Bell and Gough 1979).

19.3.1.2 Total and Spectral Gamma-Ray Logs

The GR log is a measure of the natural formation's radioactivity versus depth. Natural GRs are electromagnetic radiations emitted by unstable atomic nuclei during the process of decaying (Serra 1984). The radiations originated from uranium (U^{238}), thorium (Th^{232}), and potassium (K^{40}), which happen naturally in the formation (Adams and Weaver, 1958). The total GR log measures the radioactivity of the three radiogenic elements (U, K, and Th) collectively, while the spectral GR log helps in the recognition of the type of radiogenic element contributing to a formation's radioactivity, as well as its abundance in the reservoir. It is better used to estimate the volume of shale in the reservoir, which is derived from thorium or potassium or the sum of both (CGR). The GR log is the most recorded log and is an extremely useful tool in picking the tops of the formations during drilling, in well-to-well correlation, and in in-depth control of perforating and testing intervals. Furthermore, its evolution with depth gives a better determination of the mineralogy and grain size vertical distribution (Hassan et al. 1976). Generally, high GR readings are characteristically used to indicate the shaliness of the reservoir. It can also be due to the high potassium content in glauconite-rich sands, or in a formation that contains a high amount of potassium (evaporitic environment). Additionally, a high GR can be also due to a high uranium content in the formation which can be generated by chemical precipitation or by adsorption by organic matter or by phosphates in the clay particles (Hassan et al.1977; Swanson 1961). In addition, the ratio of these radioactive elements (Th/U, U/K, and Th/K) can be used to determine the environmental deposit, the source rock potential, and the recognition of rock types of different facies (Adams, 1958). As an example, the Th/K increases from glauconite – muscovite – illite – mixed layer clays – kaolinite – chlorite to bauxite. Quantitatively, the most significant use of the GR log is the estimation of the volume fraction of the clay in the formation. The GR index is defined as a linear rescaling of the GR log between the gamma-ray minimum (GR_{min}) and gamma-ray maximum (GR_{max}) such that:

$$V_{clay} = GR_{index} = \frac{GR - GR_{min}}{GR_{max} - GR_{min}} \tag{19.1}$$

Where,

GR: is the total measurable gamma radiation for a given interval,
GR_{min}: is the total measurable gamma radiation for a clean interval,
GR_{max}: total measurable gamma radiation for a shaly interval,
V_{clay}: volume of clay.

More importantly, in this study, in the Three Forks reservoir, high gamma signatures can be associated with high content in K-feldspars in some wells, which contain high proportions of potassium. Therefore, the clay volumes estimated from the total GR may be higher than the volume estimated from other methods or tools. In contrast, the anhydrite in the lower Three Forks reservoir provides very low readings on the GR log scale.

19.3.1.3 Electrical Properties (Resistivity)

The resistivity log is a measure of the combined resistivity of the rock matrix and the fluids inside the pores. An emitter sends an electrical current into the formation and a receiver situated at a certain distance from the emitter measures the ability of the formation to impede the flow of electric current through it (Serra 1984). There are two types of tool measurements of the resistivity of subsurface. The first type of tools are the resistivity tools which require water-based mud in the borehole, in which an alternating electrical current at a low frequency flows from the electrodes to the formation and the electrical resistivity of the formation is measured by the receiver. In contrast, the second type of tool requires oil-based mud, which induces an electromagnetic field in the formation, and the total field is detected by the receivers. The resistivity log is generally used to quantify the water saturation in a given reservoir. In explicit cases, the resistivity log can be a useful tool for lithology identification, for example, salt, anhydrite, coal, or tight formation are characterized by high resistivities. On the other hand, low resistivities may be indicative of clay or conductive minerals. In the Three Forks Formation, the oil-based mud system was used for drilling the vertical section of the wells. Therefore, the induction resistivity was recorded. In terms of thin beds, the resistivity log signature depends on the thickness of the bed layers compared to the resolution of the tools. The thin beds can lead to a large anisotropy of the resistivity in the upper and middle Three Forks. In addition, an elevated resistivity signature was observed in the lower Three Forks which indicates the presence of anhydrite.

19.3.1.4 Neutron Logs

The formation is bombarded continuously by fast neutrons (4 MeV and 6 MeV) from a chemical source and the neutron tools record the formation's reaction to the neutron's bombardment (Serra 1984). It is primarily a measurement of the apparent concentration of hydrogen atoms in the fluids existing in the pores of the formation. Then the hydrogen concentration is converted to porosity. Qualitatively, the neutron log is used in combination with other logs such as RHOB, PEF, and ECS to identify the lithology. In addition, the cross-plot of formation bulk density versus neutron porosity is an extremely useful tool for identifying the subsurface lithology and estimating the volume of clay and combined neutron-density porosity in the reservoir. Generally, the neutron log measures the porosity of the formation.

19.3.1.5 Bulk Density Log

The bulk density log measures how a formation responds to high-energy gamma-ray bombardment. Electrons in the formation scatter the gamma rays through Compton scattering, and the degree of scattering or attenuation is measured by receivers. The electron density (electrons/cc) is highly correlated with the bulk density (g/cc) and is often used as a direct measure of density (Serra 1984). The density tool also measures the density of the formation's components, including matrix and fluids in the pore space. Additionally, the litho-density tool measures the photoelectric factor, which is a direct indication of the formation's lithology. Fortunately, many minerals have distinct differences in bulk density, especially when cross-plotted against photoelectric or neutron porosity. The primary purpose of the bulk density log is to calculate porosity using the equation below, in addition to analyzing geophysical characteristics (acoustic impedance) and rock strength.

$$\phi_D = \frac{\rho_{\text{ma}} - \rho_b}{\rho_{\text{ma}} - \rho_f} \tag{19.2}$$

Where,

ϕ_D: density porosity (v/v),
ρ_{ma}: matrix density (g/cc),
ρ_f: fluid density (g/cc),
ρ_b: bulk density (g/cc).

19.3.1.6 Acoustic Logs

The acoustic logs measure the formation's capacity to transmit sound waves. This is essentially accomplished by measuring the time for a pulse of sound to travel a distance through the formation (Serra 1984). This capacity usually varies with lithology and rock texture, particularly porosity.

Three types of sound sources are used. The monopole and dipole sources are used for shear wave detection in slow formation. Therefore, three types of sound waves can be detected, the compressional wave, *P*, shear wave, *S*, and Stoneley wave, *St*. The *P* waves are longitudinal; they are caused by particles' movement in the direction of the propagation. They are the first movement to arrive at the receivers. On the other hand, the shear waves are transverse; the direction of their propagation is perpendicular to the direction of particle displacement and arrives after the compressional wave. Due to the low viscosity and rigidity of the fluids, the shear waves cannot propagate through them. Stoneley waves are a type of surface movement, they form at the surface of the borehole wall in a cylindrical environment. At low frequencies, the Stoneley waves are sensitive to the formation permeability and borehole size (Serra 1984). Consequently, high values of permeability and borehole size can cause energy loss (Serra 1984). The sonic logs primarily measure matrix porosity. They can be used to consistently identify the lithology. Sonic logs are widely used for correlation, identification of the source rock, normal compaction, and overpressure, in addition to geomechanical properties of the formation and the construction of synthetic seismograms. According to Wyllie et al. (1956), the compressional slowness measured is the sum of the interval transit time in the matrix and the interval transient time in the pores (fluids). Therefore, the porosity is estimated from Equation (19.3) as follows:

$$\phi_s = \frac{\mathrm{DT} - \mathrm{DT}_{\mathrm{ma}}}{\mathrm{DT}_f - \mathrm{DT}_{\mathrm{ma}}} \tag{19.3}$$

Where,

ϕ_S: sonic porosity (v/v),
DT: tool-measured compressional slowness (μs/ft),
DT_f: compressional slowness of interstitial fluid (μs/ft),
$\mathrm{DT}_{\mathrm{ma}}$: compressional slowness of matrix (μs/ft).

The sonic scanner tool is widely used for unconventional reservoirs. When considering representative velocities within the Three Forks Formation, it should be noted that the presence of bedded anhydrite and carbonate can have an appreciable effect on the average velocity through the interval.

19.3.1.7 Elemental Capture Spectroscopy

The formation is bombarded with high-energy neutrons, which interact with atomic nuclei and lose energy as they are scattered, primarily by hydrogen (www.slb.com/-/media/files/fe/brochure/ecs-brochure). The geochemical elemental measurements consist of detecting the GR emissions produced by the neutron reactions in the formations that can be processed to identify the chemical elements involved in the reactions and also their concentration (Galford et al. 2009). The ECS tool quantifies common elements found in sedimentary rocks, namely aluminum (Al), silicon (Si), calcium (Ca), iron (Fe), sulfur (S), gadolinium (Gd), chlorine (Ci), barium (Ba), hydrogen (H), and titanium (Ti). These relative elemental yields are converted to dry-weight elemental concentration logs for the elements Si, Al, Mg, Fe, Ca, S, Ti, and Gd (Egbe et al., 2007). Then the quantitative dry-weight lithologies are calculated using SpectroLith which is a series of proprietary empirical relationships between these elements and the commonly occurring sedimentary minerals such as clay, carbonate, and quartz. In SpectroLith, clay is estimated from Al; Carbonate (Calcite + Dolomite) from Ca and Mg, and anhydrite from S and Ca; while sand (assumed to be composed of quartz, feldspars, and micas) is obtained by subtracting clay and carbonate and other mineral fractions from 100 wt% (Egbe et al. 2007).

19.3.1.8 Nuclear Magnetic Resonance

The NMR tool was used to provide mineralogical independent total porosity and to measure the pore size distribution (Allen et al. 2000; McKenon et al. 1999). The porosity is divided into different pore size ranges. The large pores are associated with the free fluids, the small pores in which the fluids are capillary-bound, and the micropores are associated with clay-bound fluids (Dunn et al. 2002). The principles and applications of the tool have been summarized by Dunn et al. (2002) and Johnson et al. (2017). The NMR tool generates an external magnetic field, B_0. With radiofrequency pulses, the applied magnetic field gets disrupted. It modifies the alignment of molecular magnetic dipoles. The proton of the hydrogen nuclei (typically protons associated with the brine, oil, and gas phases in pores) around the tool aligns with this magnetic field (the process of alignment, called relaxation). The intensity of the NMR signal of each fluid is related to its volume fraction and its specific hydrogen index (Allen et al., 2008). The frequency dampens exponentially over time in relation to the fluid types and pore sizes of the formation and this dampening results in the decay time (Bloembergen et al., 1948). The rate at which the protons align to B_0 is defined by a decay time constant T_1 (longitudinal relaxation time). In disparity, the transverse relaxation time, T_2, is described by the time it takes hydrogen nuclei to lose alignment to the external magnetic field when a small electromagnetic pulse with a defined frequency and duration is applied to the polarized protons (Moss et al., 2001). The energy of the pulse is adsorbed by the proton when it is applied but at the end of the pulse, the proton is radiated back again (Rider and Kennedy, 2011). The T_2 distribution comprises a continuous range of relaxation times from hydrogen atoms in multiple pore sizes contributing to different signal amplitudes (Moss et al., 2001). According to Allen et al. (2008), three principal fluid and rock interactions exist that control T_2 distribution of hydrocarbon and water in pores: spin-spin interactions of neighboring hydrogen nuclei in the fluid, the diffusion gradient of the fluid type, and interactions of spinning protons coming into contact with grain surfaces due to Brownian motion. The T_2 distribution is decreased by the fluid viscosity (Allen et al. 2008). The surface relaxivity parameter (ρ), and the surface area-to-volume of a pore ratio (S/V) quantify the interaction of the molecules with the rock matrix and they have been related to T_2 of a single-fluid phase by the following relationship (Mitra et al. 1993, Simpson et al. 2015):

$$T_2 \approx V/\rho S \tag{19.4}$$

According to Xu et al. (2013), the distribution of the pore sizes is not the only controlling factor on the T_2 distribution, but in some rocks, T_2 frequency will also reflect changes in pore shape, surface relaxivity, and fluid type. T_2 has been associated with fluid volumes in three pore-size types. T_2 cutoff time for the free fluid index (FFI) is fixed at 33 milliseconds (Prammer et al., 1996). The T_2 times less than 3 msec has been associated with clay-bound water (CBW) and viscous hydrocarbons. The fluid contributing T_2 greater than 3 msec and less than 33 msec has been associated with capillary-bound water, BFV (Chen et al. 1998).

19.3.1.9 Multifrequency Array Dielectric Measurements

The dielectric logging tool measures the speed of electromagnetic waves sent to the formation (Bateman, 1940). According to Hizem et al. (2008), the dielectric constant of a material affects the way in which electromagnetic wave passes through it. It is a measure of the relative ability of electrically charged particles in a substance to be polarized by an electric field (dielectric permittivity). It measures the dielectric dispersion and the variation of formation dielectric properties as a function of the frequency (Hizem et al., 2008). The dielectric dispersion measurements convey information on rock properties and fluid distribution for advanced petrophysical analysis and to

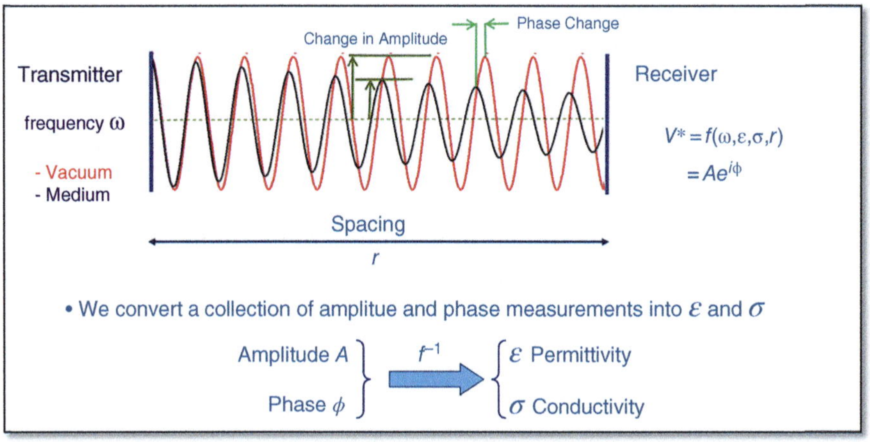

Figure 19.3 Phase shift and attenuation. *Source:* Schlumberger (2011).

Table 19.1 The permittivity of some minerals.

Minerals and fluids	Relative permittivity
Sandstone	4.65
Dolomite	6.8
Limestone	7.5–9.2
Oil	2–2.2
Gas	1

better characterize the reservoir through the matrix textural, water-filled porosity, and water saturation independent of the resistivity (Simpson et al., 2015).

The tool transmits electromagnetic waves with a specific frequency into the formation, the interaction with fluids and matrix results in the attenuation of the amplitude, and changes in wave velocity-phase shift. The changes in amplitude and the phase shift of the wave are functions of initial frequency, formation dielectric permittivity, and conductivity, and the transmitter-receiver spacing (Hizem et al., 2008). These changes are measured at the receiver and then inverted to output permittivity, conductivity, and water-filled porosity (see Figure 19.3). Table 19.1 gives the permittivity of some abundant minerals and fluids in the geological formations.

19.3.2 Petrophysical Analysis Challenges

In the Three Forks interlaminated complex reservoir, a higher level of reservoir knowledge is required. It is mainly needed during the initial phase of the petrophysical analysis for assessing the reservoir characteristics in terms of mineralogy, pore size distribution, and S_w estimation. The primary complexity concerning the petrophysical analysis of the Three Forks reservoir is the presence of thin beds, which can result in large anisotropy of resistivity and high-water saturation. In addition, the Three Forks reservoir is mineralogically complex containing different components with different volumes from well to well. Two substitute methods for thin beds analysis were described in this chapter. First, a robust solution was applied in the Quanti.Elan program to resolve ambiguity from conventional logs by integrating the advanced logging in conjunction with the conventional logs into the workflow. Both probabilistic and deterministic methods were used to quantify the fluid and mineral volumes. The second approach to thin beds analysis consists of the estimation of the water saturation from dielectric dispersion measurements. It is used to better estimate the water saturation independent of the resistivity and Archie parameters (m and n) (Hizem et al. 2008). The difference between the total porosity and the water-filled porosity from array dielectric measurements supplied a straightforward estimation of the S_w in the invaded zone (Simpson et al. 2015).

The petrophysical evaluation was performed on a well-by-well basis and each member was examined individually. The overall objective in doing petrophysical work was to first build a detailed petrophysical model and quantify reservoir parameters in areas where advanced logging and core data are available. After building databases from several widely spaced wells, typical petrophysical parameters for each reservoir zone in each area were defined. These parameters were then used in analyzing wells that are remote from any advanced logs and core measurements. The petrophysical analysis was carried out for 52 wells of the total wells selected, and the same methodology was applied to characterize the output component and fluid volumes for all the interpreted wells. However, two example wells (Anderson Federal 152-96-9-4-11H and Pumpkin 148-93-14C-13H-TF) are chosen and illustrated in detail in this chapter to characterize the Three Forks reservoir. The logging suite for both wells included triple-combo and core analysis. However, the first well also included ECS, NMR, and array dielectric measurements.

19.3.2.1 Formation Components and Volumes

One of the challenges in the petrophysical analysis of the Three Forks reservoir is related to the complexity of the reservoir, being composed of illite, chlorite, quartz, dolomite, calcite, K-feldspars-plagioclase, pyrite, hematite, ankerite, halite, and anhydrite as depicted in Figure 19.4. A multimineral solver, as a probabilistic approach, was used to establish the lithological model based on the input components with simultaneous optimization of different equations being used in the model. The results are shown in Figure 19.5.

First, the conceptual lithological model was created based on multi cross-plots (bulk density versus neutron porosity, thorium versus potassium, and bulk density versus photoelectric factor with GR indicating the color) as presented in Figure 19.6 and on the mineralogy from XRD analysis for 16 wells (Figures 19.17, 19.20, and 19.22). Then, the mineral volumes were quantified from the elemental dry weight fractions computed by ECS, which were used in combination with neutron porosity, bulk density, and photoelectric factor to establish the detailed multimineral model (illite, chlorite, quartz, dolomite, calcite, K-feldspars, pyrite, and anhydrite) and to quantify the content of various matrix components. The output of the mineral volumes is compared with XRD analysis. Estimating the total clay volume using conventional logs such as total GR would give inaccurate results. This is because GR activity is higher due to the abundance of K-feldspars in the reservoir. The integration of the elemental dry weights fraction from ECS allows more accurate mineralogical determination and calculation. The lithology including the volume of clay was derived by combining ECS results with neutron porosity, bulk density, and photoelectric factor. In the second step, the permittivity (1-inch vertical resolution) of the array dielectric measurements were integrated as an input into the model.

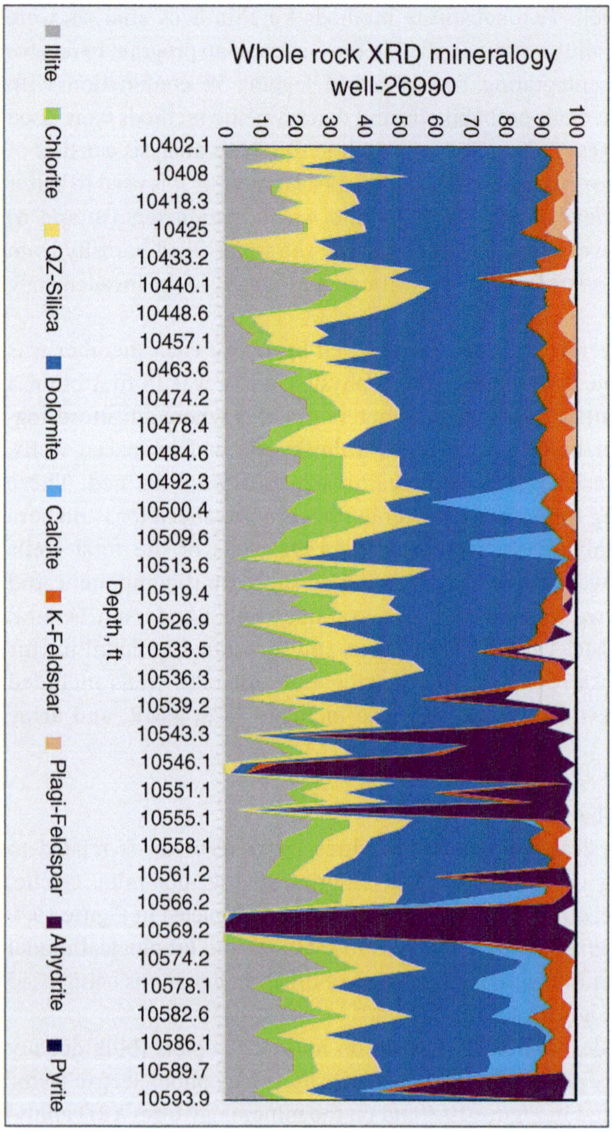

Figure 19.4 Well Hognose 152-94-18B-19H. X-ray diffraction mineralogy (XRD).

In the post-processing phase, after adding the fluids' components (oil and water) to the workflow, the effective porosity was corrected for the effect of clay minerals, and the water saturation was calculated and compared to the core analysis data. The combined neutron porosity and bulk density technique was used to estimate the total and effective porosity. In the case of large washouts affecting the accuracy of the density readings, the sonic log combined with neutron porosity was used to determine the effective porosity. The final log porosities were calibrated to the core porosities. In addition, the true formation resistivity was taken from the invasion-corrected deep resistivity data. On the other hand, the modified Simandoux equation was used for water saturation calculations with Archie parameters m and n equal to 1.7 and 2, respectively. It is observed that the modified Simandoux equation model gave better results than the Archie equation model or other complex models (Dual Water or Waxman-Smits) compared to Dean-Stark water

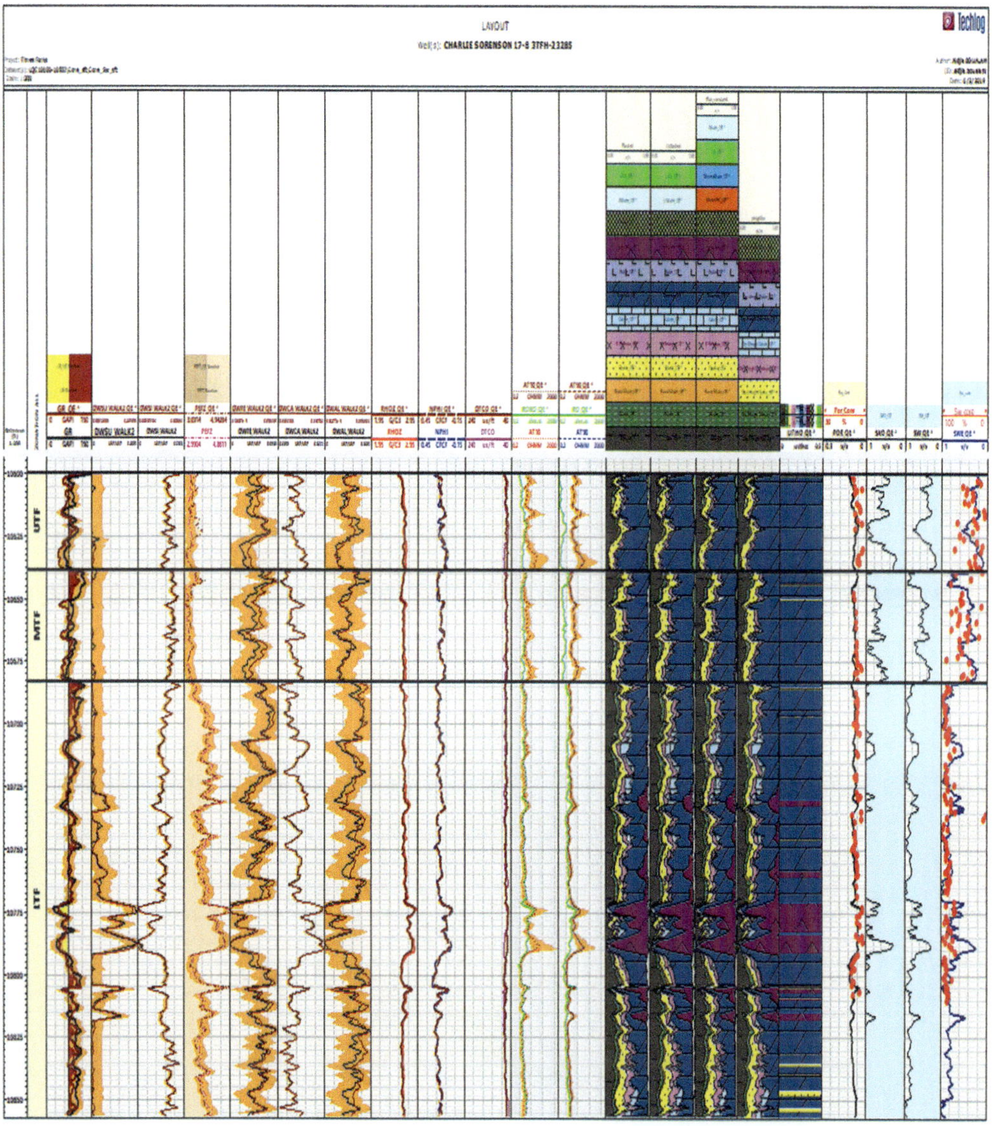

Figure 19.5 Well Charlie Sorenson 17-8 – 3 TFH. Montage showing Elan processing template. Track 1, Total GR and GR elan model. Track 2, DWSU, and DWSU elan model. Track 3, DWSI, and DWSI elan model. Track 4, PEF, and PEF elan model. Track 5, DWFE, and DWFE elan model. Track 6, DWCA, and DWCA elan model. Track 7, DWAL, and DWAL elan model. Track 8, RHOZ, and RHOZ elan model. Track 9, NPHI, and NPHI elan model. Track 10, DTCO, and DTCO elan model. AT10 and AT10 elan model. Track 12, AT9,0, and AT90 elan model. Track 13, Elan model in the flushed zone. Track 14, Elan model in the unflushed zone. Track 15, The cumulative elan. Track 16, Elemental weight fraction. Track 18, effective porosity compared to core porosity. Track 19, Flushed zone water saturation. Track 20, unflushed zone water saturation. Track 21, effective water saturation compared to Dean-Stack water saturation.

saturation. At this step, the nonreservoir facies which are interbedded with porous dolomitic-silt facies were better identified. Consequently, the resulting petrophysical model was extrapolated to the wells with a limited set of data (conventional logs). However, the input components were rescaled to the minimum components to be solved considering the fact that the minimum

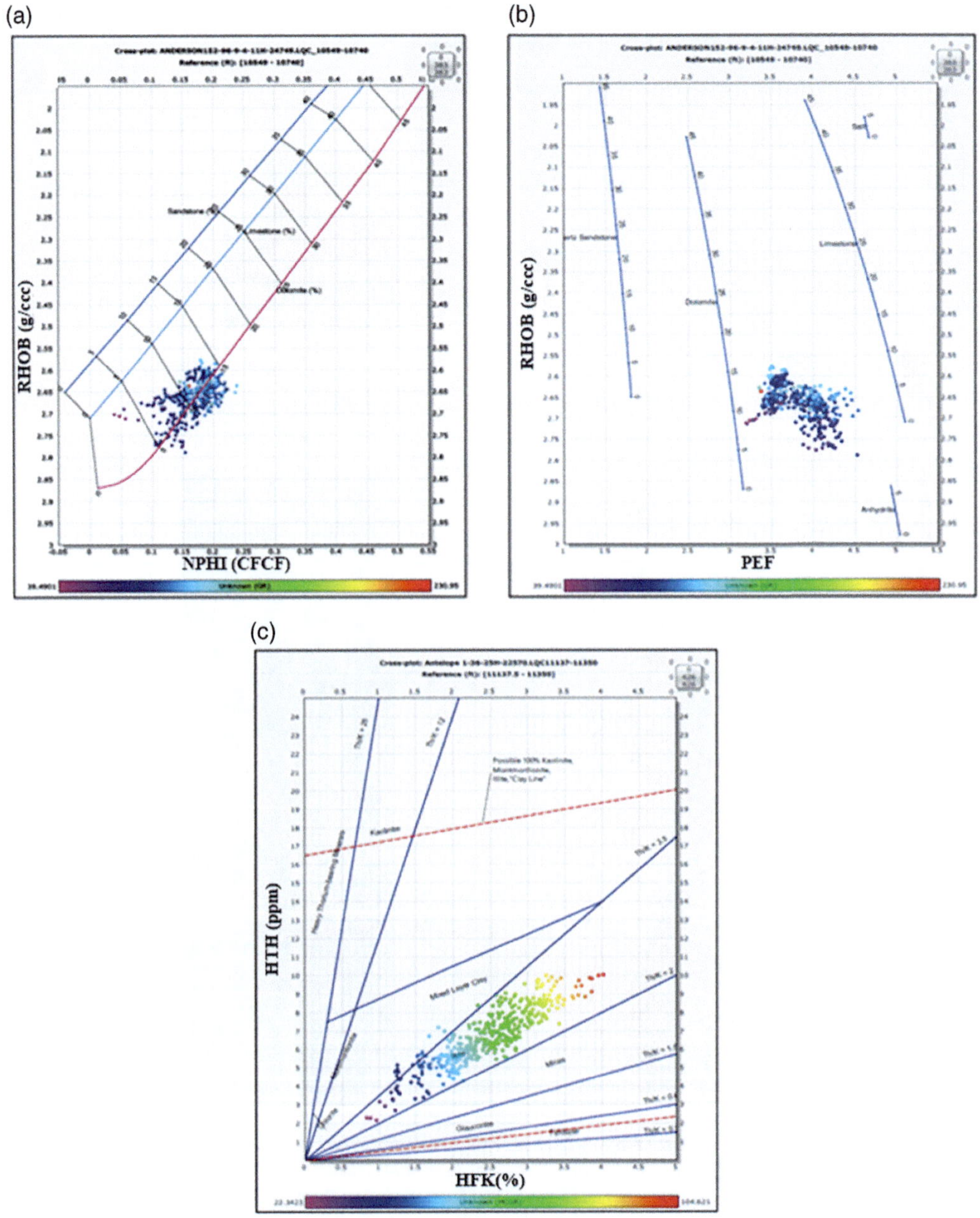

Figure 19.6 Well Anderson Federal 152-96-9-4-11H cross-plots: RHOZ versus NPHI and RHOZ versus PEF show the dolomite is the most abundant mineral with calcite and anhydrite. From HTHO versus HFK, cross-plot illite is the abundant clay mineral in addition to chlorite and mixed layer clay.

component cannot exceed the total number of equations in use. During this phase, the clay minerals were rescaled to illite, while quartz, K-feldspars, and plagioclase were rescaled to quartz and K-feldspars, the carbonate group rescaled to dolomite, and the anhydrite with pyrite and halite was rescaled to anhydrite as depicted in Figure 19.7. The appropriate uncertainties, matrix parameters, and weight multipliers for each equation must be set. Also, additional constraints, based on XRD analysis, were necessary to supply the model with more information.

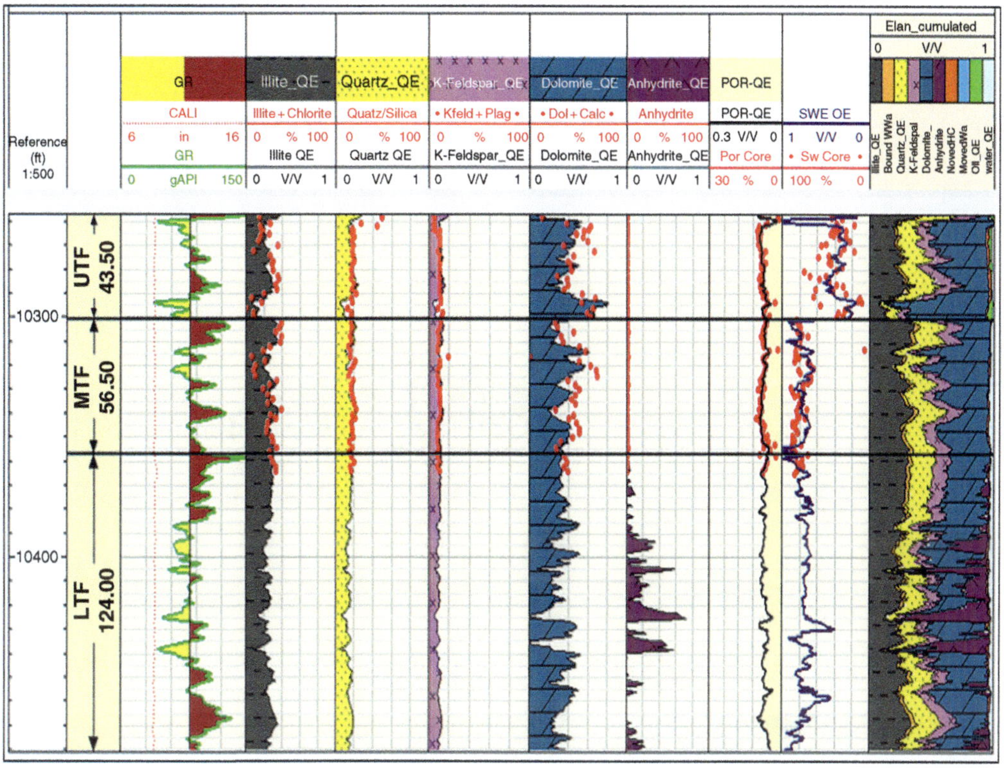

Figure 19.7 The advanced petrophysical model is extrapolated to the Well Pumpkin 148-93-14C-13H-TF with triple combo data set. Track 1, GR. Tracks 2 to 5, mineral volumes calibrated with XRD analysis. Track 6, Total porosity estimated from the combination between NPHI and RHOZ compared to the core porosity. Track 7, modified Simandoux calculated saturation is compared to Dean-Stark S_w. Track 8, lithological model.

19.3.2.2 Water Saturation Model

Another challenge in the characterization of the Three Forks Formation is associated with the development of the S_W model, which lies in the estimation of the Archie parameters: cementation factor (m) and saturation exponent (n). These two parameters are highly variable in carbonate reservoirs due to the high variation of carbonate texture (Focke and Munn 1987). It is important to note that many authors have examined the errors involved in the water saturation equation model and have presented different conclusions. Rosepiler (1981) mentioned that the errors in water saturation are essentially due to the uncertainties in estimating porosity, whereas Dorfman (1984) and Walstrom (1967) concluded that the error is related to the values used for saturation exponent (n) and cementation factor (m), respectively. In the Three Forks Formation, based on the petrophysical analysis, the effective porosity is consistent across the reservoir with a good agreement between the core and NMR porosities. On the other hand, the Archie parameters m and n are better quantified in the laboratory on the core samples, but the laboratory data was not available in this study. However, Johnson et al. (2017) found out, based on the energy dispersive spectrograms in Three Forks, that illite-filled pores contain a hydrocarbon residue. Nevertheless, they did not observe the carbon peaks on the dolomite crystal faces, suggesting that the upper Three Forks (UTF) is a water-wet formation. Therefore, the saturation exponent (n) was taken equal to 2 in the modified Simandoux S_W calculation. It should be noted that in the study area, the oil-based mud system was used for drilling the vertical section of the wells, which prevents the use of

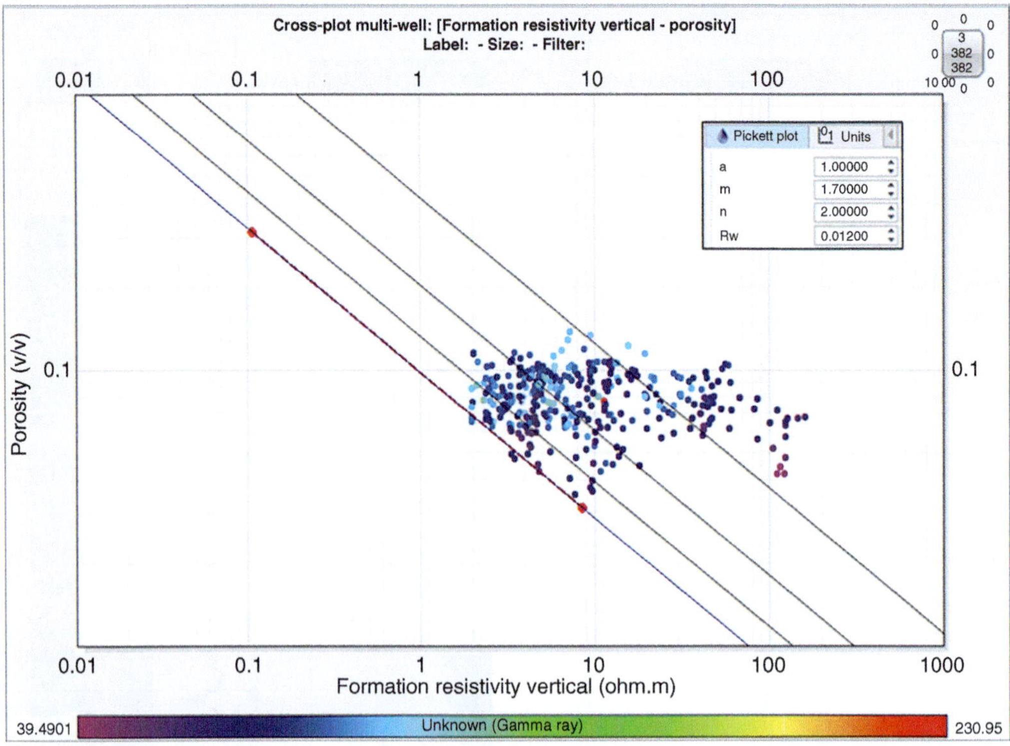

Figure 19.8 Pickett plot for the well Anderson Federal 152-96-9-4-11H.

spontaneous potential measurements that can be an alternative way for water formation resistivity calculation. Figure 19.8 shows the Pickett plot for well Anderson Federal 152-96-9-4-11H over the Three Forks Formation. The vertical resistivity from tri-axial resistivity measurements was used instead of deep resistivity. The slope of the S_w equal to the 100% line is inversely associated with the Archie cementation factor (m), which leans toward 1.7, and the equivalent values of 2 and 1 were considered for the Archie exponent (n) and for the tortuosity factor (a), respectively. The formation water resistivity was equal to 0.012 ohms.m.

Another attempt was made to identify the formation of water resistivities using a multiwell Pickett plot. However, the lower values of the deep resistivities of Three Forks, which resulted from the thin beds' characteristics of the reservoir, made the results insignificant and poorly defined. Formation water resistivity, generated from the Pickett plot, is affected not only by the low resistivity of the thin beds but also by the anhydrite content in the lower Three Forks. Therefore, the formation water resistivities were computed using salinities and temperature profiles. The salinity of the Three Forks Formation is very high, averaging 240 kppm NaCl. This is confirmed by the XRD analysis where the weight volume of halite can reach 3% at well Short-Fee 31-3 in lower Three Forks. The equivalent formation water resistivity corresponding to approximately 0.012 ohm.m to 0.014 ohm.m was estimated and used in the modified Simandoux equation. This confirms the results of the Pickett plot for the well Anderson Federal 152-96-9-4-11H. According to Nordeng (2010), the water in the Bakken formation is close to the saturation of sodium chloride. Thus, the formation water resistivity may be confidently assumed to be close to the resistivity of a salt-saturated brine (0.04 ohm.m at 75 °F). Plotting the latter on Gen-9, Schlumberger chart, and extrapolating the salinity line to the Three Forks reservoir temperature which varies between

180 and 250 °F for the studied wells, it can be seen that the formation water resistivity ranges from 0.012 to 0.016 ohm.m (see Figure 19.9). In addition, Warren (2006) stated that “the evaporates in the Three Forks reservoir commonly follow carbonate precipitation in the mineral series forming from a fluid concentrated by evaporation which indicates a high salinity.”

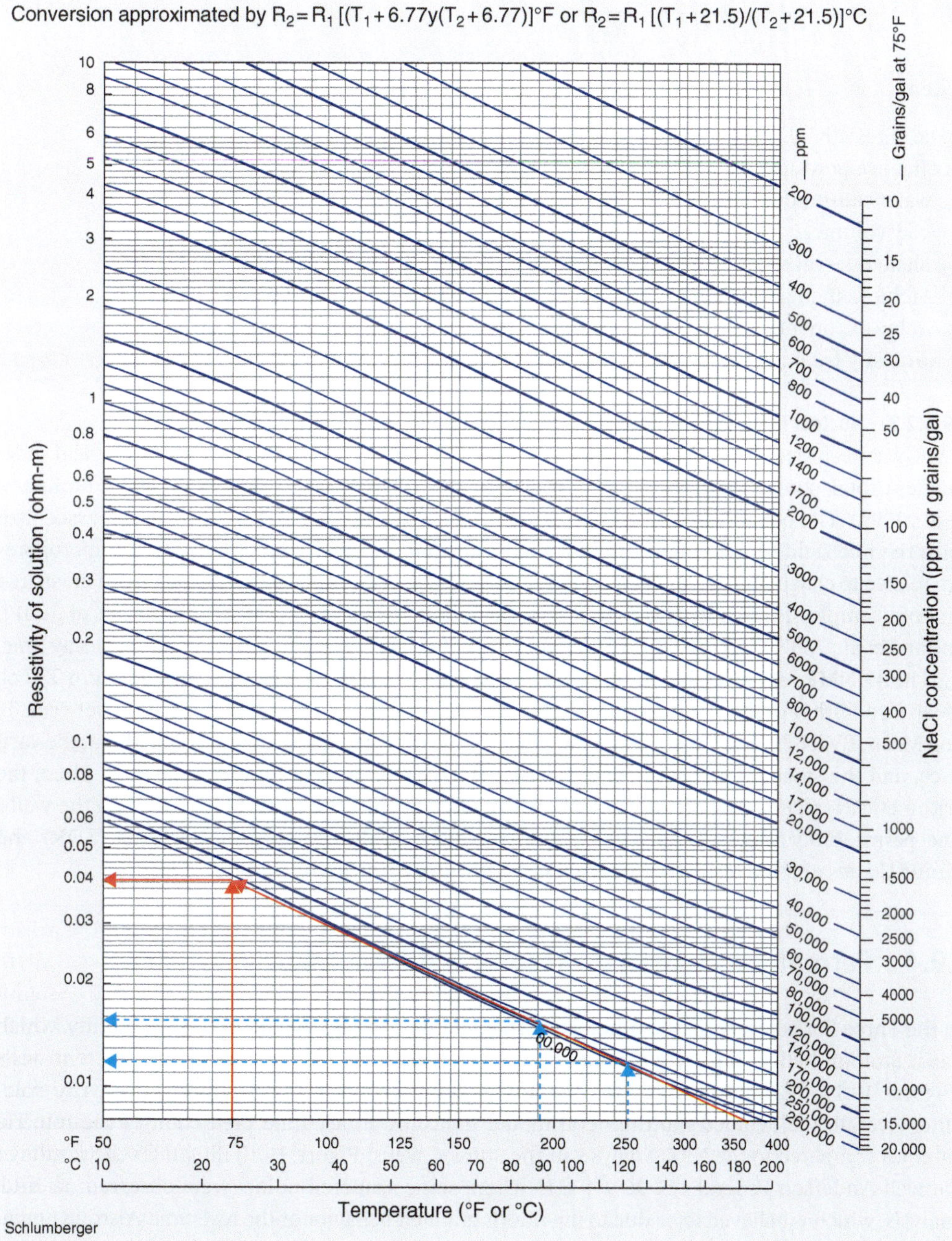

Figure 19.9 GEN 9, Schlumberger chart.

Water saturation was then calculated using the modified Simandoux equation. The “m” values were either obtained from the Pickett plot analysis or were standard values for the type of reservoir under consideration. The “n” values and “a” parameter used were always the standard values of 2 and 1, respectively. After an initial set of log output curves were produced, the results were displayed in a graphical log plot. The equation used for water saturation was:

$$\left(\frac{1}{R_t}\right)=\frac{\phi^m * S_w^n}{a * R_w *\left(1-V_{\text{cl}}\right)}+\frac{V_{\text{cl}} * S_w}{R_{\text{cl}}} \tag{19.5}$$

Where,

R_t: deep resistivity,
Φ: effective porosity,
S_W: water saturation,
V_{cl}: clay volume,
R_{cl}: shale interval resistivity,
m: Archie cementation factor,
n: Archie exponent;
a: tortuosity factor.

19.3.2.3 Nuclear Magnetic Resonance

NMR was recorded in most of the selected wells and was used to provide a mineralogical independent total porosity and to measure the pore size distribution (Allen et al. 2000; McKenon et al., 1999). The porosity is divided into different pore size ranges. The large pores are associated with the free fluid, small pores represent the fluids that are capillary-bound, and the micropores are related to clay-bound fluids (Dunn et al., 2002). Many authors have used T_2 Log Mean measured on core samples to identify the pore size distribution (Allen et al., 2000; Johnson et al., 2017; Sørland et al., 2007; Peterson, 2017; Xu et al., 2013). The larger the pores, the longer the decay time, T_2. The 41 NMR logs available in this study were used to show the changes with depth in T_2 Log Mean in addition to the pore size distribution for each member. Also, the correlation between T_2 Log Mean, BVI, FFI, and CBW was investigated. In the first step, the average T_2 Log Mean, the variance, and the standard deviation were calculated for each member and for each well. Then, the arithmetic average T_2 Log Mean, per well, was plotted against the Dean-Stark S_w for all the wells. The permeability was calculated using two methods: Schlumberger-Doll-Research (SDR) and Timur-Coates and the results were compared with core permeability.

19.4 Petrophysical Analysis Results and Discussion

In the Three Forks reservoir, the lithology is characterized by significant vertical variability, which has a profound effect on reservoir quality. It can be seen from XRD analysis for different wells (Figures 19.17, 19.20, and 19.22) that the dominated lithological composition is primarily dolomite. Overall, the detailed multimineral model indicates an accurate prediction of the mineral volumes compared to the XRD analysis in the studied wells. Figure 19.10 illustrates the results of the well Anderson Federal 152-96-9-4-11H where some scattered points were observed on XRD analysis, which is believed to be due to the interlaminated behavior of the reservoir. Also, an agreement was observed between the estimated porosity, core porosity, and NMR total porosity. The water saturation was accurately reproduced by the two applied models except for the lower interval

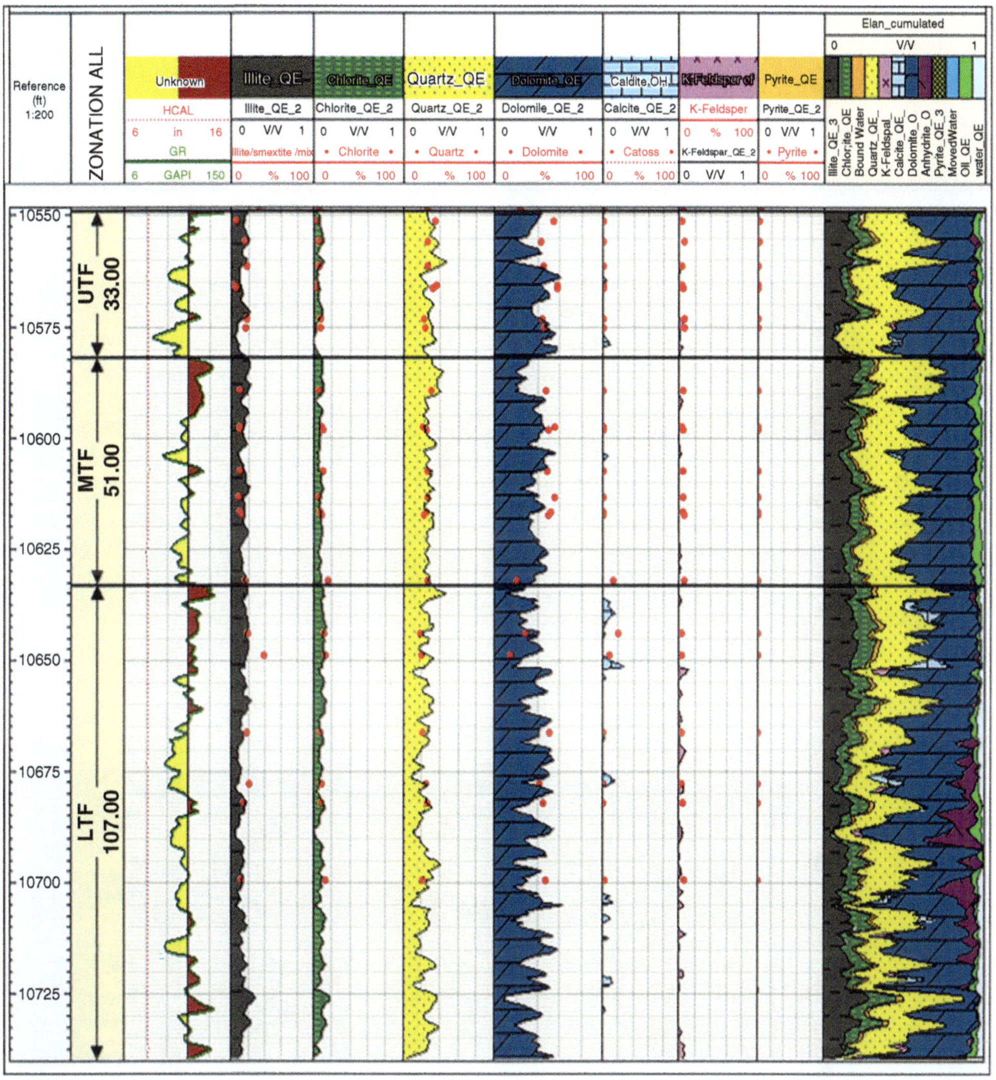

Figure 19.10 Anderson Federal 152-96-9-4-11H Well. Complex mineralogy model determined using ECS, combined with RHOZ, NPHI, and PEF using Quanti. Elan multimineral solver. In special core analysis, XRD (red points) are overlaid for comparison.

where the dielectric model gave better results compared to the Dean-Stark core data. This is due to the abundance of the anhydrite, which leads to an increase in resistivity (see Figure 19.11, track 6). Also, there is a discrepancy between Dean-Stark water saturation and the resistivity-based water saturation model in the thin dolo-mudstone intervals. Indeed, the dielectric measurements provide accurate water saturation results in these intervals.

Figure 19.12 displays the average T_2 Log Mean, per well, versus the Dean-Stark water saturation for a set of wells. The results show that the varied reservoir characteristics are related to the presence of multiple units. Two disparate clouds of points are illustrated on the graph. The first group has a water saturation ranging between 26% and 60% with an average of T_2 Log Mean per well greater than 8 msec. The second group has a water saturation value greater than 60% with an

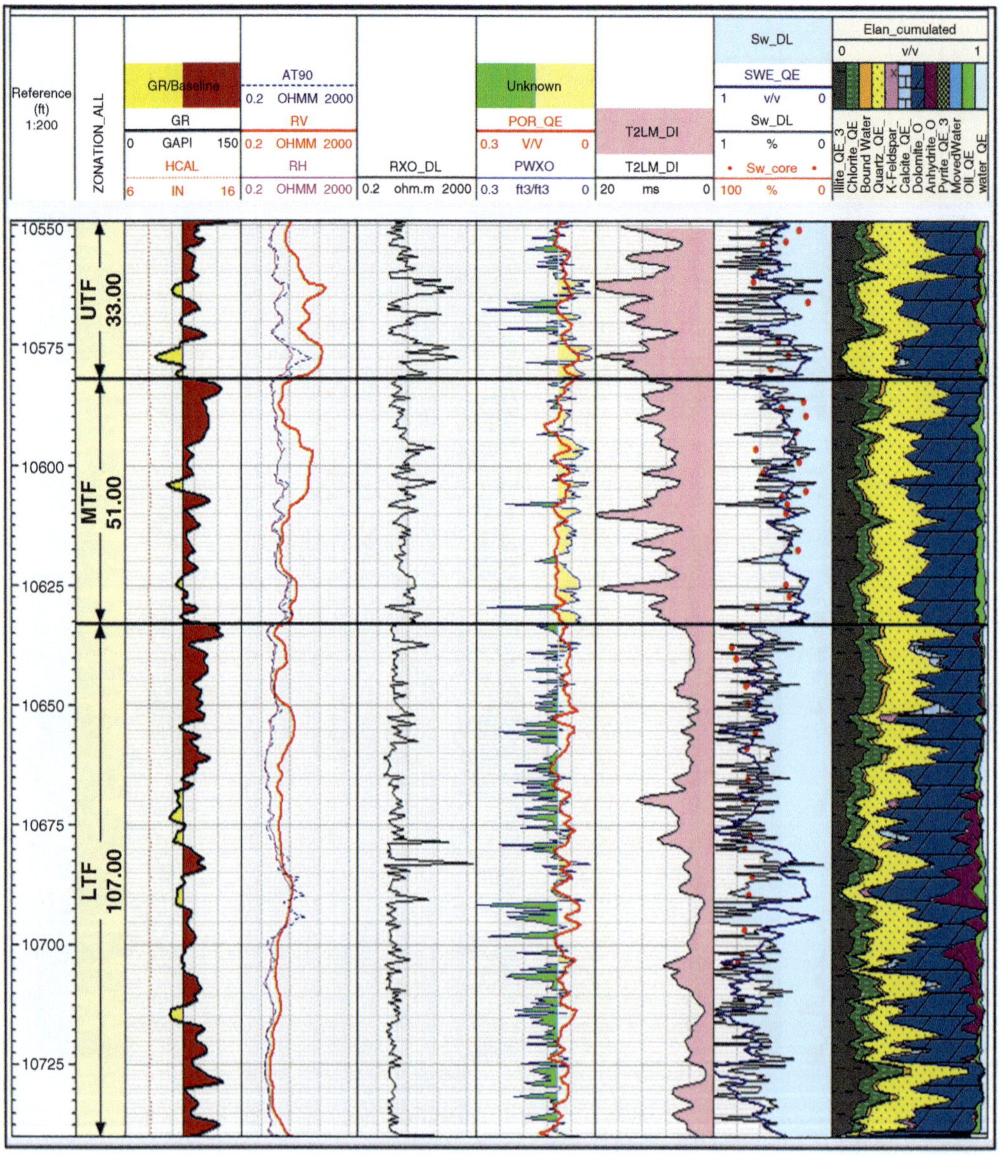

Figure 19.11 Anderson Federal 152-96-9-4-11H Well. Track 2, Total GR. Track 3, Tri-axial induction resistivity measurements (R_V & R_H) versus deep resistivity from conventional induction log. Track 4, shallow resistivity from dielectric measurements. Track 5, effective porosity versus water-filled porosity. Track 6, S_w models derived from petrophysical analysis using modified Simandoux equation (S_{WE}_QE) and array dielectric measurements (S_w_DL) versus Dean Stark S_w (S_w_core). Track 7, complex mineralogical model and fluids model.

average of T_2 Log Mean less than 8 msec (for the lower Three Forks, the T_2 Log Mean is even lower, less than 4 msec). It can be seen that in spite of a wide variability, the UTFs is characterized mainly by large pores with an average T_2 Log Mean of 12.79 msec for all the studied wells (Table 19.2). In contrast, the middle and lower members have an average T_2 Log Mean of 5.95 msec and 2.59 msec, respectively (Tables 19.3 and 19.4). These results are confirmed by the increase of the capillary and

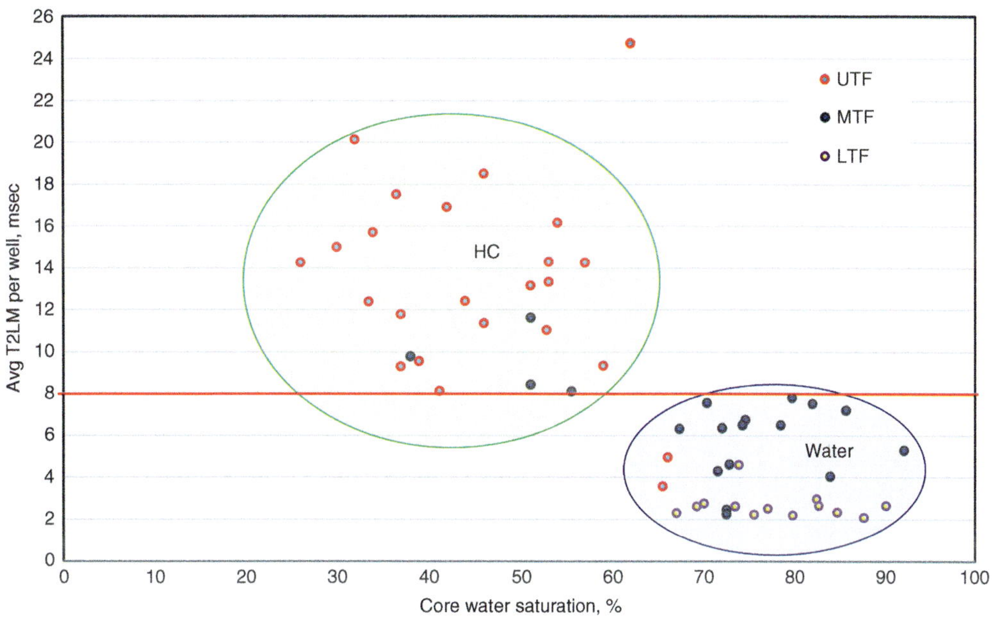

Figure 19.12 Average T_2 Log Mean (Avg T_2LM), per well, versus Dean-Stark S_w cross-plot for UTF, MTF, and LTF for 24 wells. The first group with $S_w < 60$ % and an average of T_2 Log Mean per well greater than 8 msec (oil-bearing zone). The second group with $S_w > 60$ % and an average of T_2 Log Mean less than 8 msec (water-bearing zone).

Table 19.2 T_2 Log mean from NMR and Dean-Stark S_w for the studied wells, UTF.

			T_2LM Min	T_2LM Max	Avg. T_2LM	Avg. $S_w$core
Well name	**NDIC No**	**Variance**	**ms**	**ms**	**ms**	**%**
SARATOGA 12-1-161-92H	22572	323.5	4.06	71.72	24.74	62
ROSENCRANS 44-21H	17193	348.83	2.55	80.27	20.16	32
BARTLESON 44-1-2TFH	27026	259.11	3.41	84.26	18.52	46
FORT BERTHOLD148-95-23D-14-1H	20172	686.68	3.41	180.15	18.67	
FAIRBANKS 1-20H	21966	186.327	5.78	76.99	17.53	36.5
JB 11-6TFH	29062	185.12	3.08	69.93	16.93	42
ROLF 1-20H	20183	79.6	6.6	53.39	16.18	54
KUBAS 11-13 TFH	18837	90.31	3.12	36.46	16.1	
RYSTEDT 4-11H	17109	209.12	1.17	84.33	15.83	
CHARLIE SORENSON 17-8 3TFH	23285	72.87	5.85	42.49	15.72	34
EN-PERSON OBS_2-32	20539	145.59	5.28	60.32	15.65	
JERICHO 2-5H-TF	18792	216.89	3.52	81.44	15.31	

(Continued)

Table 19.2 (Continued)

Well name	NDIC No	Variance	T_2LM Min ms	T_2LM Max ms	Avg. T_2LM ms	Avg. $S_w$core %
NESSON STATE 42X-36	17015	58.5	6.64	44.53	15.03	30
GO-BIWER 157-98-2635H-1	21932	53.76	4.82	40.85	14.52	
ERIE 44-19H	17289	157.69	5.03	77.52	14.41	
THORLAKSEN 11-14H-17325	17325	81.83	3.7	43.41	14.34	
JANE FEDERAL	17430	246.22	3.41	107.04	14.29	53
COMFORD 9-12H	19060	35.96	5.8	29.2	14.28	26
WAYZETTA 46-11M	26661	120.57	4.36	60.72	14.27	57
BARENTHSEN 11-20H	17194	73.54	3.9	41.08	13.36	53
HAWKINSON 14-22H2	24456	345.17	4.22	135.34	13.17	51
PARSHALL 408-15M	27850	416.09	3.18	131.54	13.04	
LAKEWOOD 1-20H	19799	42.06	3.81	31.22	12.43	
ANDERSON FEDERAL 152-96-9-4-11H	24749	15.81	5.21	22.79	12.42	44
FORT BERTHOLD 147-94-3B-10-3H	24272	29.98	2.77	28.68	12.41	33.5
SHELL CREEK 1-01	17058	88.39	3.36	51.39	12.007	
MC. D. TRUST FEDERAL 31-3PH	26269	190.45	1.83	114.85	11.79	37
BRAAFLAT 11-11H	17023	62.87	2.98	45.05	11.38	46
GULLIKSON 44-34H	32044	25.37	4.44	33.52	11.06	52.8
FARHART 11-11H	17096	27.2	1.43	32.28	9.55	39
EN-PERSON OBS_2-24	20442	10.66	3.56	18.65	9.53	
TRACKER HOVDEN 15-1H	20457	95.49	1.32	81.94	9.35	59
CHARLOTTE 1-22H	19918	31.71	3.39	31.01	9.31	37
KOSTELECKEY 31-6H	19264	44.88	2.16	42.49	9.042	
MARIANA TRUST 12X-20G2	24123	6.96	3.56	16.53	8.14	41.2
FAIRFIED STATE 21-16-1H	21947	9.55	3.86	14.75	7.26	
LIND 2-1H	18450	9.47	1.9	18.34	6.87	
BONNIE DIVIDE 16-1H	18976	15.21	2.02	33.5	6.45	
GRAVOS 42-13-14H	24118	6.41	1.19	11.76	4.97	66
EDNA 11-2#1H	22022	3.06	2.62	11.67	4.81	
DEBRECEN 1-3H	20034	3.77	0.97	10.36	3.59	65.5
Average					**12.79**	

Table 19.3 T_2 Log mean from NMR and Dean-Stark S_w for the studied wells, MTF.

Well name	NDIC No	Variance	T_2LM Min	T_2LM Max	Avg. T_2LM	Avg. S_w core
			ms	ms	ms	%
CHARLIE SORENSON 17-8 3TFH	23285	43.22	3.76	36.96	11.61	51
ANDERSON FEDERAL 152-96-9-4-11H	24749	9.64	5.37	21.54	9.76	38
GO-BIWER 157-98-2635H-1	21932	28.17	2.3	29.11	8.7	
HAWKINSON 14-22H2	24456	50.03	2.92	31.57	8.42	51
LAKEWOOD 1-20H	19799	29.48	1.84	34.07	8.17	
FORT BERTHOLD 147-94-3B-10-3H	24272	10.75	1.94	17.88	8.1	55.5
ROLF 1-20H	20183	16.1	2.5	27.91	7.78	79.7
JANE FEDERAL	17430	20.14	2.45	25.24	7.54	70.3
COMFORD 9-12H	19060	3.04	3.96	10.72	7.5	82
BARENTHSEN 11-20H	17194	18.64	2.6	24.21	7.21	85.7
NESSON STATE 42X-36	17015	3.12	3.75	11.34	6.96	
FORT BERTHOLD148-95-23D-14-1H	20172	21.65	2.15	29.7	6.84	
FAIRBANKS 1-20H	21966	5.37	3.23	16.8	6.75	74.5
EN-PERSON OBS_2-24	20442	5.67	2.88	13.03	6.65	
EN-PERSON OBS_2-32	20539	5.71	2.9	18.48	6.64	
CHARLOTTE 1-22H	19918	5.64	3.51	16.24	6.5	78.46
SARATOGA 12-1-161-92H	22572	12.75	1.42	17.82	6.47	74.2
JB 11-6TFH	29062	9.13	2.24	21.68	6.42	
FARHART 11-11H	17096	9.17	2.07	19.9	6.36	72
MARIANA TRUST 12X-20G2	24123	5.17	2.74	11.71	6.32	67.3
ROSENCRANS 44-21H	17193	7.57	2.37	15.58	6.088	
THORLAKSEN 11-14H-17325	17325	6.44	2.83	17.08	6.024	
PARSHALL 2-36H	16324	16.4	1.78	15.38	6.02	
ERIE 44-19H	17289	7.27	2.17	15.67	5.81	
JERICHO 2-5H-TF	18792	11.98	2.18	19.26	5.61	
RYSTEDT 4-11H	17109	7.9	1.49	13.79	5.56	
WAYZETTA 46-11M	26661	6.59	2.24	15.23	5.28	92
FAIRFIED STATE 21-16-1H	21947	7.39	1.3	13.94	4.89	
KUBAS 11-13 TFH	18837	9.14	1.89	12.59	4.67	
GULLIKSON 44-34H	32044	9.64	1.32	16.77	4.62	72.8
BARTLESON 44-1-2TFH	27026	3.93	1.03	11.88	4.52	
SHELL CREEK 1-01	17058	2.63	1.74	9.19	4.37	
BRAAFLAT 11-11H	17023	3.67	1.3	11.77	4.3	71.5
TRACKER HOVDEN 15-1H	20457	5.19	1.36	9.64	4.01	84
BONNIE DIVIDE 16-1H	18976	5.8	1.01	12.55	3.58	
KOSTELECKEY 31-6H	19264	3.43	0.94	10.81	3.45	

(Continued)

Table 19.3 (Continued)

Well name	NDIC No	Variance	T_2LM Min ms	T_2LM Max ms	Avg. T_2LM ms	*Avg.* S_w core %
EDNA 11-2#1H	22022	3.33	1.007	10.15	3.39	
LIND 2-1H	18450	2.79	1.05	7.73	3.1	
MC. D. TRUST FEDERAL 31-3PH	26269	1.52	1.5	6.93	3.02	
GRAVOS 42-13-14H	24118	4.15	0.93	14.82	2.44	72.5
DEBRECEN 1-3H	20034	1.28	0.92	6.89	2.23	72.5
Average					**5.94**	

Table 19.4 T_2 Log Mean from NMR and Dean-Stark S_w for the studied wells, LTF.

Well name	NDIC No	Variance	T_2LM Min ms	T_2LM Max ms	Avg. T_2LM ms	*Avg.* S_w core %
KOSTELECKEY 31-6H	19264	88.23	0.9	91.98	4.84	
ANDERSON FEDERAL 152-96-9-4-11H	24749	3.83	1.23	12.84	4.6	73.8
JERICHO 2-5H-TF	18792	31.15	0.9	40.87	3.75	
GO-BIWER 157-98-2635H-1	21932	57.95	0.87	105.92	3.69	
LAKEWOOD 1-20H	19799	4.51	0.84	12.19	3.14	
ROLF 1-20H	20183	3.3	0.72	12.9	3	82.5
JANE FEDERAL	17430	2.34	0.92	11.86	2.86	
FORT BERTHOLD 147-94-3B-10-3H	24272	2.16	0.89	9.87	2.78	70
FORT BERTHOLD 148-95-23D-14-1H	20172		0.66	10.94	2.65	
WAYZETTA 46-11M	26661	0.98	1.16	8.08	2.68	90
CHARLOTTE 1-22H	19918	1.34	0.95	6.6	2.66	82.7
HAWKINSON 14-22H2	24456	1.78	0.91	10.17	2.64	73.42
MARIANA TRUST 12X-20G2	24123	1.42	0.79	6.84	2.63	69.2
MC. D. TRUST FEDERAL 31-3PH	26269	42.52	0.66	82.84	2.6	
FAIRBANKS 1-20H	21966	1.18	0.85	6.85	2.54	77
FAIRFIED STATE 21-16-1H	21947	2.72	0.78	14.14	2.49	
SHELL CREEK 1-01	17058	1.19	0.97	9.25	2.49	
FARHART 11-11H	17096	6.38	0.7	29.26	2.47	
BARENTHSEN 11-20H	17194	14.2	0.74	53.61	2.34	
CHARLIE SORENSON 17-8 3TFH	23285	1.47	0.77	9.15	2.34	84.8
KUBAS 11-13 TFH	18837	2.45	0.92	10.11	2.33	
GRAVOS 42-13-14H	24118	1.95	0.89	15.97	2.32	67
EN-PERSON OBS_2-24	20442	0.79	0.58	5.9	2.27	
DEBRECEN 1-3H	20034	1.72	0.73	12.5	2.24	75.5

Table 19.4 (Continued)

Well name	NDIC No	Variance	T_2LM Min ms	T_2LM Max ms	Avg. T_2LM ms	Avg. S_w core %
JB 11-6TFH	29062	1.11	0.45	10.14	2.23	
EDNA 11-2#1H	22022	1.057	0.85	6.45	2.23	
BARTLESON 44-1-2TFH	27026	0.75	0.76	6.58	2.19	79.8
ERIE 44-19H	17289	1.2	0.52	5.022	2.19	
EN-PERSON OBS_2-32	20539	0.79	0.77	6.21	2.18	
ROSENCRANS 44-21H	17193	1.16	0.87	8.2	2.18	
SARATOGA 12-1-161-92H	22572	1.24	0.82	9.1	2.17	
TRACKER HOVDEN 15-1H	20457	0.88	0.62	6.06	2.11	87.6
BONNIE DIVIDE 16-1H	18976	0.56	0.88	4.34	2.08	
THORLAKSEN 11-14H-17325	17325	1.48	0.79	13.45	1.96	
PARSHALL 2-36H	16324	1.65	0.68	16.53	1.85	
RYSTEDT 4-11H	17109	1.13	0.53	8.42	1.83	
Average					**2.599**	

clay-bound fluids with depth (see Figure 19.13, track 3; Figure 19.14, track 2). Also, the relationship between T_2 Log Mean and volume of chlorite versus Dean-Stark S_w is plotted in Figure 19.15. As can be seen from this figure, a correlation is observed between T_2 Log Mean and the chlorite content. As the T_2 Log Mean decreases, the volume of chlorite increases. This shows that the larger the volume of chlorite, the more complex the pore size structure, mainly in the middle Three Forks. This correlation is accentuated in the lower interval, where the nanopores and micropores are most abundant. Also, from this figure, a strong correlation is observed between the cutoff of T_2 Log Mean at 8 msec and the Dean-Stark S_w. The higher the values of T_2 Log Mean equal to or greater than 8 msec, the lower the S_w and the larger the pore size.

19.4.1 Upper Three Forks

The UTF interval is a drilling target by itself in the Bakken petroleum system. It displays better characteristics than the middle and the lower Three Forks, which suggests that this interval is oil-bearing. It can be seen from the well Anderson Federal 152-96-9-4-11H that the laminations are highlighted by the large anisotropy of the resistivity shown by the vertical (R_V) and horizontal (R_H) resistivities from tri-axial induction resistivity, and high variations in the shallow resistivity from dielectric measurements (Figure 19.11, tracks 2 and 3). The anisotropy arises from resistivity contrasts between dolo-mudstone, clay, and dolomitic-silt layers. In the lower zone of the UTF unit, the water-filled porosity from array dielectric measurements is lower than the effective porosity from the petrophysical analysis (Figure 19.11, track 4) which confirms the oil-bearing zone. However, in the upper part of the UTFs, large variations are observed between these two curves. This is due to the laminations, thus, the water-bearing intervals are characterized by a higher water-filled porosity compared to the effective porosity. The lower values of the effective porosity compared to water-filled porosity suggest that the interbedded dolo-mudstone facies are

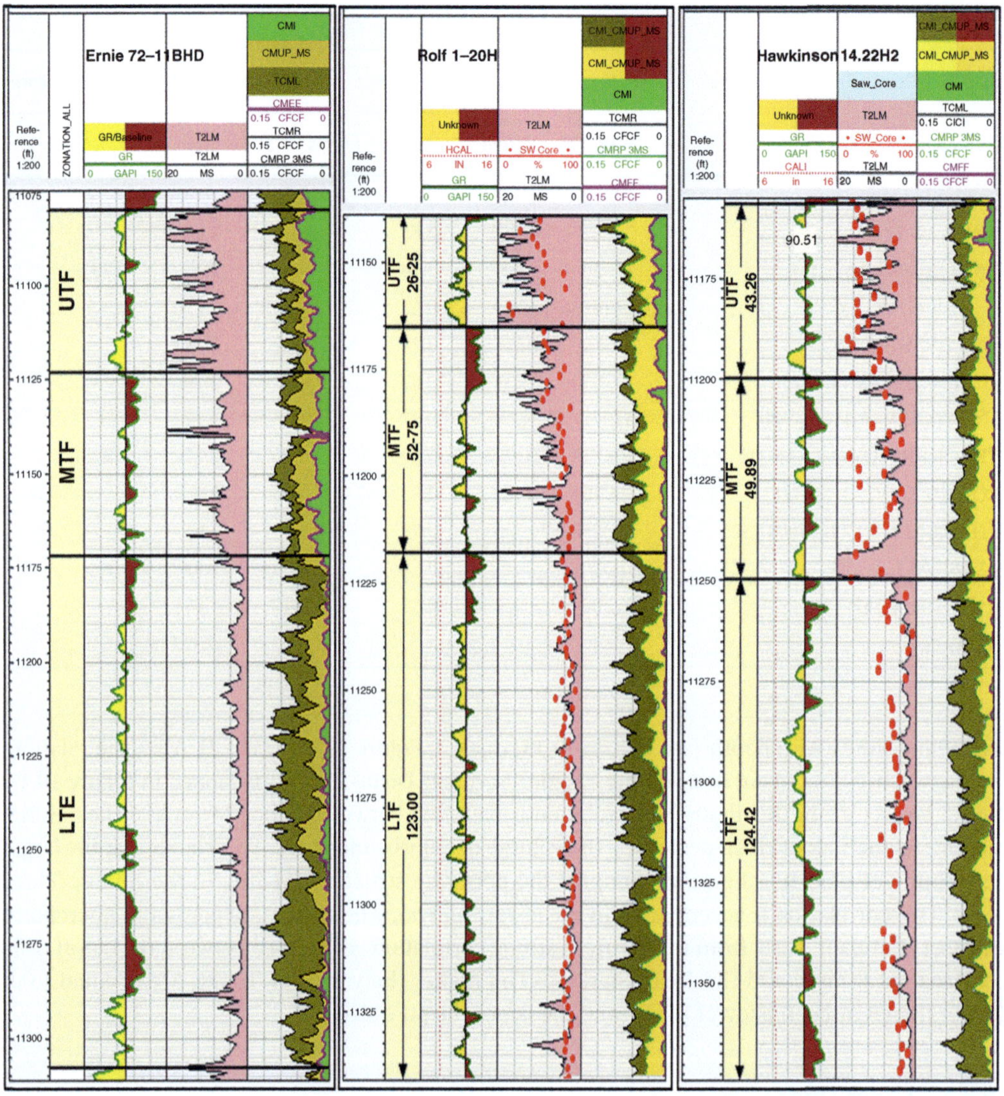

Figure 19.13 From left to right: Well Ernie 7 2-11BHD, Rolf 1-20H, and Hawkinson 14-22H2. Track 2, total GR. Track 3, T_2LM. Track 4, Free fluid index-CMFF (T_2 cutoff > 33 msec), capillary bound fluid-BFV (3 msec < T_2 cutoff < 33 msec), clay bound fluid-CBW (T_2 cutoff < 3 msec).

water-saturated (Figure 19.11, track 4). The corresponding Dean-Stark S_w is significantly higher in these intervals, averaging at 68% as depicted in Figure 19.11, track 6.

Figure 19.16 illustrates the core porosity and permeability distribution at a reference net confining stress (NCS) varies between 2500 and 3000 psi. It can be seen from the graph that UTFs exhibit the classic exponential relationship. The average porosity is between 6% and 7.5%, with maximum values reaching up to 11%. However, the permeabilities (Klinkenberg) are less than 1 mD.

From lithological models combined with XRD analysis, it is seen that the UTF unit is composed principally of dolomite, quartz, K-felspars, plagioclase, pyrite, and clay minerals mainly illite. Dolomite is the most abundant mineral ranging between 25% and 65%. However, the volume of quartz and feldspars are approximately constant, averaging at 25% and 15%, respectively

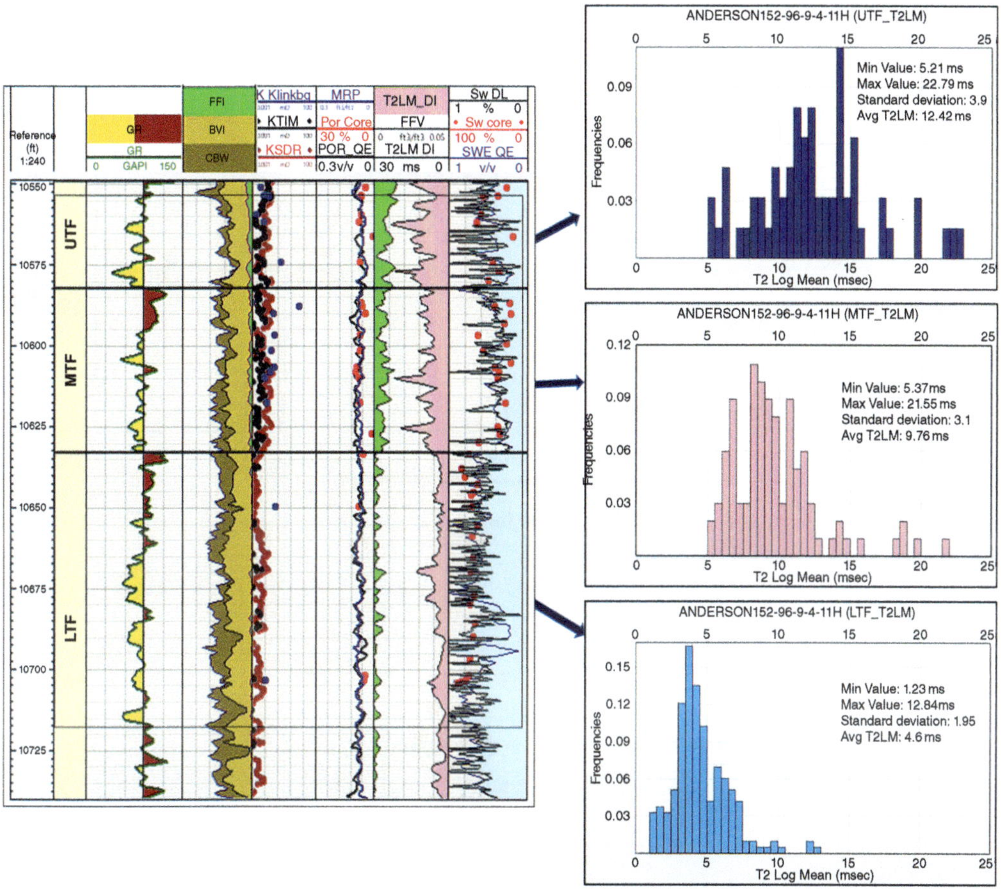

Figure 19.14 Anderson Federal 152-96-9-4-11H Well. Track 2, Total GR. Track 3, Free fluid volume-FFI (T_2 cutoff > 33 msec), capillary bound fluid – BFV (3 msec < T_2 cutoff < 33 msec), clay bound fluid-CBW (T_2 cutoff < 3 msec). Track 4, a good agreement between core permeability (blue points) with permeability from SDR and Timur-Coats. Track 5, a good agreement between effective porosity, NMR total porosity, and core porosity (red points). Track 6, FFI and T_2 Log Mean (T_2LM) decrease with depth, as can be seen from the histograms, the average T_2LM for UTFs is 12.42 msec, middle Three Forks (MTF) is 9.76 msec, and lower Three Forks (LTF) is 4.6 msec. Track 7, a good agreement between the based resistivity saturation model (S_{WE}_QE), S_w model from dielectric measurements (S_w_DL), and to Dean-Stark S_w (red points). Overall, an excellent correlation between FFI, T_2 Log Mean, and the S_w. Higher FFI and T_2LM in the UTF. T_2LM and FFI are lower in the MTF compared to the previous unit. There is a significant decrease in both T_2LM and FFI in the LTF accompanied by an increase in the S_w and CBW.

(Figure 19.17). The volume of quartz can reach 40% in some wells. The volume of chlorite is lower than 5%, whereas the content of illite varies between 5% and 20% in most samples. The clean intervals with a high dolo-siltite content and low content of clay minerals of less than 10% present better reservoir characteristics. A good relationship can be seen between reservoir characteristics, mineralogical composition, and their contents. In addition, the water saturation from Dean-Stark is significantly higher in laminated dolo-mudstone averaging at 65%.

Overall, T_2 Log Mean distribution is highly variable in the UTFs (standard deviation: 2.5 to 20.4). The average T_2 Log Mean values, by well, is ranging from 3.59 msec to 24.74 msec as shown in Figure 19.18. The maximum value reaches up to 180.15 msec at well Fort Berthold 148-95-23D-14-1H

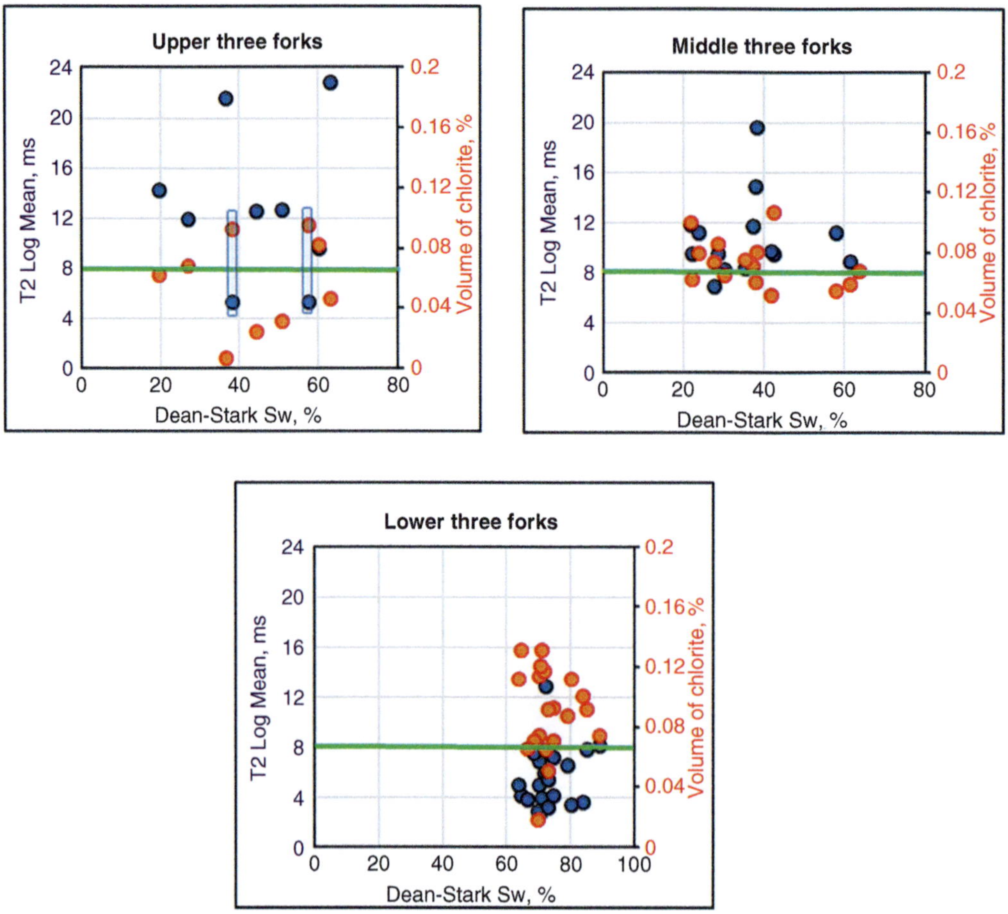

Figure 19.15 Anderson Federal 152-96-9-4-11H Well. T_2 Log Mean (blue points) and chlorite volume (orange points) versus Dean-Stark S_w. The lower the chlorite content, the higher the T_2 Log Mean and the lower S_w.

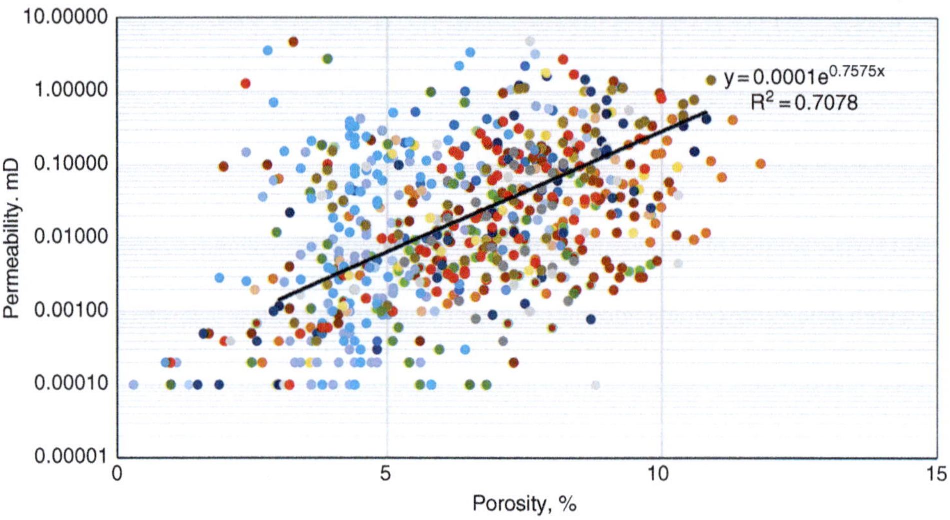

Figure 19.16 Permeability (Klinkenberg) versus porosity core cross-plot, UTF, at NCS between 2500 and 3000 psi. The porosity is between 1% and 11.5% and the permeability is generally less 1 mD.

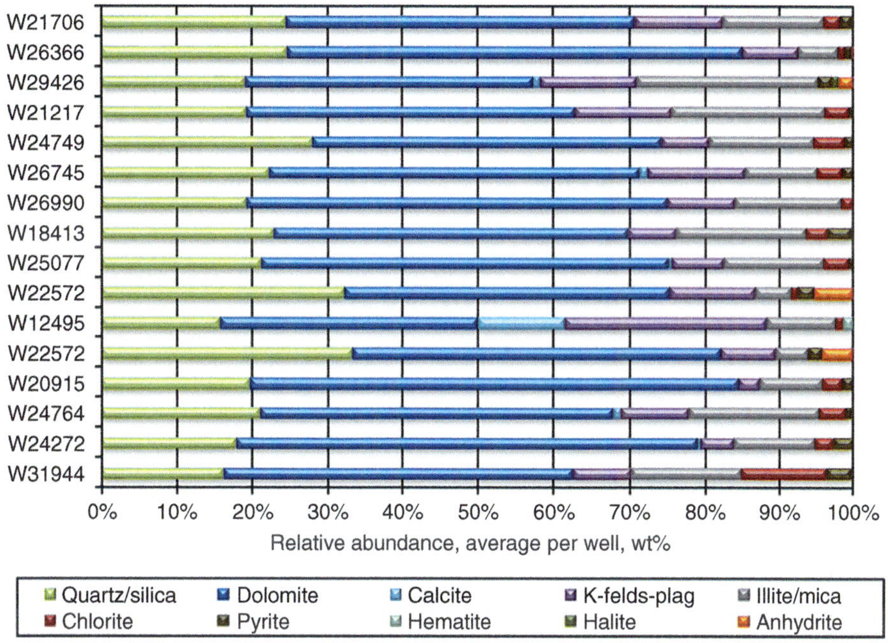

Figure 19.17 Average relative abundance of minerals per well for UTF from XRD experiments. Dolomite is the dominant component. The dominant clay mineral is illite.

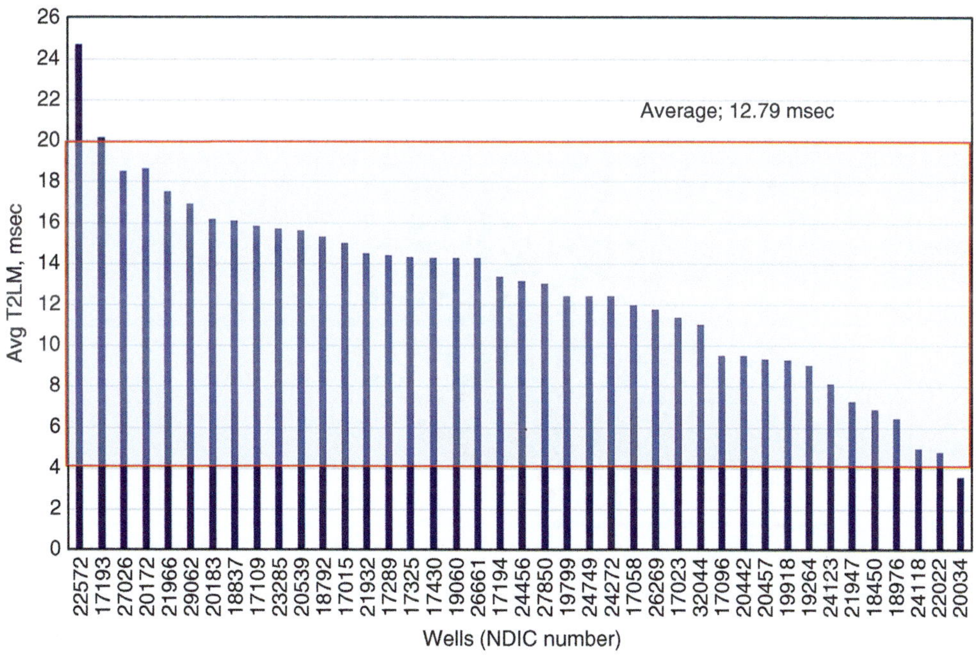

Figure 19.18 Average T_2 Log Mean (T_2LM), per well, for UTF. T_2LM is greater than 8 msec for most of the wells.

with a large variance of 686.32 msec as can be seen from Table 19.2. These variations are generally attributed to the interlaminated nature of the reservoir. The lowest T_2 Log Mean values are recorded by the well Debrecen 1-3H ranging between 0.97 msec and 10.36 msec with an average S_W of 65.7%. This well is located in the south part of the studied area where 40 ft of Pronghorn overlay the UTF member. The FFI, BVI, and CBW from NMR remain fairly consistent in most of the wells across this unit. It is characterized by lower BVI and CBW compared to middle and lower Three Forks members. One can see that the higher FFI corresponds to the large pores ($T_2LM > 8ms$) and lower S_W (see Figure 19.14). Therefore, it appears that in this unit, the water occupies the small pores and the oil fills the large pores. A relationship is also seen between T_2 Log Mean, FFI, BVI, and CBW with S_W and mineralogical composition. The dolo-siltite with a low volume of clay is characterized by high T_2 Log Mean (large pores), high FFI, and low S_W. The dolo-mudstone facies have higher S_W and lower T_2 Log Mean.

19.4.2 Middle Three Forks

The Middle Three Forks is more complex than the upper unit. The hydrocarbon storage capacity of this unit depends on many factors, including its characteristics and the thermal maturity of the black Bakken shales (Sonnenberg, 2015). Also, according to Rice (2000) and Al Duhailan (2014), due to the poor characteristics of the Three Forks reservoir, the Bakken organic shales' forces of expulsion can drain only 50 ft to 75 ft through the Three Forks interval. The results of the studied wells show that the reservoir characteristics of the middle unit are lower than the upper unit. The core porosities range between 2% and 11%, with an average of 6.5% and the permeability varies from 0.0001 to 10 mD based on the 1300 studied core samples (Figure 19.19). An agreement can be seen in Figure 19.11 (track 6) among the water saturation models, averaging 35%. In addition, the water-filled porosity reads much lower than the effective porosity which indicates a clear oil-bearing zone except at the top of this unit where the shale content is higher. It can be seen from a

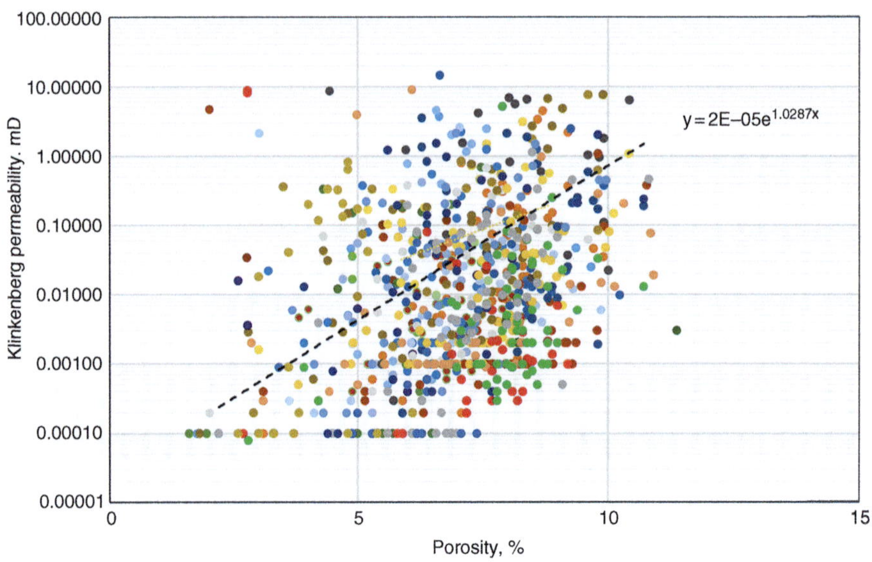

Figure 19.19 Permeability (Klinkenberg) versus core porosity cross-plot, MTF. The porosity is between 2 and 11% and the permeability is generally less 10 mD.

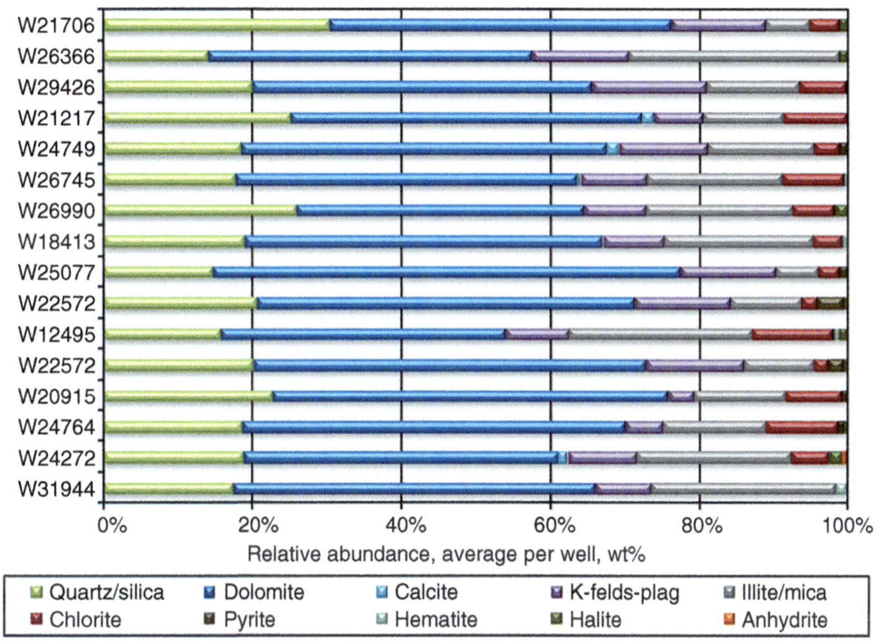

Figure 19.20 Average relative abundance of minerals per well for MTF from XRD experiments. It is predominantly composed of dolomite, the content of clay minerals increases (chlorite), and the volume of the dolomite is lower than the previous unit.

petrophysical analysis and XRD analysis (Figure 19.20) that the MTF reservoir is composed primarily of dolomite, quartz, K-feldspars, plagioclase, pyrite, calcite, and clay minerals. Dolomite is the most abundant mineral (45%) in this unit. While the quartz content ranges from 15 to 30%, averaging at 20%, K-feldspars with plagioclase contents vary from 3 to 15%. Nevertheless, the content of the clay minerals is higher (9% to 35%) mainly in chlorite content. The abundance of non-reservoir intercalated dolo-mudstone layers and clay with dolomitic-silt is the origin of the deterioration of the reservoir properties. The average T_2 Log Mean, per well, is plotted against the Dean-Stark S_w in Figure 19.21. It can be seen that the average T_2 Log Mean, per well, is lower than the upper unit ranging from 2 msec to 11.6 msec (refer to Table 19.3). However, most of the wells have an average T_2 Log Mean lower than 8 msec. The middle Three Forks pore structure complexity is highlighted by high variability in S_w for the same value of T_2 Log Mean. For example, the average T_2 Log Mean observed at wells Hawkinson 14-22-H2 and Rolf 1-20H is around 8 msec. However, the S_w values measured for these wells are 51% and 79.7%, respectively (see Figure 19.13). This variation in S_w can be related to the high capillary-bound water at well Rolf 1-20H.

19.4.3 Lower Three Forks

The lower Three Forks member, in contrast to the previously described units, is characterized by the abundance of dolo-mudstone facies and anhydrite (Figure 19.22). Additionally, a high variation happens in clay minerals (16% to 40%) and calcite content (0% to 35%). The amount of quartz ranges from 9% to 17% and K-feldspars with plagioclase contents vary from 3% to 11%. XRD analysis also shows an insignificant amount of pyrite (0.8% to 1.7%), hematite (0% to 3%), and halite (0% to 3.7%). The water zone is easily identified in the totality interval of the lower Three Forks

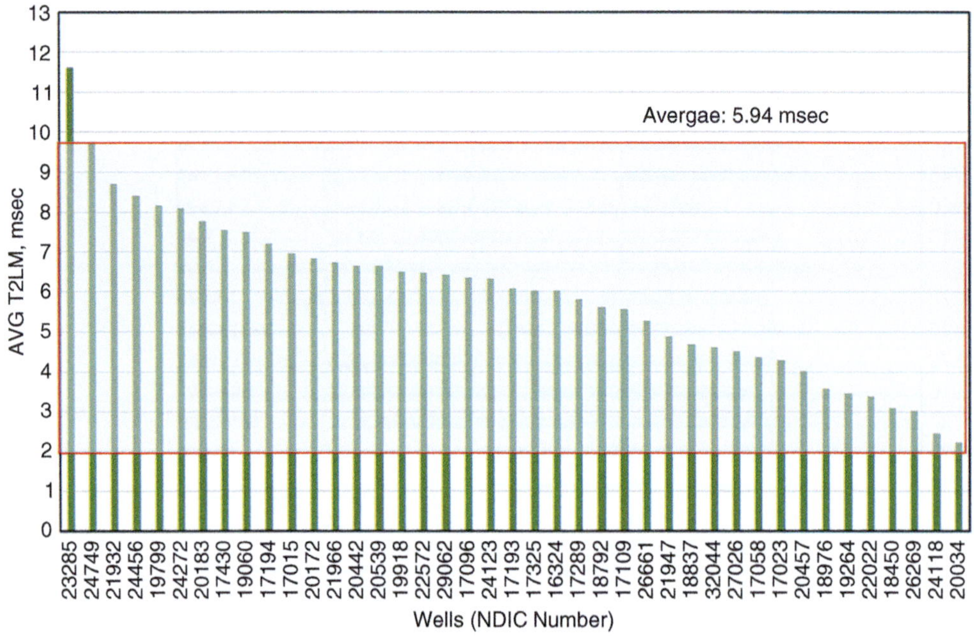

Figure 19.21 Average T_2 Log Mean (T_2LM) per well for MTF. T_2LM is less than 8 msec for most of the wells.

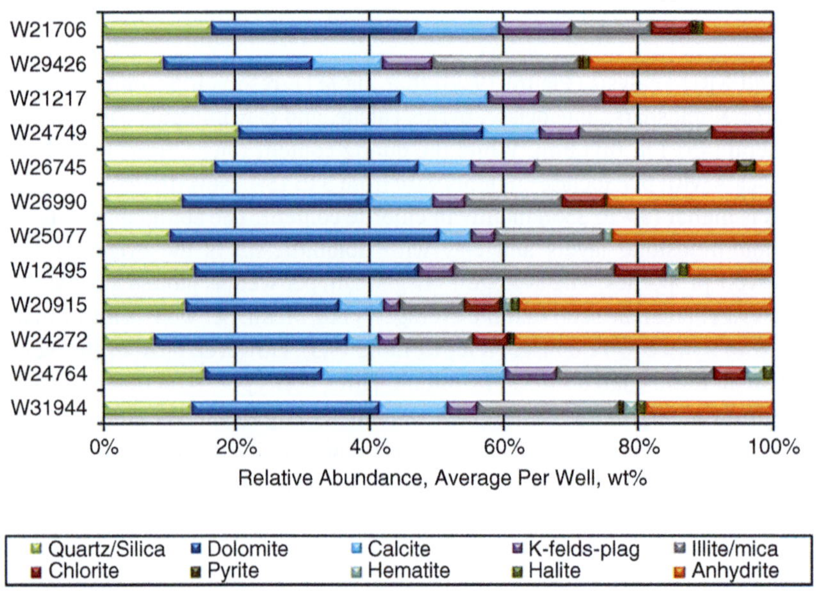

Figure 19.22 Average relative abundance of minerals per well for LTF from XRD experiments. It is composed predominantly of dolomite and anhydrite. It has the highest amount of calcite and clay minerals.

unit where the effective porosity is much lower than water-filled porosity (Figure 19.11, track 4), in addition to the high-water saturation. The water saturation computed from array dielectric measurements is compared to Dean Stark analysis and to the water saturation from the petrophysical analysis (Figure 19.11, track 6). An excellent agreement can be seen between the water saturation model from array dielectric measurements and Dean-Stark analysis mainly in the lower interval where the petrophysical analysis resulted in low values of water saturation, due to the abundance of the anhydrite. Figure 19.23 illustrates the permeability versus porosity distribution. The core porosities range from 0.5% to 11%, averaging at 7%. The permeability varies between 0.0001 and 10 mD. The higher permeabilities are usually related to the fractured samples. Furthermore, the low values of T_2 Log Mean are observed in Figure 19.24 throughout most of the interval, ranging from 2 msec to 4 msec (standard deviation 1.48). It is obvious from the well Anderson Federal 152-96-9-4-11H (Figure 19.14, track 5) that the average T_2 Log Mean decreases by 47% from the

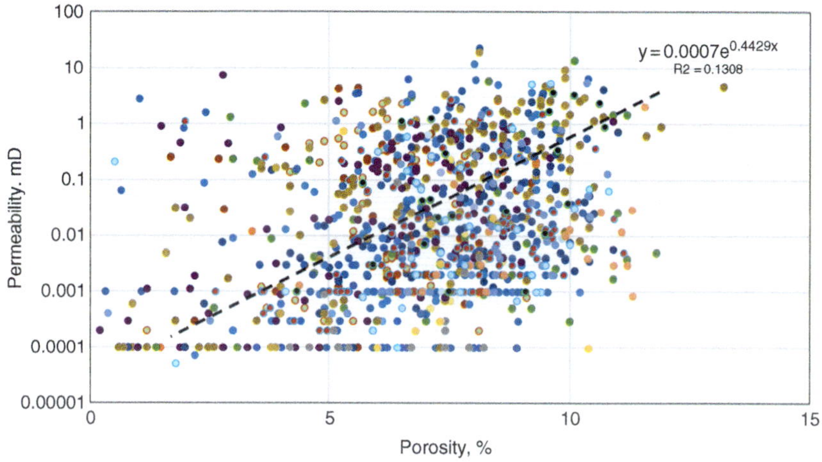

Figure 19.23 Permeability (Klinkenberg) versus porosity core cross-plot, LTF. The porosity can reach 11% and the permeability is generally less than 10 mD.

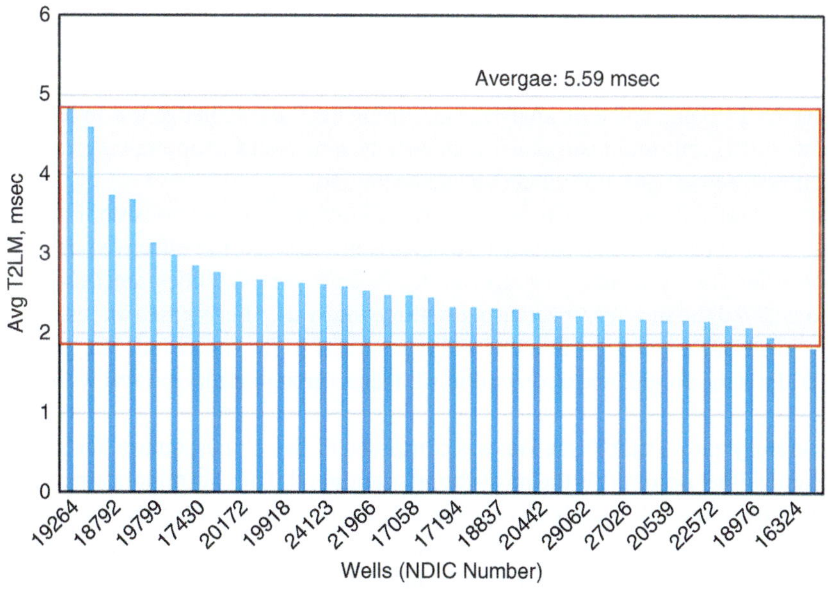

Figure 19.24 Average T_2 Log Mean (T_2LM), per well, for LTF. The average T_2LM, per well, is between 2 and 4 msec for most of the wells.

middle Three Forks, from 9.76 msec to 4.6 msec. This is confirmed by the increase in BVI and CBW, which are associated with small and micropores, respectively. On the other hand, the large pores are completely absent, resulting from the variable lithology of this unit, with the average T_2 Log Mean value of 2.59 msec. The dolo-mudstone facies characteristics combined with the abundance of anhydrite and clay contents deteriorate the reservoir quality. These characteristics make this unit nonreservoir with any petroleum interest. A consistent correlation is found between T_2 Log Mean and S_w in this unit (see Figure 19.13).

19.5 Conclusion

This chapter represents the alternative approaches to thin beds petrophysical analysis for the water saturation estimation to gain a better understanding of the reservoir characterization. Two methods were described in this chapter:

A complex petrophysical model was developed for the Three Forks reservoir by the integration of the advanced logging into the workflow including (1) ECS for the mineralogy and grain density, (2) NMR for the porosity, CBW, and free fluid, (3) DL for water saturation. A probabilistic method was used to assess the output component and fluid volumes. The conceptual lithological model was created based on multicross plots and on the mineralogy from XRD analysis. The integration of the elemental dry weights fraction with conventional logs allows more accurate mineralogical determination and calculation. The combined neutron porosity and bulk density technique was used to estimate the effective porosity. In addition, the modified Simandoux equation is used for water saturation calculation with Archie parameters m and n equal to 1.7 and 2, respectively. The formation of water resistivity was calculated using salinities and temperature profiles. The accuracy of the model was compared to the routine and special core analysis. The second approach consists of the estimation of the water saturation from array dielectric measurements independent of the resistivity. An agreement was seen in the water saturation estimation between the two methods. This agreement confirms the Archie parameters and the formation water resistivity used as an input into the modified Simandoux equation.

In the second step, the resulting petrophysical model was extrapolated to the wells that are remote from any advanced logging and core analysis measurements. The challenge was to rescale the input components to the minimum components to be solved and set the appropriate matrix parameters, uncertainties, and weight multipliers for each equation.

The petrophysical analysis and core porosity versus permeability plots show a consistent porosity across the Three Forks interval with a satisfactory agreement between the effective porosity, core, and NMR porosities. The porosities average 6% to 7% and permeabilities are less than 1 mD. The UTF is considered the most prospective of the three intervals. This is based on reservoir parameters, pore size distribution, and the amount and type of clay in the formation. The prominent feature of the UTFs is the abundance of dolomitic-silt facies throughout its interval with low clay content.

Overall, XRD mineralogy shows that Three Forks is composed of dolostones, quartz, feldspars, and clay minerals mainly illite. Dolomite is the most abundant mineral. The clay minerals increase with depth predominantly in the volume of chlorite. The calcite and hematite contents are higher in the middle and in the lower Three Forks in some wells. The pyrite content increases at the top of the UTFs. However, the anhydrite content significantly increases in abundance in lower Three Forks.

The clean dolomitic-silt facies within the reservoir are characterized by high values of T_2 Log Mean and FFI. In contrast, the dolo-mudstone facies are characterized by high water saturation. The petrophysical analysis becomes very challenging when the thinly dolo-mudstone facies are interlaminated with oil-bearing dolomitic-silt facies. That is why, a detailed multimineral analysis is one of the significant solutions in Three Forks reservoir characterization.

The average T_2 Log Mean equal to or greater than 8 msec is defined as the cutoff corresponding to the oil-bearing zone within the Three Forks interval. The two uppermost sequences (UTF and MTF) exhibit a range of pore size distribution with significantly varying T_2 Log Mean, due to the laminations. Also, the chlorite contents have an influence on T_2 Log Mean. The larger the volume of chlorite, the more complex the pore size structure and the lower the T_2 Log Mean.

Furthermore, in the UTFs, higher values of T_2 Log Mean equal to or greater than 8 msec and FFI are visible for almost all of the studied wells, suggesting the existence of large pore sizes in this unit. However, in the lower Three Forks, T_2 Log Mean remains almost constant, where the small and micropores are the most dominant pores.

There is a significant difference between the middle Three Forks and the previously described units (UTF and LTF) where a large variation occurs in capillary-bound water, indicating a high heterogeneity of the pore size distribution. Overall, through the entire Three Forks interval, the capillary-bound water, and CBW are observed to increase with depth. Relatively, a reasonable relationship can be seen between the water saturation, the free fluid, T_2 Log Mean, capillary-bound water, and CBW.

References

Adams, J.A. and Weaver, C.E. (1958). Thorium-uranium ratios as indicators of sedimentary processes: example of concepts of geochemical facies. *Bell. Am. Assoc. Petrol. Geol.* 42 (2): 387–430.

Adeniran, A., Elshafei, M., and Hamada, G. (2009). Functional network soft sensor for formation porosity and water saturation in oil wells. *2009 IEEE Instrumentation and Measurement Technology Conference*, Singapore, pp. 1138–1143. http//:doi.org/10.1109/IMTC.2009.5168625.

Aguilera, R. (1990b). A new approach for analysis of the nuclear magnetic log-resistivity log combination. *Journal of Canadian Petroleum Technology* 29 (1): 67–71.

Aguilera, R. (2002). Incorporating capillary pressure, pore aperture radii, height above free water table, and Winland r 35 values on Pickett plots. *AAPG Bulletin* 86 (4): 605–624.

Aguilera, R. (2003a). Determination of matrix flow units in naturally fractured reservoirs. *Journal of Canadian Petroleum Technology* 42 (12): 54–61.

Aguilera, R. and Aguilera, M.S. (2002). The integration of capillary pressures and Pickett plots for determination of flow units and reservoir containers. *Society of Petroleum Engineers Reservoir Evaluation and Engineering* 5 (6): 465–471.

Al Duhailan, M. (2014). Petroleum-expulsion fracturing in organic-rich shales: genesis and impact on unconventional pervasive petroleum systems. Ph.D dissertation. Colorado School of Mines, 206p.

Al-Bulushi, N., King, P.R., Blunt, M.J., and Kraaijveld, M. (2009). Development of artificial neural network models for predicting water saturation and fluid distribution. *J. Pet. Sci.Eng.* 68 (3–4): 197–208.

Alexeyev, A., Ostadhassan, M., Bubach, B. et al. (2017). Integrated reservoir characterization of the Middle Bakken in the Blue Buttes field, Williston Basin, North Dakota. *Society of Petroleum Engineers* https://doi.org/10.2118/185664-MS.

Allen, D., Crary, S., Freedman, S. et al. (1997). How to use borehole nuclear magnetic resonance. *Oilfield Review* 9 (3): 34–57.

Allen, D., Flaum, C., Ramakrishnan, T.S. et al. (2000). Trends in NMR logging. *Oilfield Review* 12 (3): 2–19.

Allen, D.C., Flaum, T.S., Ramakrishnan, J. et al. (2008). Trends in NMR logging. *Oilfield Review* 12 (3): 2–19.

Allwein, E.L., Schapire, R.E., and Singer, Y. (2000). Reducing multiclass to binary: a unifying approach for margin classifiers. *J.Mach.Learn.Res.* 1: 113–141.

Al-Wardy, W. (2002). Measurement of the poroelastic parameters of reservoir sandstones. Ph.D. thesis. Imperial College, London.

Basak, D., Pal, S., and Patranabis, D.C. (2007). Support vector regression. *Neural Information Processing – Letters and Reviews* 11 (10): 21 pp.

Bateman, R.M. (1940). Open hole log analysis and formation evaluation, 668 pp. Softcover. ISBN: 978-1-61399-156-5.

Bell, J.S. and Gough, D.I. (1979). Northeast-southwest compressive stress in Alberta- evidence from oil wells. *Earth and Planetary Science Letters* 45: 475–482.

Bernabé, Y. (1991). Pore geometry and pressure dependence of the transport properties in sandstone. *Geophysics* 56: 436–446.

Bernabe, Y., Brace, W.F., and Evans, B. (1982). Permeability, porosity and pore geometry of hot-pressed calcite. *Mech. Maters.* 1982 (1): 173–183.

Bernabé, Y., Mok, U., and Evans, B. (2003). Permeability-porosity relationships in rocks subjected to various evolution processes. *Pure Appl. Geophys.* 160: 937–960.

Berryman, J.G. (1992). Effective stress for transport properties of inhomogeneous porous rock. *J. Geophys. Res.* 97 (17): 409–424.

Berwick, B.R. (2008). Depositional environment, mineralogy, and sequence stratigraphy of the late devonian sanish member (upper three forks formation), Williston Basin, North Dakota. Master's thesis, Colorado School of Mines, Golden, CO, 262p.

Bottjer, R., Sterling, R., Grau, A., and Dear, P. (2011). Stratigraphic relationships and reservoir quality at the three forks–bakken unconformity, Williston basin, North Dakota. In: *The Bakken–Three Forks Petroleum System in the Williston Basin* (ed. J.W. Robinson, J.A. LeFever, and S.B. Gaswirth), 173–228. Rocky Mountains Associations of Geologists.

Boualam, A. (2019). Impact of stress on the characterization of the flow units in the complex three forks reservoir, Williston Basin. Theses and Dissertations. 2839. https://commons.und.edu/theses/2839.

Boualam, A., Rasouli, V., Dalkhaa, C., and Djezzar, S. (2020a). Advanced petrophysical analysis and water saturation prediction in three forks, Williston basin. *SPWLA 61st Annual Logging Symposium*. Virtual Online Webinar (June 2020). https://doi.org/10.30632/SPWLA-5104.

Boualam, A., Rasouli, V., Dalkhaa, C., and Djezzar, S. (2020b). Stress-dependent permeability and porosity in three forks carbonate reservoir, Williston Basin. *54th U.S. Rock Mechanics/Geomechanics Symposium*.

Bush, D.C. and Jenkins, R.E. (1970). Proper hydration of clays for rock property determinations. *J. Pel. Tech.* 800–804.

Chen, S., Ostroff, G., and Georgi, D.T. (1998). Improving estimation of NMR log T2 cutoff value with core NMR and capillary pressure measurements. *SCA International Symposium*, The Hague (14 September 1998), pp. 14–16

Coalson, E.B., Hartmann, D.J., and Thomas, J.B. (1985). Productive characteristics of common reservoir porosity types. *Bulletin of the South Texas Geological Society* 25 (6): 35–51.

Cristianini, C. and Shawe-Taylor, J. (2000). *An Introduction to Support Vector Machine and Other Kernel-based Learning Methods*, 1e, 189. New York, NY: Cambridge University Press.

Cui, A., Wust, R., Nassichuk, B. et al. (2013). A nearly complete characterization of permeability to hydrocarbon gas and liquid for unconventional reservoirs: a challenge to conventional thinking. *SPE Paper 168730 Presented at the SPE Unconventional Resources Technology Conference*, Denver (12–14 August 2013).

David, C. and Darot, M. (1989). Permeability and conductivity of sandstones. In: *Rock at Great Depth* (ed. V. Maury and D. Fourmaintraux), 203–209. Balkema.

Davies, J.P. and Davies, D.K. (1999). Stress-dependent permeability: Characterization and modeling. *SPE Annual Technical Conference and Exhibition*. Houston: SPE 56813.

Dennis, D. (2008). Measuring porosity and permeability from drill cuttings. *Journal of Petroleum Technology* 60 (8): https://doi.org/10.2118/0808-0051-JPT.

Djebbar, T. and Donaldson, E.C. (2012). *Petrophysics: Theory and Practice of Measuring Reservoir Rock and Fluid Transport Properties*, 3e. Elsevier.

Dong, J.J., Hsu, J.Y., Wu, W.J. et al. (2010). Stress-dependence of the permeability and porosity of sandstone and shale from TCDP Hola-A. *International Journal of Rock Mechanics and Mining* 47: 1141–1156.

Dorfman, M.H. (1984). Discussion of reservoir description using well logs. *Journal of Petroleum Technology* 36: 2196–2197.

Doveton, J.H., Guy, W., Watney, L.W. et al. (1996). Log analysis of petrofacies and flow units with microcomputer spreadsheet software: Kansas Geological Survey. University of Kansas 66047, 10p.

Droege, L.A. (2014). Sedimentology, facies architecture and diagenesis of the middle Three Forks Formation – North Dakota, U.S.A. Master's thesis. Colorado State University, Fort Collins, CO, 14p.

Drucker, H., Chris J.C., Burges, H.J.C et al. (1997). Support Vector Regression Machines. ARPA contract number N00014-94-C-1 086.

Dumonceaux, G.M. (1984). Stratigraphy and depositional environments of the Three Forks formation (upper Devonian), Williston basin, North Dakota. Master thesis.

Dunham, R.J. (1962). Classification of carbonate rocks according to depositional texture. In: *Classification of Carbonate Rocks* (ed. W.E. Ham), 108–121. AAPG Memoir.

Dunn, K.J., Bergman, D.J., and Lattoraca, G.A. (2002). *Nuclear Magnetic Resonance Petrophysical and Logging Applications (Handbook of Geophysical Exploration, Seismic Exploration)*, vol. 32. Amsterdam: Pergamon an Imprint of Elsevier Science. 293p.

Ebanks, W.J. Jr. (1987). Flow unit concept – integrated approach to reservoir description for engineering projects. *AAPG Bulletin* 71 (5): 551–552.

Efron, B. and Tibshirani, R. (1994). *An Introduction to the Bootstrap*. New York: Chapman & Hall.

Egbe, S.J., Omole, O., Diedjomahor, J., and Crowe, J. (2007). Calibration of the elemental capture spectroscopy tool using the Niger Delta formation. *31st Nigeria Annual International Conference and Exhibition Held in Abuja*. SPE 111910, Nigeria (6–8 August 2007).

El-sebakhy, E.A., Asparouhov, O., Abdulraheem, A. et al. (2010). Data mining in identifying carbonate litho-facies from well logs based from extreme learning and support vector machines. *AAPG GEO, Middle East Geoscience Conference & Exhibition*, Manama, Bahrein (7–10 March 2010).

Farber, D.L., Bonner, B.P., Balooch, M. et al. (2001). Observations of water induced transition from brittle to viscoelastic behavior in nano-crystalline swelling clay. *EOS* 82: 1189.

Focke, J.W. and Munn, D. (1987). Cementation exponents in middle eastern carbonate reservoirs. *SPE Formation Evaluation*. SPE, Qatar General Petroleum Corp *2* (02): 155–167.

Franklin, A. (2017). Deposition, stratigraphy, provenance, and reservoir characterization of carbonate mudstone: The Three Forks Formation, Williston Basin. Ph.D thesis. Colorado School of Mines.

Galford, J., Quirein, J., Shannon, S. et al. (2009). Field test results of a new neutron induced gamma-ray spectroscopy geochemical logging tool. *SPE Annual Techinical and Exibition Held in New Orleans*. SPE 123992.

Ganpule, S.V., Srinivasan, K., Izykowski, T. et al. (2015). Impact of geomechanics on well completion and asset development in the Bakken formation. *Society of Petroleum Engineers* https://doi.org/10.2118/173329-MS.

Gaswirth, S.B. and Marra, K.R. (2015). U.S. Geological Survey 2013 assessment of undiscovered resources in the Bakken and Three Forks Formations of the U.S. Williston Basin Province.

Gerhard, L.C., Anderson, S.B., and Fischer, D.W. (1990). Petroleum geology of the Williston Basin. *American Association of Petroleum Geologists Memoir* 51: 507–559.

Ghanizadeh, A., Amann-Hildenbrand, A., and Gasparik, M. (2014). Experimental study of fluid transport processes in the matrix system of the European organic rich shales: II Posidonia Shale (Lower Toarcian, Noerthern Germany). *International Journal of Coal Geology* 123: 20–33. https://doi.org/10.1016/j.coal.2013.06.009.

Gottlib Zeh, S., Briqueu, L., and Veillerette, A. (1999). Indexed Self-Organizing Map: a new calibration system for a geological interpretation of logs. *The Fifth Annual Conference of the International Association for Mathematical Geology, Proc IAMG*, Trondheim, Norway (6–11 August 1999), pp. 183–188.

Gunn, S.R. (1998). Support Vector Machine for Classification and Regression. Tech. Rep., Univ. Southampton, Southampton, U.K.

Gunter, G.W., Finneran, J.M., Hartmann, D.J., and Miller, J.D. (1997). Early determination of reservoir flow units using an integrated petrophysical method: Society of Petroleum Engineers. *SPE-38679-MS Presented at the SPE Annual Technical Conference and Exhibition*, San Antonio, TX (5–8 October 1997), pp. 373–380. http://doi.org/10.2118/38679-MS.

Hashem, S. (1997). Optimal linear combinations of neural networks. *Neural Net-works* 10 (4): 599–614.

Hassan, M., Selo, M., and Combaz, A. (1975). Uranium distribution and geochemistry as criteria of diagenesis in carbonate rocks. *9th International Sedimentological Congress,* Nice, France, 7pp.

Hassan, M., Hossin, A., and Combaz, A. (1976). Fundamentals of the differential gamma-ray log. Interpretation technique. SPWLA. *17th Ann. Log. Sym. Trans., Paper H,* Denver, CO (June 1976).

Havens, J.B. and Batzle, M.L. (2011). Minimum horizontal stress in the bakken formation. *45th U.S. Rock Mechanics/Geomechanics Symposium*, San Francisco, CA (June 2011). Paper Number: ARMA-11-322.

He, J., Ling, K., Wu, X. et al. (2019). Static and dynamic elastic moduli of bakken formation. *International Petroleum Technology Conference.* https://doi.org/10.2523/IPTC-19416-MS.

Heck, T.J., LeFever, R.D., Fischer, D.W., and LeFever, J.A. (2002). Overview of the petroleum geology of the North Dakota Williston Basin: North Dakota Geological Survey Report, Bismarck, ND, USA.

Herron, S., Herron, M., Pirie, I. et al. (2014). Application and quality control of core data for the development and validation of elemental spectroscopy log interpretation. *Petrophysics* 55 (5): 392–414.

Hizem, M., Budan, H., Devillé, B. et al. (2008). Dielectric dispersion: A new wireline petrophysical measurement. *Paper SPE-116130 Presented at the SPE Annual Technical Conference and Exhibition*, Denver, CO, USA (21–24 September 2008).

Holubnyak, Y.I., Bremer, J.M., Mibeck, B.A.F. et al. (2011). *Understanding the sourcing at Bakken oil reservoirs.* Woodlands, TX, USA: Society of Petroleum Engineers. http://doi.org/10.2118/141434-MS.

Isleyen, E., Demirkan, D. C., Duzgun, H.S., and Rostami, J. (2019). Lithological classification of limestone with self-organizing maps. *53rd U.S. Rock Mechanics/Geomechanics Symposium*, New York, NY, USA (23–26 June 2019). Paper Number: ARMA-2019-1791.

Jin, H., Sonnenberg, S.A., and Sarg, J.F. (2015). Source rock potential and sequence stratigraphy of bakken shales in the williston basin. *Unconventional Resources Technology Conference*. http://doi.org/10.15530/URTEC-2015-2169797.

Johnson, R., Longman, M., and Ruskin, B. (2017). Petrographic and petrophysical characteristics of the upper Devonian three forks formation, southern Nesson anticline, North Dakota. *The Mountain Geologist* 54 (3): 181–201.

Jones, S.C. (1988). Two-point determinations of permeability and PV vs net confining stress. *SPE Formation Evaluation* (March 1988), pp. 235–241. ISSN: 0885-923X. EISSN: 2469-8512.

Jones, F.O. and Owens, W.W. (1980). A laboratory study of low-permeability gas sand. *Journal of Petroleum Technology* 32: 1631–1640.

Kamalyar, K., Sheikhi, Y., and Jamialahmadi, M. (2012). Using an artificial neural network for predicting water saturation in an Iranian oil reservoir. *Petroleum Science and Technology* 30 (1): 35–45.

Kausik, R., Fellah, K., Feng, L. et al. (2016). High and low field NMR relaxometry and diffusometry of the Bakken petroleum system. *SPWLA 57th Annual Logging Symposium*. Society of Petrophysicists and Well-Log Analysts, Reykjavik, Iceland (25–29 June 2016), 7p.

Kenyon, W., Day, P., Straley, C., and Willemsen, J. (1988). A three-part study of NMR longitudinal relaxation properties of water-saturated sandstones. *SPE Formation Evaluation* 3 (3): 622–636.

Kerans, C., Lucia, F.J., and Senger, R.K. (1994). Integrated characterization of carbonate ramp reservoirs using Permian San Andres Formation outcrop analogs. *AAPG Bulletin* 78: 181–216.

Kingma, D.P. and Lei Ba, J. (2015). *Adam: A Method for Stochastic Optimization*. ICLR.

Klenner, R., Braunberger, J., Sorensen, J. et al. (2014). A formation evaluation of the middle Bakken member using multimineral petrophysical analysis approach. *Unconventional Resources Technology Conference. SPE/AAPG/SEG Unconventional Resources Technology Conference*. URTeC:1922735, Denver, CO, USA (25–27 August 2014).

Kwon, O., Kronenberg, A.K., Gangi, A.F., and Johnson, B. (2001). Permeability of Wilcox shale and its effective pressure law. *Journal of Geophysics Research* 106 (19): 339–353.

LeFever, J.A. and Nordeng, S.H. (2009). The three forks formation–North Dakota to sinclair field. Manitoba. *North Dakota Geological Survey Geological Investigations* 76: 1.

LeFever, J.A. (2008). *Isopach of the Three Forks Formation*. North Dakota Geological Survey Geological Investigations. No. 64, 1 plate.

LeFever, J.A., Le Fever, R.D., and Nordeng, S.H. (2011). Revised nomenclature for the Bakken Formation (Mississippian-Devonian), North Dakota. The Bakken-Three Forks petroleum system in the Williston basin. Chapter 1, pp. 11–26.

LeFever, J.A., LeFever, R.D., and Nordeng, S.H. (2014). *Reservoirs of the Bakken Petroleum System: A Core-based Perspective*. North Dakota Geological Survey Geologic Investigations. GI. 171, 3 Posters.

Leon, M., Lafournere, A.N., Bourge, J.P. et al. (2015). Rock typing mapping methodology based on indexed and probabilistic self-organized map in shushufindi field. *Society of Petroleum Engineers* https://doi.org/10.2118/177086-MS.

Loucks, R.G., Reed, R.M., Ruppel, S.M., and Jarvie, D.M. (2009). Morphology, genesis, and distribution of nanometer-scale pores in siliceous mudstones of the Mississippian Barnett Shale. *Journal of Sedimentary Research* 79: 848–861.

Loucks, R.G., Reed, R.M., Ruppel, S.C., and Hammes, U. (2012). Spectrum of pore types and networks in mud-rocks and a descriptive classification for matrix-related mud-rock pores. *AAPG Bulletin* 96: 1071–1098.

Loucks, R.G., Ruppel, S.C., Zhang, T., and Peng, S. (2017). Origin and characterization of eagle ford pore networks in the south Texas upper cretaceous shelf. *AAPG Bulletin* 101 (3): 387–418.

Lucia, F.J. (1983). Petrophysical parameters estimated from visual description of carbonate rocks: a field classification of carbonate pore space. *Journal of Petroleum Technology* 35 (3): 629–637.

Lucia, F.J. (1995). Rock fabric and petrophysical classification of carbonate pore space for reservoir characterization. *AAPG Bulletin* 79 (9): 1275–1300.

Lucia, F.J., Kerans, C., and Senger, R.K. (1992). Defining flow units in dolomitized carbonate-ramp reservoirs: proceedings. *Society of Petroleum Engineers*, paper SPE 24702: 399–406.

Mardi, M., Nurozi, H., and Edalatkhah, S. (2012). A water saturation prediction using artificial neural networks and an investigation on cementation factors and saturation exponent variations in an Iranian oil well. *Petroleum Science and Technology* 30 (4): 425–434.

Martin, A.J., Solomon, S.T., and Hartmann, D.T. (1997). Characterization of petrophysical flow units in carbonate reservoirs. *AAPG Bulletin* 81: 734–759.

Mayergoyz, I.D. (1986). Mathematical models of hysteresis and their application. *IEEE Transactions on Magnetics* 22: https://doi.org/10.1109/TMAG.1986.1064347.

McKenon, D., Cao Minh, C., Freedman, R. et al. (1999). An improved NMR tool design for faster logging, *paper CC, transactions. SPWLA 40th Annual Logging Symposium*, Oslo, Norway (30 May–3 June).

Merkel, R., Machesney, J., and Tompkins, K. (2018). Calculated determination of variable wettability in the middle Bakken and Three Forks, Williston basin, USA. *SPWLA 59th Annual Logging Symposium*, London, UK (June 2018).

Milijkovic, D. (2017). Brief review of self-organizing maps. *40th International Convention on Information and Communication Technology, Electronics and Microelectronics (MIPRO)*, Opatija, Croatia (22–26 May 2017), 6pp.

Millard, M. and Brinkerhoff, R. (2016). The integration of geochemical, stratigraphic, and production data to improve geological models in the bakken-three forks petroleum system, williston basin, North Dakota. In: *Hydrocarbon Source Rocks in Unconventional Plays* (ed. M.P. Dolan, D.K. Higley, and P.G. Lillis), 190–211. Denver: Rocky Mountain Region: RMAG.

Mitchell, R. (2013). Sedimentology and reservoir properties of the Three Forks dolomite, Bakken Petroleum System, Williston Basin, U.S.A: AAPG. Search and Discovery Articles, no. 120079.

Mitra, P.P., Sen, P.N., and Schwartz, L.M. (1993). Short-time behavior of the diffusion coefficient as a geometrical probe of porous media. *Physical Review B* 47: 8565–8574.

Mohaghegh, S., Arefi, R., Ameri, S. et al. (1996). Petroleum reservoir characterization with the aid of artificial neural networks. *Journal of petroleum Science and Enigineering* 16 (1996): 263–274.

Mollajan, A., Memarian, H., and Jalali, M.R. (2013). Prediction of reservoir water saturation using support vector regression in an Iranian carbonate reservoir. *ARMA* 13–311.

Moss, A.K and Jing, X.D. (2001). An investigation into the effect of clay type, volume and distribution on NMR measurements in sandstones. *Paper SCA2001-29 Presented at the SCA International Symposium*, Edinburgh, Scotland, UK (17–19 September 2001).

Newman, J., Edman, J., Howe, J., and LeFever, J. (2013). The bakken at parshall field: inferences from new data regarding hydrocarbon generation and migration. *Unconventional Resources Technology Conference* https://doi.org/10.15530/URTEC-1578764-MS.

Nordeng, S. H. and LeFever, J.A., 2009. Three Forks Formation Log to Core Correlation. North Dakota Geological Survey Geologic Investigations No. 75,. 1p.

Passey, Q.R., Dahlberg, K.E., Sullivan, K.B. et al. (2004). A systematic approach to evaluate hydrocarbons in thinly bedded reservoirs.

Peters, K.E., Walters, C., and Moldowan, J.M. (2005). *The Biomarker Guide*, vol. 2. Cambridge, UK: Cambridge University Press, 490 p.

Peterson, K.J. (2017). Pore-size distributions from nuclear magnetic resonance and corresponding hydrocarbon saturations in the Devonian Three Forks Formation, Williston Basin, North Dakota (abstract). *Rocky Mountain Section Annual Meeting, Billings*, Montana (25–28 June 2017).

Petty, D. (2014). Mineralogy and petrology controls on hydrocarbon saturation in the Three Forks reservoir, North Dakota. AAPG Search and Discovery Article-10623.

Pittman, E.D. (1992). Relationship of porosity and permeability to various parameters derived from mercury injection-capillary pressure curves for sandstone. *AAPG Bulletin* 76: 191–198.

Prammer, M.G., Drack, E.D., Bouton, J.C. et al. (1996). Measurements of clay-bound water and total porosity by magnetic resonance logging: SPE- 36522-MS, *Presented at the SPE Annual Technical Conference and Exhibition*, Denver, pp. 311–320. http://dx.doi.org/10.2118/36522-MS (accessed 14 April 2017).

Price, L.C. (2000). Origins and characteristics of the basin-centered continuous reservoir unconventional oil-resource base of the Bakken source system, Williston Basin, paper. Available at: http://www.unddeerc.org/Price/.

Qi, L. and Carr, T.R. (2006). Neural network prediction of carbonate lithofacies from well logs, Big Bow and Sand Arroyo Creek fields, South west Kansas. *Computures and Geosciences* 32 (7): 947–964.

Ramakrishna, S., Balliet, B., Miller, D., and Sarvotham, S. (2010). Formation evaluation in the Bakken Complex using laboratory core data and advanced logging technologies. *SPWLA 51st Annual Logging Symposium,* Perth, Australia (19–23 June 2010).

Rezaee, M., Ilkhchi, A., and Alizadeh, P. (2008). Intelligent approaches for the synthesis of petrophysical logs. *Journal of Geophysics and Engineering* 5: 12–26.

Rider, M. and Kennedy, M. (2011). *The Geological Interpretation of Well Logs*. Rider-French Consulting Limited. 3rd Revised edition.

Roland, L. and Olivier, F. (2007). Advances in measuring porosity and permeability from drill cuttings. *SPE/EAGE Reservoir Characterization and Simulation Conference* (28–3 October), Abu Dhabi, UAE. https://doi.org/https://doi.org/10.2118/111286-MS.

Rosepiler, M.J. (1981). Calculation and significance of water saturation in low porosity shaly gas sands. *SPWLA 22nd Annual Symposium*, Mexico City, Mexico (June 1981).

Saffarrzadeh, S. and Shadizadeh, S.R. (2012). Reservoir rock permeability prediction using support vector regression in an Iranian oil field. *Journal of Geophysics and Engineering* 9 (3): 336–344.

Sandberg, C.A. and Hammond, C.R. (1958). Devonian system in Williston basin and central Montana. *AAPG Bulletin* 42: 2293–2334.

Saneifar, M., Skalinski, M., Theologou, P. et al. (2015). Integrated petrophysical rock classification in the McElroy Field, West Texas, USA. *Petrophysics* 56: 493–510.

Schmoker, J.W. and Halley, R.B. (1982). Carbonate porosity versus depth: a predictable relation for south Florida. *AAPG Bull* 1982 (66): 2561–2570.

Serra, O. (1984). *Fundamentals of Well-Log Interpretation. Developments in Petroleum Science 15A*. Elsevier.

Simpson, G., Hohman, J., Pirie I., and Horkowitz, J. (2015). Using advanced logging measurements to develop a robust petrophysical model for the Bakken petroleum system. *SPWLA 56th Annual Logging Symposium,* Long Beach, CA, USA (18–22 July 2015).

Sloss, L.L. (1984). Comparative anatomy of cratonic unconformities. In: *Interregional Unconformities and Hydrocarbon Accumulation* (ed. J.S. Schlee), 7–36. American Association of Petroleum Geologist Memoir 36.

Smola, A.J. and Schölkopf, B. (1998). A Tutorial on Support Vector Regression, NeuroCOLT. Technical Report NC-TR-98-030, Royal Holloway College, University of London, UK.

Soeder, D.L. (1986). Laboratory Drying Procedures and the Permeability of Tight Sandstone Core. SPE, Inst. of Gas Technology.

Soeder, D.L. and Doherty, M.G. (1983). The effects of laboratory drying techniques on the permeability of tight sandstone core. *SPE 11622 Presented at the SPEIDOE*, Denver (13–16 March 1983).

Sonnenberg, S.A. (2015). Keys to production, Three Forks, Williston Basin. SPE-178510-MS/URTeC:2148989.

Sonnenberg, S.A. (2017). Sequence stratigraphy of the Bakken and Three Forks Formations, Williston basin, USA. *AAPG Rocky Mountain Section Annual Meeting, Billings*, Montana (25–28 June 2017).

Sonnenberg, S.A., Gantyno A., and Sarg R. (2011). Petroleum potential of the upper Three Forks formation, Williston basin, USA. *AAPG Annual Convention and Exhibition*, Houston, TX (10–13 April 2011).

Sørland, G.H., Djurhuus, K., Widerøe, H.C. et al. (2007). Absolute pore size distributions from NMR. *Diffusion Fundamentals* 5: 4.1–4.15.

Swanson, V.E. (1960). Oil yield and Uranium Content of black shales. *Geol. Survey* Prof. Paper 356-A.

Teklu, T., Zhou, Z., Li, X., and Abass, H. (2016). *Cyclic Permeability and Porosity Hysteresis in Mudrocks-Experimental Study. 50th US Rock Mechanics/Geomechanics Symposium*, vol. 1, pp. 1–12.

Teklu, T., Zhou, Z., Li, X., and Abass, H. (2016b). Experimental investigation on permeability and porosity hysteresis in low-permeability formations. *The SPE Low Perm Symposium Held in Denver, Colorado, USA*, https://doi.org/10.2118/180226-MS.

Terzaghi, K., 1943. *Theoretical Soil Mechanics.*

Thrasher, L. (1987). Macrofossils and stratigraphic subdivisions of the Bakken Formation (Devonian-Mississippian), Williston Basin, North Dakota. In: *Fifth International Williston Basin Symposium* (ed. D.W. Fischer), 53–67.

Vapnik, V.N. (1995). *The Nature of Statistical Learning Theory*. Springer-Verlag New York, Inc. 99-39803.

Vapnik, V.N. and Chervonenkis, A. (1964). A note on one class of perceptron. *Automation and Remote Control* 25.

Vapnik, V.N. and Lerner, A. (1963). Pattern recognition using generalized portrait method. *Automation and Remote Control* 24: 774–780.

Walls, J. and Nur, A. (1979). Pore pressure and confining pressure dependence of permeability in sandstone. *7th Form. Eval. Symp. Can. Well Logging Soc.*, Calgary.

Walstrom, J.E., Mueller, T.D., and McFarlane, R.C. (1967). Evaluating uncertainty in engineering calculations. *Journal of Petroleum Technology* 1595–1599.

Wang, W., Ren, X., Zhang, Y., and Li, M. (2018). Deep learning based lithology classification using dual-frequency Pol-SAR data. *Applied Sciences* 8: 1513.

Wang, X., Yang, S., Zhao, Y., and Wang, Y. (2018b). Lithology identification using an optimized KNN clustering method based on entropy-weighed cosine distance in Mesozoic strata of gaoqing field, jiyang depression. *Journal of Petroleum Science and Engineering*. 166: 157–174.

White, F.M. (1974). *Viscous Fluid Flow*. New York: McGraw-Hill.

Xu, J. and Sonnenberg, S.A. (2017). An SEM study of porosity in the organic-rich lower Bakken lower Member and Pronghorn member, Bakken Formation, Williston Basin. *Unconventional Resources Technology Conference*, Austin, TX, USA. http://doi.org/10.15530/URTEC-2017-2697215.

Xu, C. and Torres-Verdin, C. (2013). Quantifying fluid distribution and phase connectivity with a simple 3D cubic pore network model constrained by NMR and MICP data. *Computers and Geosciences* 61: 94–103.

Yu, L., Porwal, A., Holden, E., and Dentith, M.C. (2012). Towards automatic lithological classification from remote sensing data using support vector machines. *Computers & Geosciences* 45: 229–239.

Zoback, M.D. (1975). High pressure deformation and fluid flow in sandstone, granite, and granular materials. Ph.D. thesis. Stanford Univ., Stanford, Calif.

Zoback, M.D. and Byerlee, J.D. (1975). Permeability and effective stress. *American Association of Petroleum Geologists Bulletin* 154–158.

Zoback, M.D. and Byerlee, J.D. (1976). Effect of high-pressure deformation on permeability of Ottawa sand. *American Association of Petroleum Geologists Bulletin* 1976 (60): 153–142.

20

Water Saturation Prediction Using Machine Learning and Deep Learning. Application to Three Forks Formation in Williston Basin, North Dakota, USA

Aldjia Boualam and Sofiane Djezzar

University of North Dakota, Grand Forks, ND, USA

KEYPOINTS

- This chapter proposes three machine learning algorithms for water saturation prediction in thin beds reservoir using conventional logs.
- The pre-processing of the dataset is performed using PCA to identify the patterns and redundancy in the dataset.
- The hydrocarbon column has a strong relationship with the output target Sw, leading to an increase in accuracy by 17%.
- Support vector regression (SVR) with RBF kernel function is used to model water saturation, yielding a target output prediction with a correlation factor of 0.78.
- The results show a small difference in performance between SVR, backpropagation neural network, and random forest regression algorithms.
- The complexity of the Three Forks Formation with a complex pore size distribution is captured, and the study concludes that increasing the input variables would improve the prediction accuracy of the models on water saturation prediction in thin beds analysis.

20.1 Introduction

The complexity associated with the water saturation (S_w) estimation in Three Forks Formation, as explained in the previous chapter, is due to the existence of the thin beds where the oil-bearing dolomitic-silt facies are interbedded with dolo-mudstone and shaly facies with high water saturation. When the thickness of the layers is less than the resolution of the conventional tools, it can result in large anisotropy and erroneously higher water saturation estimation (Passey et al. 2006). The practical impact is an underestimation of the net thickness with a higher potential of hydrocarbon (HC) that is amenable to hydraulic fracturing in this type of unconventional reservoir.

The thin bed formation evaluation has been studied by many authors by using advanced tools and techniques. Several models are widely used to estimate the S_W such as Modified Simandoux, Dual Water, Waxman-Smits, in addition to the Thomas-Steiber method developed for thin sand-shale beds analysis. However, the incoherent response of conventional logs in thinly carbonate

Unconventional Hydrocarbon Resources: Prediction and Modeling Using Artificial Intelligence Approaches, First Edition. Edited by Sid-Ali Ouadfeul.

layers makes the estimation of S_w ineffective and may lead to erroneous interpretation due to the wide range of responses for each layer. A key challenge is to accurately predict the water saturation from conventional well logs in a heterogeneous carbonate reservoir. Thus, a powerful approach is needed such as the application of machine learning and deep learning algorithms, which can be generalized across the entire extent of the Three Forks in Williston Basin at a lower cost compared to the advanced logging and coring. Since the discovery of the Three Forks reservoir, there have been more than three thousand wells drilled and completed with conventional logs. Due to the complexity of the reservoir, full cores were acquired across the Three Forks Formation and advanced logging was recorded in some wells.

Several intelligent techniques have successfully been applied for predicting S_W and other petrophysical properties such as porosity and permeability in conventional and carbonate reservoirs (Karimian et al. 2013; Kenari and Mashohor 2013; Mardi et al. 2012; Kamlyar et al. 2012; Adeniran et al. 2009; Bulushi et al. 2009; Rezaee et al. 2008; Helle and Bhatt 2002; Yan 2002; Hachem 1997; Mohaghegh et al. 1996). However, there are yet challenges to accurately predicting S_W in thin beds carbonate reservoirs.

In this chapter, we examine the capability of support vector regression (SVR) combined with the grid searching algorithm and kernel functions to predict S_w in Three Forks Formation. We first analyzed and generated the input variables and the output target. Secondly, the model was built and trained based on the search for the best combination of several parameters (gamma, regularization parameter, epsilon, and kernel functions). The SVR model constructed is then tested on the test data set to compare the accuracy of the model. Additionally, the performance of SVR was compared to that of backpropagation artificial neural network and random forest (RF) regression models based on correlation coefficient, root mean square error, and mean absolute error indexes (Table 20.1). Python 3.6 with different libraries such as Scikit-learn, TensorFlow, NumPy, Matplotlib, and Pandas were used to build the models. Also, the curve reconstruction using K.Mod will be discussed in this chapter. The workflow used to predict water saturation is illustrated in Figure 20.1.

Table 20.1 Loss formulas for accuracy assessment.

Root Mean Square Error (RMSE)	$\sqrt{\frac{1}{n}\sum_{i=1}^{n}\left(y_i - y_{\text{ip}}\right)^2}$
Mean Absolute Error (MAE)	$\frac{1}{n} * \max\left\|y_i - y_{\text{ip}}\right\|$
Correlation Coefficient (R^2)	$100 * \left[\frac{\sum_{i=1}^{n}\left(y_{\text{ip}} - \bar{y}_{\text{ip}}\right)\left(y_i - \bar{y}_i\right)}{\sqrt{\sum_{i=1}^{n}\left(y_{\text{ip}} - \bar{y}_{\text{ip}}\right)^2 \sum_{i=1}^{n}\left(y_i - \bar{y}_i\right)^2}}\right]^2$

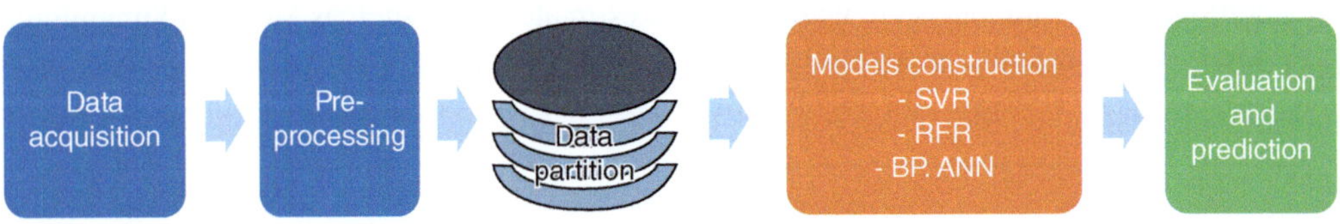

Figure 20.1 Workflow for water saturation prediction.

20.2 Experimental Procedure and Methodology

20.2.1 Support Vector Machine Concepts

The support vector machine (SVM) has been developed by Vapnik in 1995. It is a supervised machine learning algorithm, which was developed first to solve classification problems (SVC), but it was extended to regression problems by the introduction of an alternative loss function, Ɛ-insensitive, by Vapnik et al. (1997). SVR tries to minimize the generalization error bound, which is represented by a combination of the training error and a regularization term, instead of minimizing the observed training error (Basak et al. 2007). The SVR algorithm was developed from a linear regression function in a high dimensional feature space where the input variables are mapped via a nonlinear function (Basak et al. 2007). For a given training dataset represented by $\{(\vec{x}_1, y_1)\ldots\ldots(\vec{x}_n, y_n)\}$, where $\vec{x}_i$ is an input vector and y_i is a corresponding output scalar (i = 1, 2, 3. . .*n*), the general linear SVR function can be expressed as:

$$f(\vec{x}_i) = \vec{\omega}\cdot\vec{x}_i + b \qquad \vec{\omega}\in X,\ b\in R \tag{20.1}$$

where $\vec{\omega}$ is a weight vector, b is a scalar called bias value. The objective of SVR is to find a function $f(\vec{x})$ that can predict the output target y_i whose deviation is less than a tolerance value ε and flat as possible (Smola and Schölkopf 2004). In order to set a reasonable tolerance value ε, the ε-insensitivity loss function (Vapnik 1995) is defined as follows:

$$L_\varepsilon(y) = \begin{cases} 0, & \text{for } |f(\vec{x}) - y| < \varepsilon \\ |f(\vec{x}) - y| - \varepsilon, & \text{otherwise} \end{cases} \tag{20.2}$$

According to Drucker et al. (1997), this equation defines an ε tube and if the predicted value is outside the tube, the loss is the magnitude of the difference between the predicted value and the radius ε of the tube, whereas if the predicted value is within the tube, then the loss is zero (Figure 20.2). The ε-insensitivity loss function can stabilize the estimation and control the accuracy into a reasonable range. A small ε represents a lower tolerance for inaccurate results, thus the model is trying to fit every extreme value and catch all the trends inside the dataset. A larger ε represents a higher tolerance of the variance, the model will prioritize simplicity, soft margin of SVR will become larger. To ensure the flatness of the function f(x), Cortes and Vapnik (1995) introduced the convex optimization formula with an ε-insensitivity loss function to minimize as follows:

$$\varphi(\vec{\omega},\varepsilon) = \frac{1}{2}\|\vec{\omega}\|^2 + C\sum_{i=1}^{N}(\xi_i^+ + \xi_i^-)$$
$$\text{Subject to}\begin{cases} y_i - (\vec{\omega}\cdot\varphi(\vec{x}_i) + b) \le \varepsilon + \xi_i^+ \\ -y_i + (\omega\cdot\varphi(x_i) + b) \ll \varepsilon + \xi_i^-, i = 1,2,\cdots,n \\ \xi_i^+, \xi_i^- \ge 0 \end{cases} \tag{20.3}$$

where *C* is a positive regularization parameter determining the tradeoff between minimizing the flatness of the function *f*(*x*) and the amount up to which deviations larger than ε are tolerated (Smola 2004). In other words, the C parameter allows the algorithm to decide how much it wants to penalize incorrect data points (Drucker et al. 1997). If C is large, more emphasis is placed on the error. Otherwise, more emphasis is placed on the norm of weight (Drucker et al. 1997).

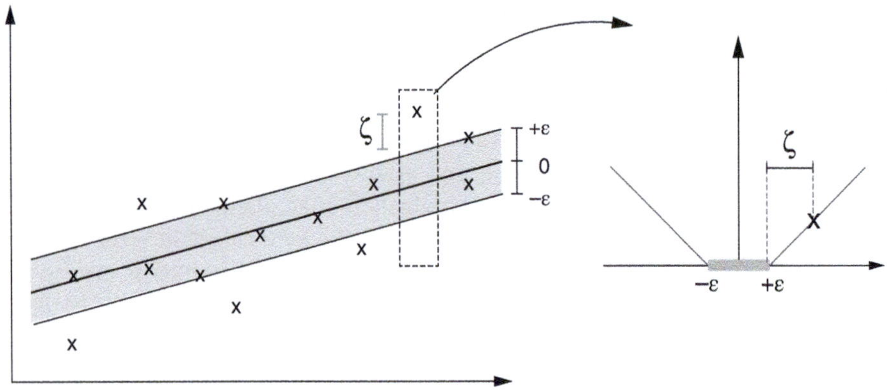

Figure 20.2 The soft margin loss setting for a linear SVR. *Source:* Schölkopf and Smola (2002).

The variables ξ_i^+, ξ_i^- are slack variables representing upper and lower constraints on the output of the system (Figure 20.2). They penalize training error by the loss function for the chosen error tolerance ε (Misra et al. 2009).

There are four types of loss functions commonly used in the SVR. Quadratic loss function corresponds to the conventional least-squares error criterion, Laplace loss function, Huber loss function, and Vapnik proposed the function called ε-insensitive loss function as an approximation to Huber's loss function.

The input space in nonlinear regression is mapped into a high dimensional space (feature space) using a kernel function. According to Basak et al. (2007), a linear algorithm in feature space is equivalent to a nonlinear algorithm in the input space. The nonlinear SVR solution, using an ε-insensitive loss function, is given by:

$$\max_{\alpha,\alpha'} W(\alpha,\alpha') = \max_{\alpha,\alpha'} \sum_{i=1}^{n} \alpha'_i(y_i - \varepsilon) - \alpha_i(y_i + \varepsilon) - \frac{1}{2}\sum_{i=1}^{n}\sum_{j=1}^{n}(\alpha'_i - \alpha_i)(\alpha'_j - \alpha_j)K(\vec{x}_i, \vec{x}_j)$$
$$\text{Subject to}\begin{cases} 0 \le \alpha'_i, \alpha_i \le C,\ i = 1,2,\ldots,n \\ \sum_{i=1}^{n}(\alpha_i - \alpha'_i) = 0 \end{cases} \tag{20.4}$$

$$f(\vec{x}) = \vec{\omega}\cdot\vec{x} + b$$
$$\text{in which}\begin{cases} \vec{\omega}\cdot\vec{x} = \sum_{i=1}^{n}(\alpha_i - \alpha'_i)\cdot K(\vec{x}_i, \vec{x}_j) \\ b = -\frac{1}{2}\sum_{i=1}^{n}(\alpha_i - \alpha'_i)\left(K(\vec{x}_i, \vec{x}_r) + K(\vec{x}_i, \vec{x}_s)\right) \end{cases} \tag{20.5}$$

where $\vec{x}_r$, $\vec{x}_s$ are any support vectors satisfying $\alpha_r, \alpha_s > 0,\ y_r = -1,\ y_s = 1$

By Lagrangian formulation, solving Equation (20.4) with constraints, the Lagrange multipliers α_i and α_i' can be determined and the regression function is obtained as follows:

$$f(x, \alpha_i, \alpha'_i) = \sum_{i=1}^{n}(\alpha_i - \alpha'_i)\cdot K(x, x_i) \tag{20.6}$$

where $K(x, x_i)$ is the kernel function. In the feature space, the dot product φ(xi).φ(xj) is introduced instead of the kernel function (Smola 2004; Gunn 1998). The linear kernel function expressed in Equation (20.7) is the simplest function:

$$K(\vec{x}, \vec{x}_i) = \vec{x}^T \cdot \vec{x}_i + C \tag{20.7}$$

where $\vec{x}^T \cdot \vec{x}_i$ is the inner product and C is an optional constant.

The polynomial kernel function, expressed in Equation (20.8), is a nonstationary kernel. It is appropriate for problems with a normalized training dataset. The parameters in the polynomial kernel function are kernel coefficient γ, the constant term C, and the polynomial degree D.

$$K(x, x_i) = \left(\gamma x^T \cdot x_i + C\right)^D \tag{20.8}$$

$$K(x, x_i) = \tanh\left(\gamma x^T x_i + C\right) \tag{20.9}$$

The sigmoid kernel function presented in Equation (20.9) comes from the artificial neural networks field. There are two variable parameters in the sigmoid kernel function, the kernel coefficient γ, and the intercept constant C.

The Gaussian kernel function presented in Equation (20.10) is an example of a radial basis function kernel, which is used in this study as follows:

$$K(x, x_i) = e^{-\left(\gamma \| x - x_i \|^2\right)} \tag{20.10}$$

in which γ is the kernel coefficient.

20.2.2 Preprocessing of the Dataset

To accurately predict S_W in thin beds reservoir, a large number of experimental S_W data points from the core analysis was collected in addition to the well logs. A total of 47 wells with both Dean-Stark S_W (2509 samples) and 282 conventional logs were collected from the NDIC website. Since some of the logs underwent different runs and core data were shifted, the pre-processing of the dataset was performed in the previous chapter, including environmental correction, depth shifting, data transformation, and petrophysical analysis to generate input variables to the model construction.

A total of six conventional logs were selected as input features. The six logs are gamma-ray (GR), bulk density (RHOB), compensated neutron (NPHI), compressional slowness (DT), deep resistivity (R_t), and shallow resistivity (R_{xo}). These variables were analyzed by applying a principal component analysis (PCA) technique. This approach allows us to recognize the patterns and redundancy in the data set and highlight the differences and similarities between the input variables. Understanding the relationship between the variables is really important in helping determine the significant logs with the most information needed as input to the model. In addition, when we are dealing with a large number of wells, it can reduce the number of input data and lower the computation time. First, PCA transforms the data into a new set of coordinate systems in multivariable space. So, PCA reorganizes the variance in the data which are arranged in the decreasing order of variance in accordance with the transformed principal components and each component is constrained by the orthogonality to other components. This means that PCA rotates the axis until it finds the maximum variance direction which corresponds to the first principal axis. Then, it searches for the second-highest variance in the data set and its direction must be perpendicular to the first axis because the variances in data in orthogonal directions are independent of one another. After that, it looks for the third direction, and so on.

The input variables (GR, R_t, R_{XO}, NPHI, RHOB, and DT) are plotted on the principal components cross-plot and are normalized between −1 and +1 scale. Figure 20.3 illustrates the projection of the input variables on the normalized principal axis. It shows the amount of influence that each variable exerts on a different axis. It can be seen that all the variable points are close to the circumference of the circle which indicates the importance and the contribution of each variable to the learning process. The NPHI and DT are at the maximum distance from the origin of the circle and contribute significantly to the variance in the first principal axis. Whereas the GR, R_t, R_{XO}, and RHOB contribute to the variances of both the first and second principal axes.

As a result, these 6 variables are important as an input to the learning process. On the other hand, the deep resistivity and medium resistivity being on top of each other means redundancy in the data set. Table 20.2 shows the correlation matrix between the six variables. It is noticeable that the highest correlation is between deep and shallow resistivities, at 97.6%, and neutron porosity

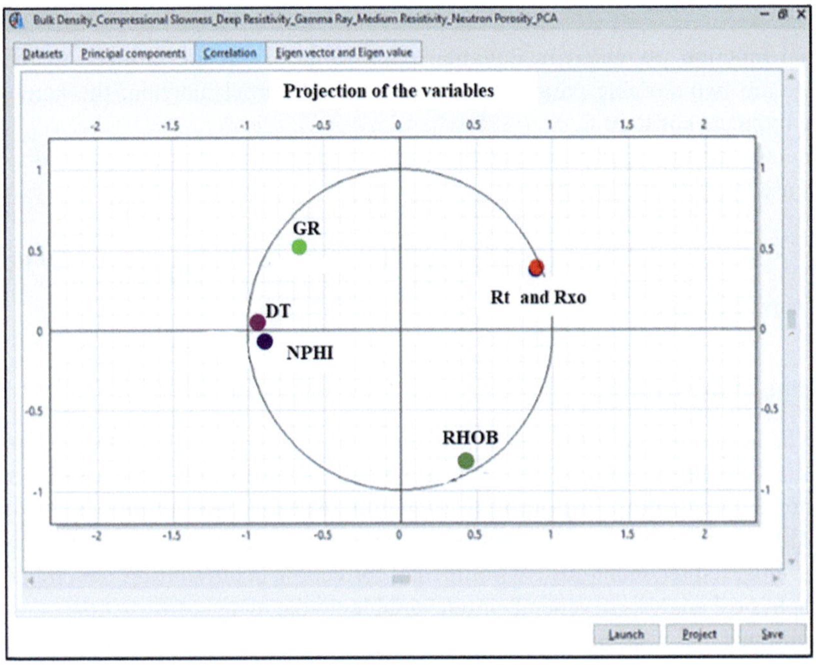

Figure 20.3 Projection of variables cross-plot.

Table 20.2 Correlation matrix.

	RHOB	DT	R_t	GR	R_{xo}	NPHI
RHOB	1	−0.6780	0.1694	−0.5671	0.2043	−0.6958
DT	−0.6789	1	−0.3656	0.6149	−0.3902	0.7324
R_t	0.1693	−0.3656	1	−0.0474	0.9767	−0.4354
GR	−0.5672	0.6149	−0.0460	1	−0.0474	0.5532
R_{xo}	0.2043	−0.3902	0.9767	−0.0474	1	−0.4903
NPHI	−0.6958	0.7324	−0.4354	0.5532	−0.4903	1

Table 20.3 The eigenvector and eigenvalue.

	Eigen Vect.1	Eigen Vect.2	Eigen Vect.3	Eigen Vect.4	Eigen Vect.5	Eigen Vect.6
RHOB	0.4144	−0.3215	0.5914	0.3573	0.4972	−0.0002
DT	−0.4703	0.1743	0.0236	0.8538	−0.1368	−0.0051
R_t	0.3462	0.5846	−0.1458	0.1016	0.1896	−0.6861
GR	−0.3473	0.4355	0.7590	−0.3180	−0.1029	−0.0411
R_{xo}	0.3620	0.5717	−0.0700	0.1015	0.0787	0.7215
NPHI	−0.4851	0.0881	−0.2175	−0.1464	0.8253	0.0830
Eigenvalue	3.3773	1.6529	0.4318	0.2709	0.2434	0.0201
Information	0.5631	0.2756	0.0720	0.0451	0.0405	0.003
Cum. Info.	0.5631	0.8388	0.9108	0.9530	0.9966	1

with compressional slowness, at nearly 74%. However, the correlation is significantly lower between gamma-ray and both resistivities, at 4.7%.

The amount of information that is being represented by each eigenvector is summarized in Table 20.3. It is obvious that the first and second eigenvectors capture nearly 84% of the information, collectively. This endorses the significance of the projection of variable cross-plot, where eigenvector 1 and eigenvector 2 are plotted.

On the basis of the inspected objective and to improve the performance of the dataset, the logarithmic of deep resistivity (Log R_t) is preferred to be used instead of R_t because Log R_t has a greater correlation with S_w than R_t. In addition, the R_t distribution seems to be skewed with the mode around 2 ohm.m (Figure 20.4). However, the logarithm of R_t shows a normal distribution (Figure 20.5).

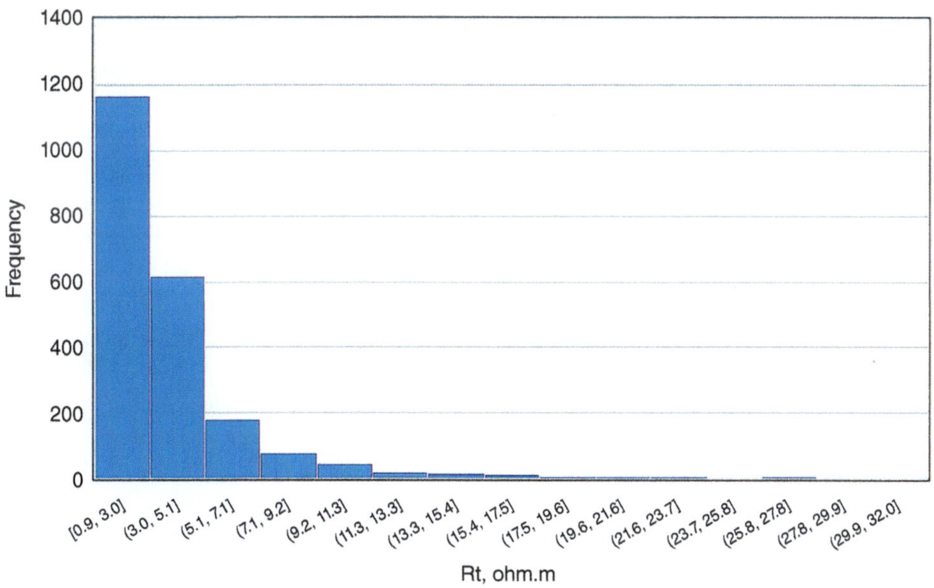

Figure 20.4 Deep resistivity histogram.

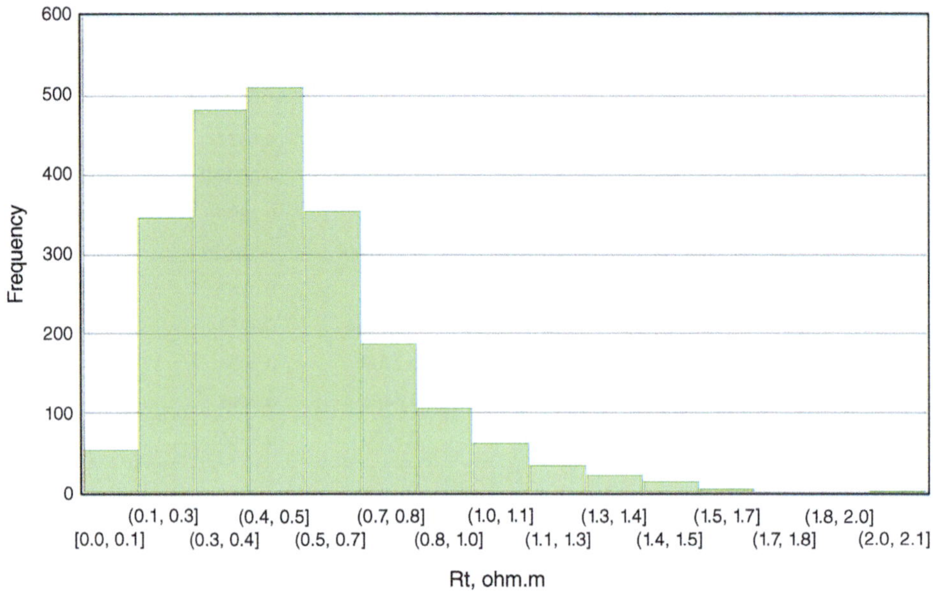

Figure 20.5 Logarithm of deep resistivity histogram.

Also, the component volumes, effective porosity, and clay volume were exported from petrophysical analysis and added to the input variable space. However, they did not improve the performance of the SVR model. Due to the complexity of the Three Forks Formation, two categorical variables were added as inputs. The first one is the HC column and the second one is the Three Forks reservoir units. Therefore, the dataset is divided into three corresponding members (UTF, MTF, and LTF). Then the formation column was expanded to 3 sub-columns to represent 3 categories (UTF: 001, MTF:010, and LTF: 100), and the HC column to 2 categories (10 and 01). Including these values into the training set, they can be represented by dummy variables which will lead to a dummy variable trap. The latter is a scenario in which one variable can be predicted from the others (multicollinear). In this case study, one column is selected to represent the HC interval and 2 columns to represent the 3 formation members.

The relationship between the variables and Dean-Stark water saturation is illustrated in Figure 20.6 and Table 20.4. One can see that the HC column has a strong relationship with the output target S_w, at 87%. It is the most significant variable for the model construction after the log R_t at 52.4% and NPHI log at 28.3%. Finally, the GR, DT, and RHOB logs have the remaining contributions at 14.2%, 10.8%, and 8.4%, respectively. The PCA analysis for the 9 variables is illustrated in Figure 20.7 which plots the proportion of variance explained by each variable. It shows a gradual increase of the cumulative variance, which means that the 9 variables are important as input features into the model construction.

After determining the number of significant input variables, each of the selected models can be built for evaluation. Two different supervised regression algorithms (SV regression and RF regression) and deep learning (backpropagation artificial neural network) were used in this study.

20.2.3 Building SVR Model

The SVR is a powerful method for regression problems with a high capability to handle small training dataset (Bruzzone et al. 2006). In addition to the preprocessed logs, the two categorical features

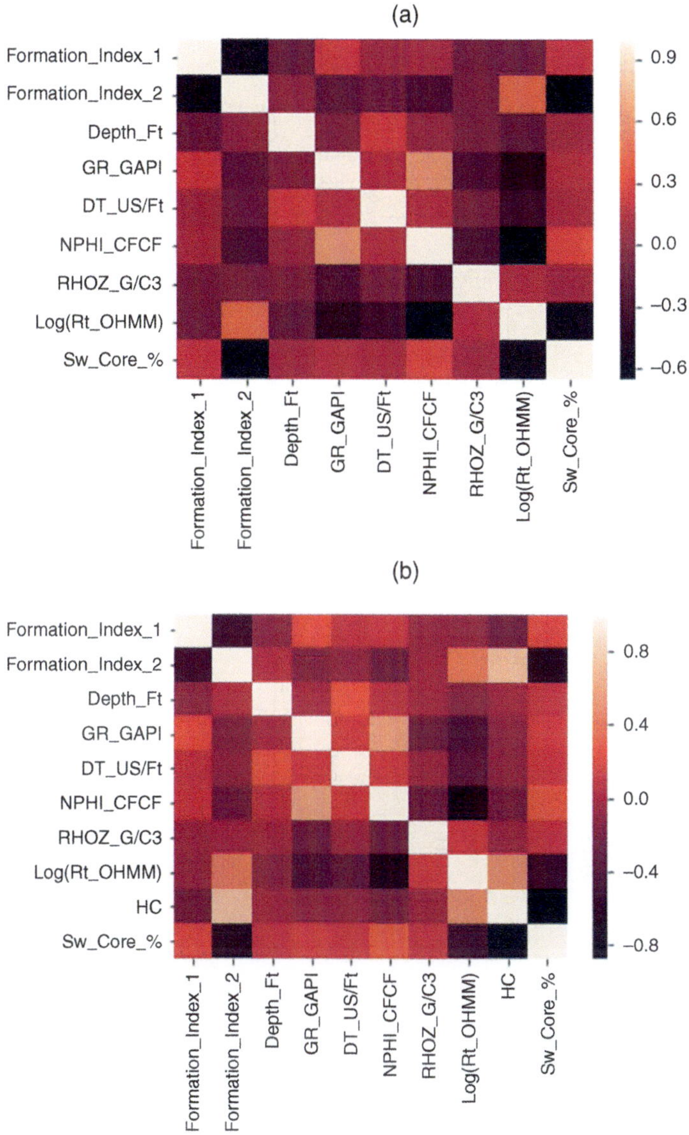

Figure 20.6 (a) Heat map of the variables used in the model without HC column. (b) Heat map of the variables without HC column. S_w_core is the output target.

are included in the training dataset. The best approach in building the SVR model is, on the basis of the training data set, to determine the appropriate kernel function and its best values of gamma, epsilon, and regularization parameter through the learning algorithms. For this case study, after a comparative investigation, the RBF kernel function was selected. The grid search was used to select the search range of the best training parameters of the SVR model. In order to optimize SVR, the initial parameters are set to default then several combinations of c, ε, γ, and n-fold values for cross-validation are examined. 10-fold cross-validation is selected and in this case, the dataset is split into 10 subsets and one subset is used as a test set each iteration (10 times). Then, the cross-validation for the training model is the average of the 10-fold iterations. The optimal SVR learning algorithm

Table 20.4 A correlation matrix with 9 input variables.

Variables	Formation_ index 1	Formation_ index 2	Depth	GR	DT	NPHI	RHOZ	Log Rt	HC_colonm	Sw_core
Formation_ index 1	1.000	−0.517	−0.131	0.218	0.070	0.097	−0.077	−0.099	−0.232	0.186
Formation_ index 2	−0.517	1.000	0.016	−0.169	−0.118	−0.247	−0.052	0.415	0.726	−0.640
Depth	−0.131	0.016	1.000	−0.031	0.242	0.041	−0.046	−0.149	−0.054	0.084
GR	0.218	−0.169	−0.031	1.000	0.126	0.591	−0.251	−0.418	−0.113	0.142
DT	0.070	−0.118	0.242	0.126	1.000	0.135	−0.061	−0.331	−0.102	0.108
NPHI	0.097	−0.247	0.041	0.591	0.135	1.000	−0.265	−0.650	−0.022	0.283
RHOZ	−0.077	−0.052	−0.046	−0.251	−0.061	−0.265	1.000	0.134	−0.073	0.084
Log Rt	−0.099	0.415	−0.149	−0.418	−0.331	−0.650	0.134	1.000	0.499	−0.524
HC_colonm	−0.232	0.726	−0.054	−0.113	−0.102	−0.220	−0.073	0.499	1.000	−0.874
Sw_core	0.186	−0.640	0.084	0.142	0.108	0.283	0.084	−0.524	−0.874	1.000

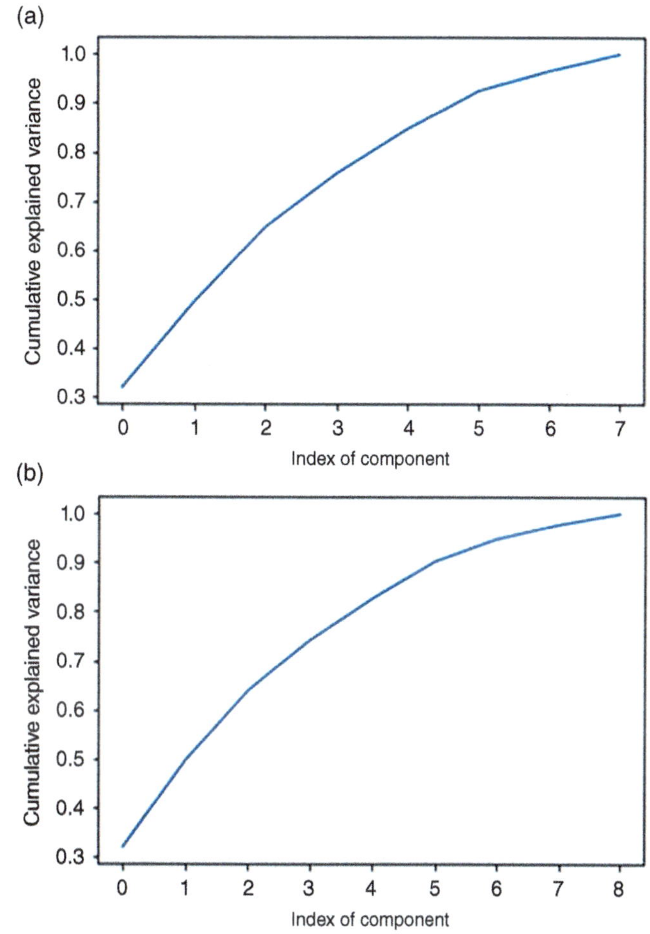

Figure 20.7 Cumulative explained variance. (a) CPA with 8 input variables, (b) with 9 input variables (with HC column).

parameters (e.g, γ, ε, and c) were set based on cross-validation score. The workflow of building the SVR model is summarized in Figure 20.8.

Once an SVR model is constructed and validated on a testing set, the program can be applied to confirmation wells that contain the same variables. There should be a good match between the SVR-generated synthetic curve and the output curve S_w in the confirmation wells. Once the model is verified using the confirmation wells, the model can be applied to wells without the output S_w with confidence.

20.2.4 Building Random Forest Regression Model

An RF is one of the powerful machine learning algorithms, which can be used for regression and classification problems (Breiman et al. 1984). Both numerical and categorical data can be used. The RF package in Patton libraries was used to build the model.

The RF algorithm combines many self-learning decision trees to parameterize models which are used to predict the target output (Mellor et al. 2013). RF helps reduce the high variance that would result from one single decision tree (Lellogg et al. 2018). RF trains a set of decision trees separately and in parallel. The algorithm randomly subsamples the initial dataset on each iteration to get a

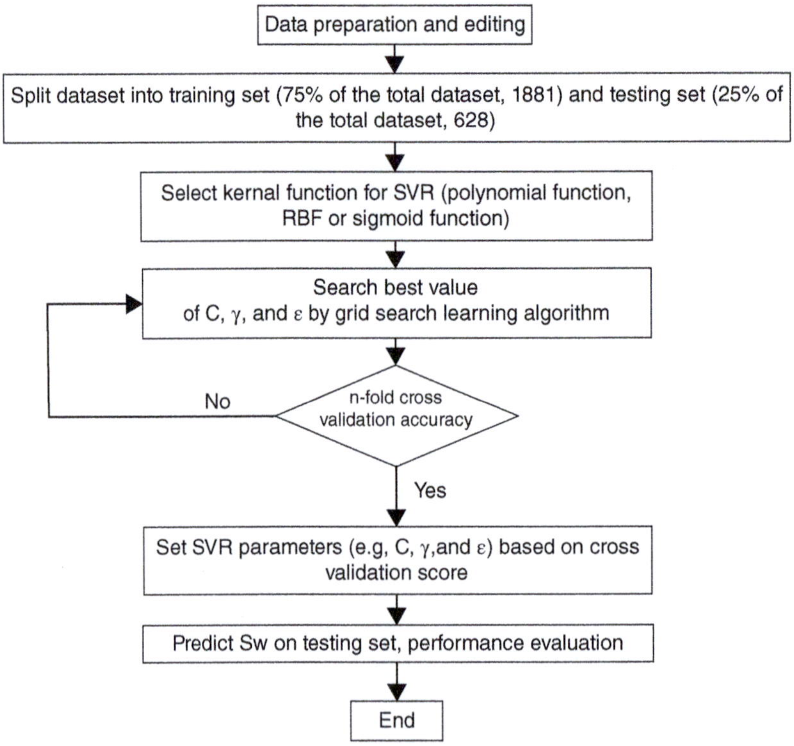

Figure 20.8 The workflow to build SVR through determining the best values of c, and γ.

different training set (bootstrap aggregation sampling) and also considers a random subset of features to split on at each decision tree node (Efron and Tibshirani 1996). The trees continue growing until there are no gains in purity (Mellor et al. 2013).

The 10-fold cross-validation is used to estimate the performance of the RFR model. Then, the average of the predictions of the trees results in the prediction of the output target. Figure 20.9 illustrates the workflow to build the RFR model by determining the best hyperparameters.

20.2.5 Building Deep Learning Model

For the construction of the deep learning (DL) model, the same dataset was used and the well logs were run through the same processing steps. A backpropagation neural network (Werbos 1974) was used in this study. It is a multihidden layer neural network technique where the initial weights are randomly chosen, then the gradient of the loss function is calculated repetitively and propagated backward through the network to adjust the initial weights (Mollajan et al. 2013).

Several network configurations were investigated to define the model structure and optimize the backpropagation neural network model in terms of the number of layers and hidden neurons. A model structure consisting of 9 input features, 3 hidden layers with 20 neurons in each hidden layer, and 1 neuron for the output layer was selected from the investigation (Figure 20.10).

The processing neurons create a sigmoid function as an activation function for each neuron to relate the output to the input. During the optimization process, the mean squared error (MSE) is

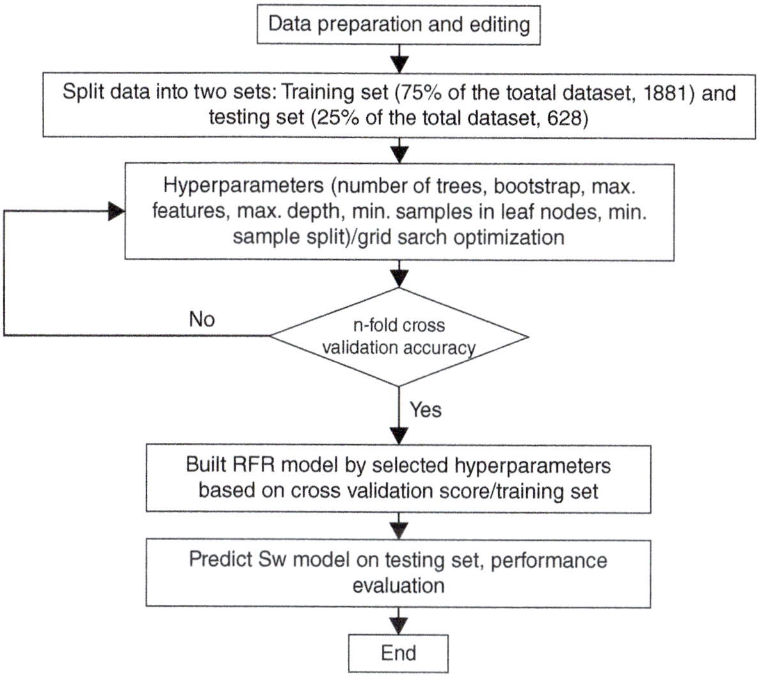

Figure 20.9 Summary of the workflow to build RFR.

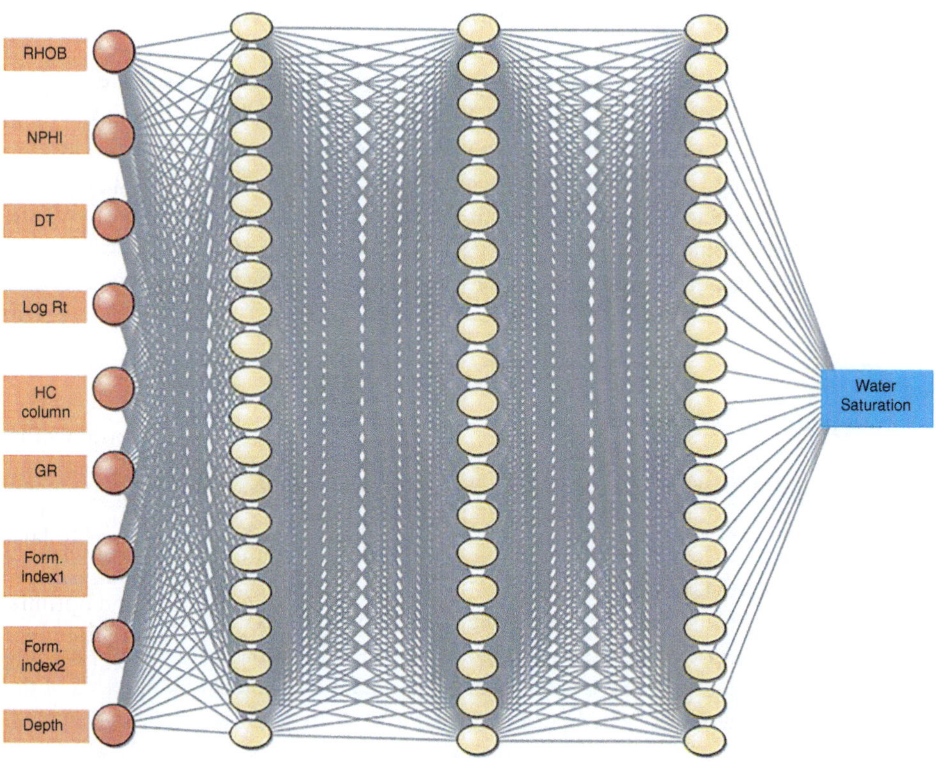

Figure 20.10 Model structure of backpropagation neural network.

used as a loss function. MSE is estimated as the average of the squared differences between the predicted and measured S_W values (see Equation 20.1):

$$MSE = \frac{1}{n}\sum_{i=1}^{n}\left(y_i - y_i'\right)^2 \tag{20.11}$$

Where n is the number of data points, y_i is the real data values, and y_i' is the predicted data value.

The loss function gives information on how the prediction model matches the actual values in the training dataset.

The training phase is performed with a learning rate of 0.01 for 200 epochs with a batch size of 32 epochs. It means that the algorithm trains the data set for 200 epochs and the loss is calculated for each epoch but the weights are updated using backpropagation of the estimated loss for every 32 runs. In addition, the Adam optimization algorithm is used to update and optimize the connection weights that the model can use for the prediction. Adam optimizer was introduced by Kingma and Lei Ba (2015). According to the authors, Adam is straightforward to implement, the hyperparameters require little tuning, and it is appropriate for problems with large data, very noisy, and/or sparse gradients. The Adam optimization algorithm calculates separately adaptive learning rates for different parameters from computing an exponential average of the gradient and the squared gradient (first and second moments of the gradients). The workflow to build backpropagation neural network model is summarized in Figure 20.11.

20.2.6 Curve Reconstruction Using K.Mod

K.Mod is generally used for curve reconstruction. The neural network modeling procedure used is a supervised artificial intelligence program that learns to build synthetic curves. It is based on the principle of multilayer perceptron which is used to separate the data in hyperspace.

The architecture of the model can be described as shown in Figure 20.12. The structure contains one input layer with 5 input curves, one layer with 10 hidden nodes which monitors the training process, and one output layer. A sigmoidal function is used by each neuron in the hidden layer to generate a sigmoid curve based on defined weights which support the matching of the output curve.

A training well is first selected that contains both the curves needed to generate the synthetic curves and the expected output curve. The input logs for 10 wells (GR, RHOB, NPHI, R_t, and DT) were trained to learn the output S_W. Another data set for 3 wells containing the same logs was used to cross-validate the model.

20.3 Results and Discussion

Due to the high concentration of laminations in the upper and middle Three Forks members which leads to the large variability of Dean-Stark S_W, it was challenging to effectively find the best techniques, kernel function, loss function, learning algorithms, and optimization parameters to accurately predict the output target S_W.

For the SVR model, the results were obtained using an ε-insensitive loss function combined with RBF kernel functions and a grid search learning algorithm. Default parameters c, ε, and γ are first used and then many combinations of c, ε, and γ were examined (see Table 20.5). In terms of the training results, RBF presented a better performance for SVR than polynomial and sigmoid functions.

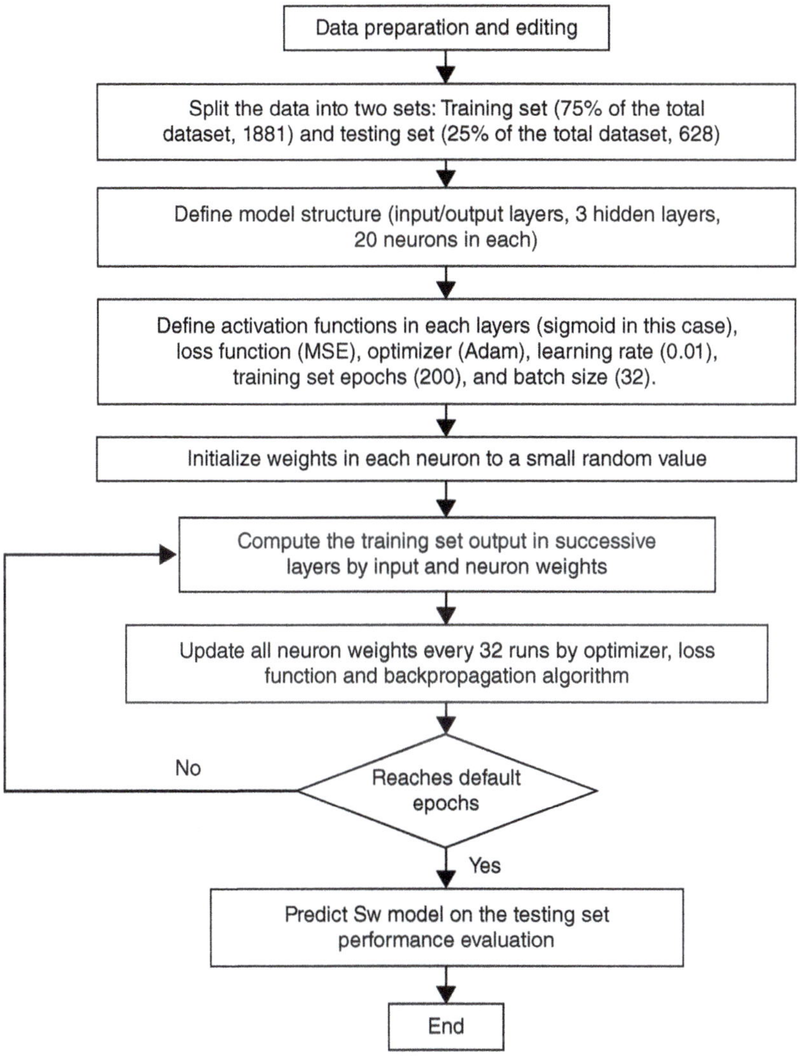

Figure 20.11 The workflow to build backpropagation artificial neural network.

The SVR displayed very stable cross-validation accuracy. Table 20.6 shows that the SVR with RBF kernel function shows the highest cross-validation accuracy at 82.95% (77.99% on average) when using grid searching as the learning algorithm and setting n-fold as 10. The best values of c, ε, and γ found by the grid searching algorithm are 2.4, 0.25, and 0.04, respectively.

The learning model of SVR with RBF kernel was successfully applied to the test data set yielding a target output prediction with a correlation factor of 0.78 (Figure 20.14). The SVR model was also constructed without the HC column as an input variable where the model yielded a lower prediction with a correlation factor of 0.65 using values of 1.9, 0.31, and 0.44 for c, ε, and γ, respectively. Figure 20.13 shows the comparison between the Dean-Stark S_w and the predicted S_w for this case. A summary of the two models and their corresponding results are illustrated in Table 20.9. Despite some differences that can be observed between the predicted S_w and Dean-Stark S_w in the

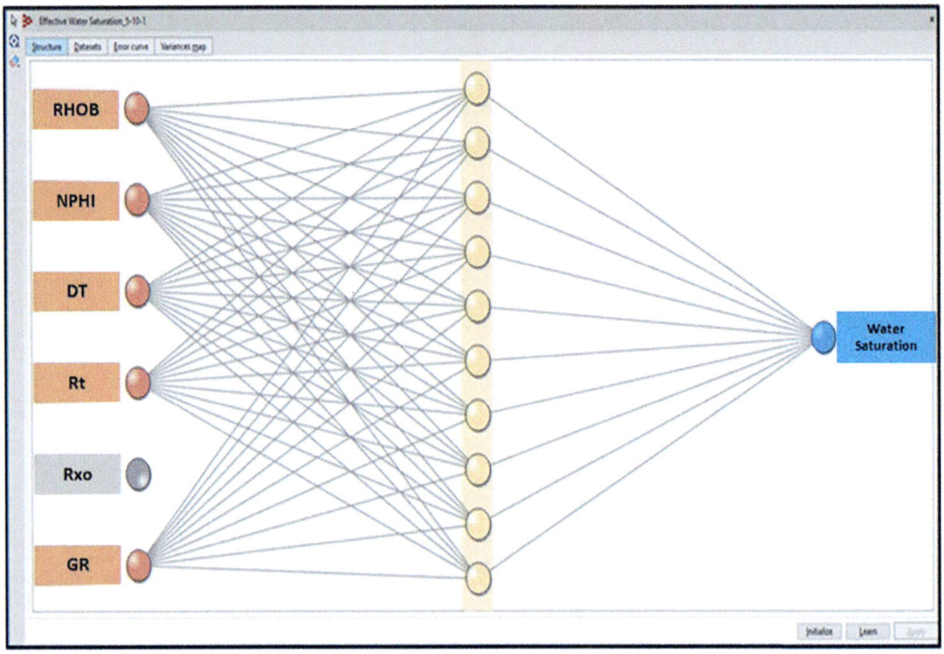

Figure 20.12 Structure of neural network, K.Mod.

Table 20.5 Search range parameters.

Parameters	Search range	Increment
c	1–10	0.1
ε	0.01–0.5	0.01
γ	0.01–0.5	0.01

Table 20.6 The cross-validation accuracy of S_w by SVR with RBF, grid searching learning algorithm, and 10-fold.

Kernel function	Learning algorithm	*n*-fold									
		$n = 1$	$n = 2$	$n = 3$	$n = 4$	$n = 5$	$n = 6$	$n = 7$	$n = 8$	$n = 9$	$n = 10$
Radial basis function (RBF)	Grid searching	80.10	77.82	80.91	76.84	78.45	82.95	77.49	72.67	73.94	78.82

upper and middle Three Forks (see Figure 20.14) where the program could not model perfectly the peaks, the agreement between the output and the measured S_W demonstrates a successful ability of the SVR to infer the complex relationship based on the conventional logs dataset between the input features and the target output.

For the second technique, RFR, the same dataset and similar steps were taken and deployed in the prediction and assessment of the RFR model. The training data set is partitioned into 9 training

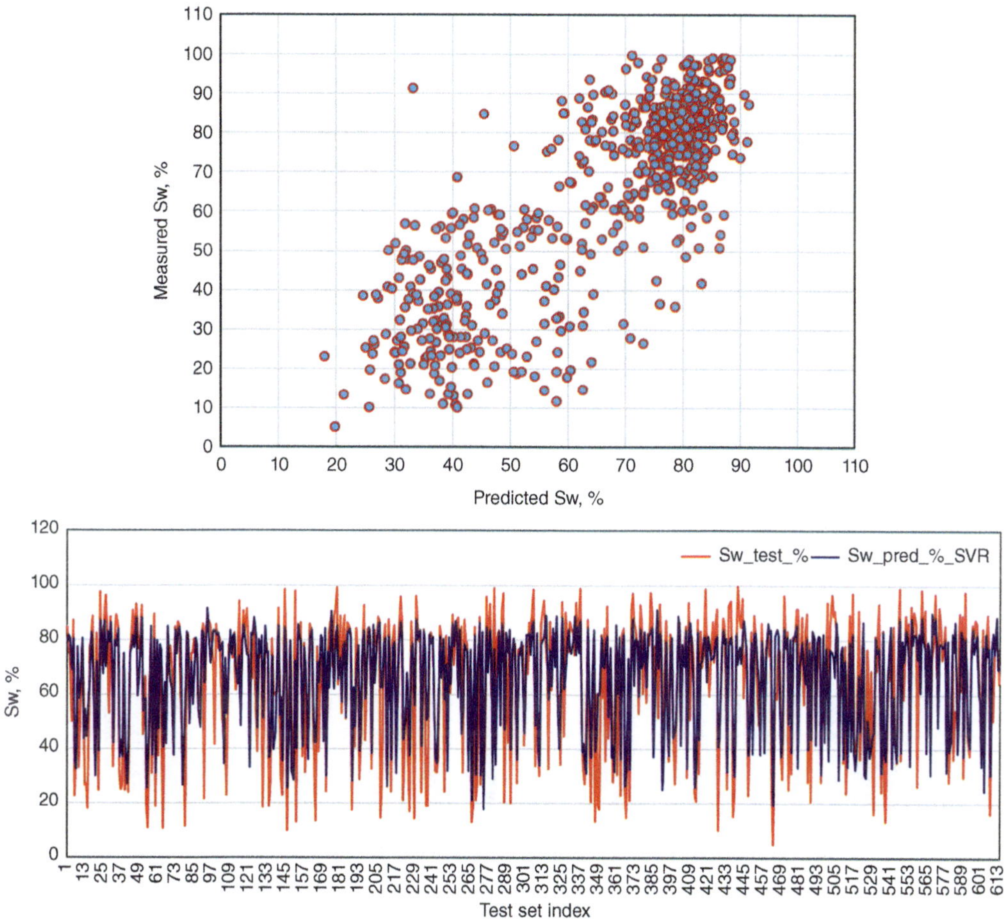

Figure 20.13 Comparison between the Dean-Stark S_w measurements and the predicted S_w from SVR model without categorical HC column as input feature.

folds and one validation fold at each iteration in addition to the grid searching learning algorithm. The grid search range is illustrated in Table 20.7.

Using the 9 features selected as shown in Figure 20.6, the results show that RFR yields a better stability cross-validation accuracy than SVR. Table 20.8 illustrates the highest cross-validation accuracy at 85% (80.58% on average) when setting n-fold as 10. In terms of the number of trees that used to generate the model, it was predicted based on 10-fold cross-validation accuracy. The depth of each tree is set to 40 and a minimum number of samples required at leaf nodes is 4. Also, the number of features to consider when looking for the best split is equal to the maximum number of features. The predicted S_W values are shown in Figure 20.16. Two separate interval units were defined which correlate to LTF with higher S_W and to UTF and MTF with an S_W less than 60% and a correlation factor of 0.786. The effect of the HC column on the RFR model performance was also investigated and the results are illustrated in Figure 20.15. The correlation factor decreases to 0.65 (Table 20.9).

Also, the prediction of the water saturation was conducted on the same dataset using a back-propagation neural network. Figure 20.11 shows the iterative deep learning workflow used to

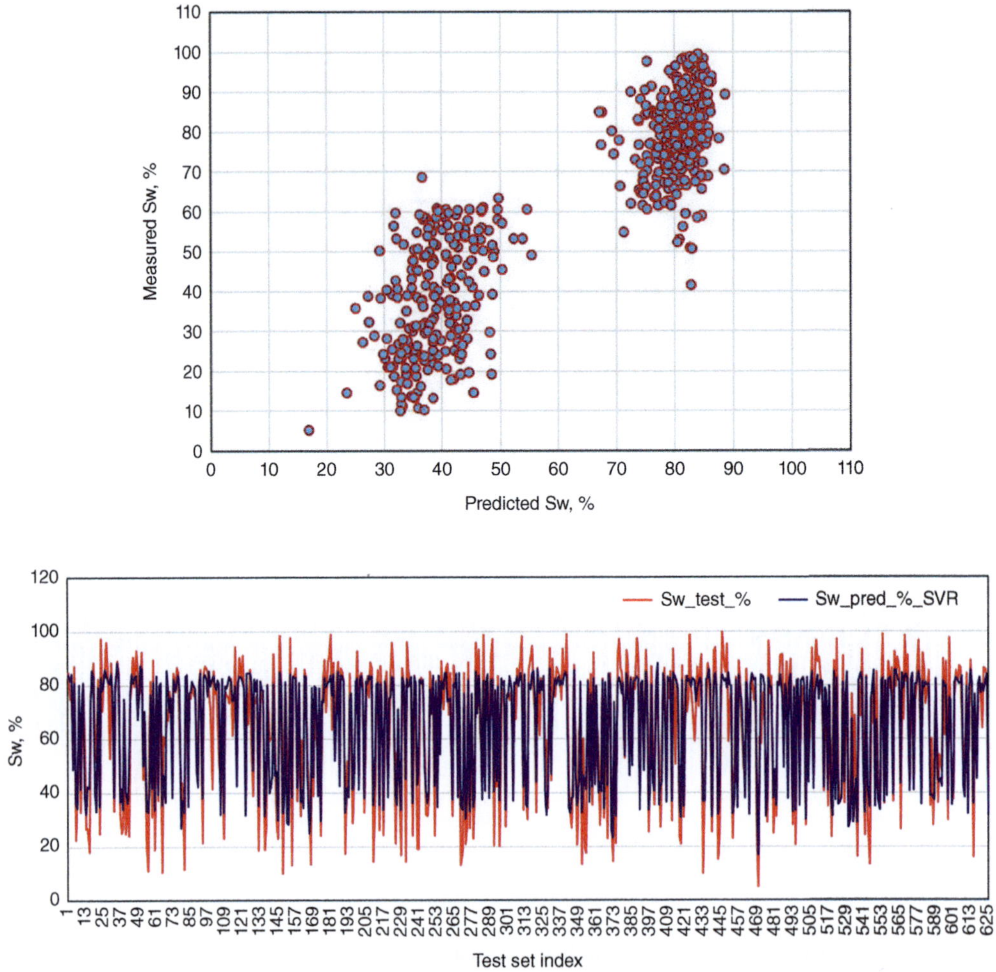

Figure 20.14 Comparison between the Dean-Stark S_w measurements and the predicted S_w from SVR model with categorical HC column as input feature.

Table 20.7 Grid search range parameters, RFR.

Key	Size	Value
bootstrap	2	[True, False]
Max. depth	12	[10, 20, 30, 40, 50, 60, 70, 80, 90, 100, . . .]
Max. features	2	[auto, sqrt]
Min. Samples leaf	3	[1, 2, 4]
Min. samples split	3	[2, 5, 10]
n. estimators	10	[200, 400, 600, 800, 1000, 1200, 1400, 1600, 1800, 2000]

Table 20.8 The cross-validation accuracy of Sw by RFR with grid searching learning algorithm, and *n*-fold values.

Learning algorithm	*n*-fold									
	n = 1	*n* = 2	*n* = 3	*n* = 4	*n* = 5	*n* = 6	*n* = 7	*n* = 8	*n* = 9	*n* = 10
Grid searching	82.66	80.82	82.80	76.38	82.65	85.03	81.08	75.62	77.89	80.87

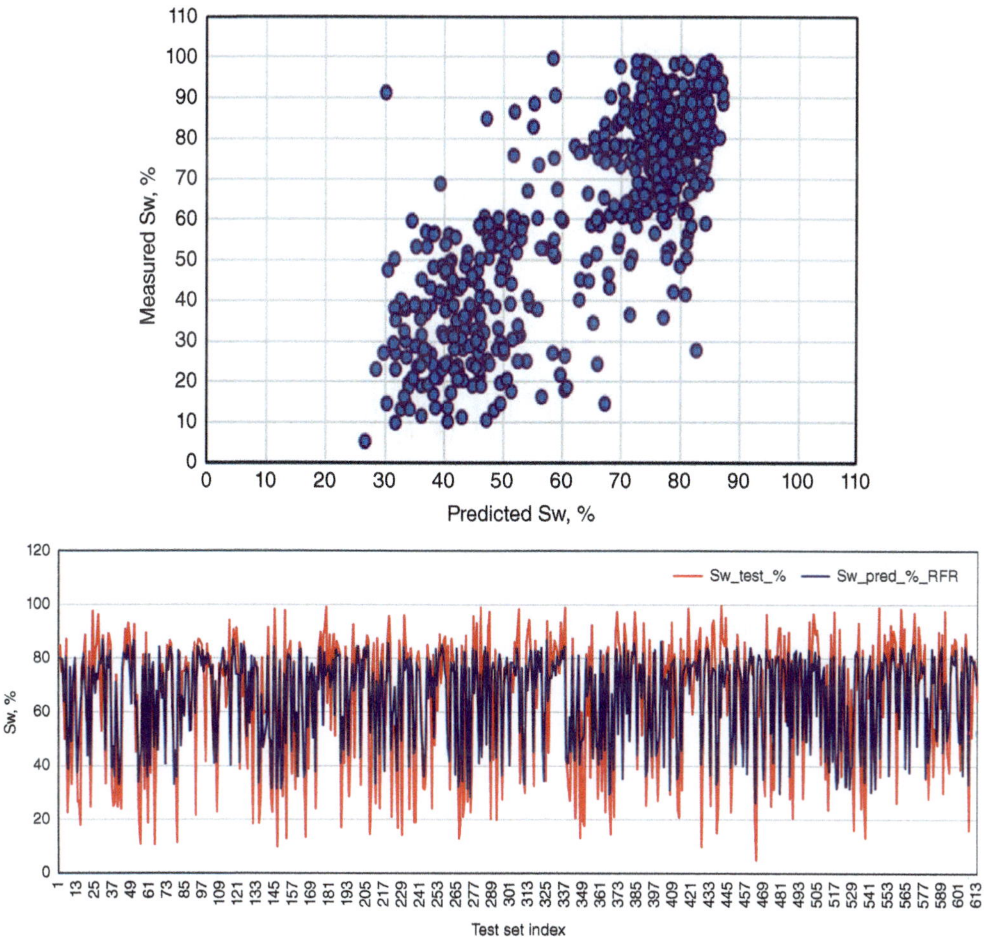

Figure 20.15 Comparison between the Dean-Stark S_w measurements and the predicted S_w from the RFR model without categorical HC column as input feature.

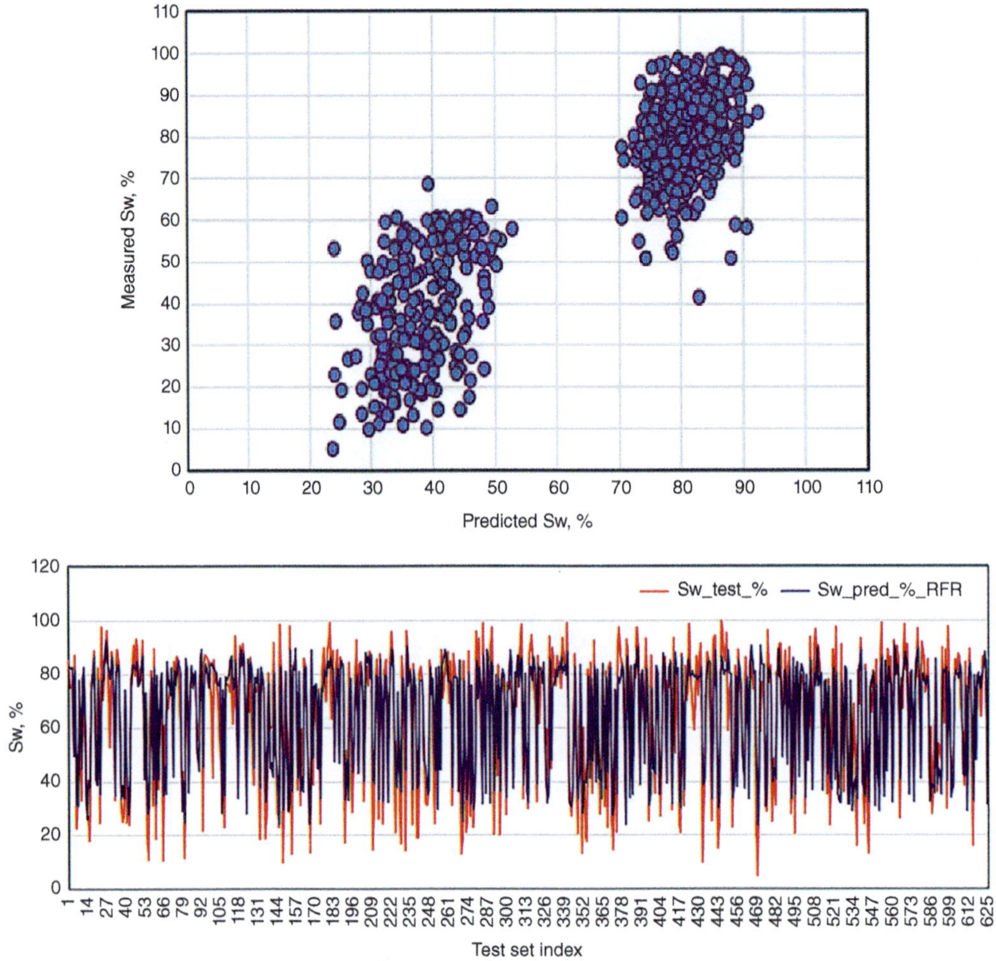

Figure 20.16 Comparison between the Dean-Stark S_w measurements and the predicted S_w from the RFR model with categorical HC column as input feature.

Table 20.9 Comparison of MAE, RMSE, and R^2 between SVR, RFR, and BP-ANN.

	Without HC column			With HC column		
Model description	**MAE**	**RMSE**	**R^2**	**MAE**	**RMSE**	**R^2**
SVR	10.524	14.052	0.648	8.640	10.897	0.788
RFR	10.557	13.996	0.650	8.709	10.941	0.786
BP-ANN	10.563	14.059	0.647	8.746	10.893	0.788

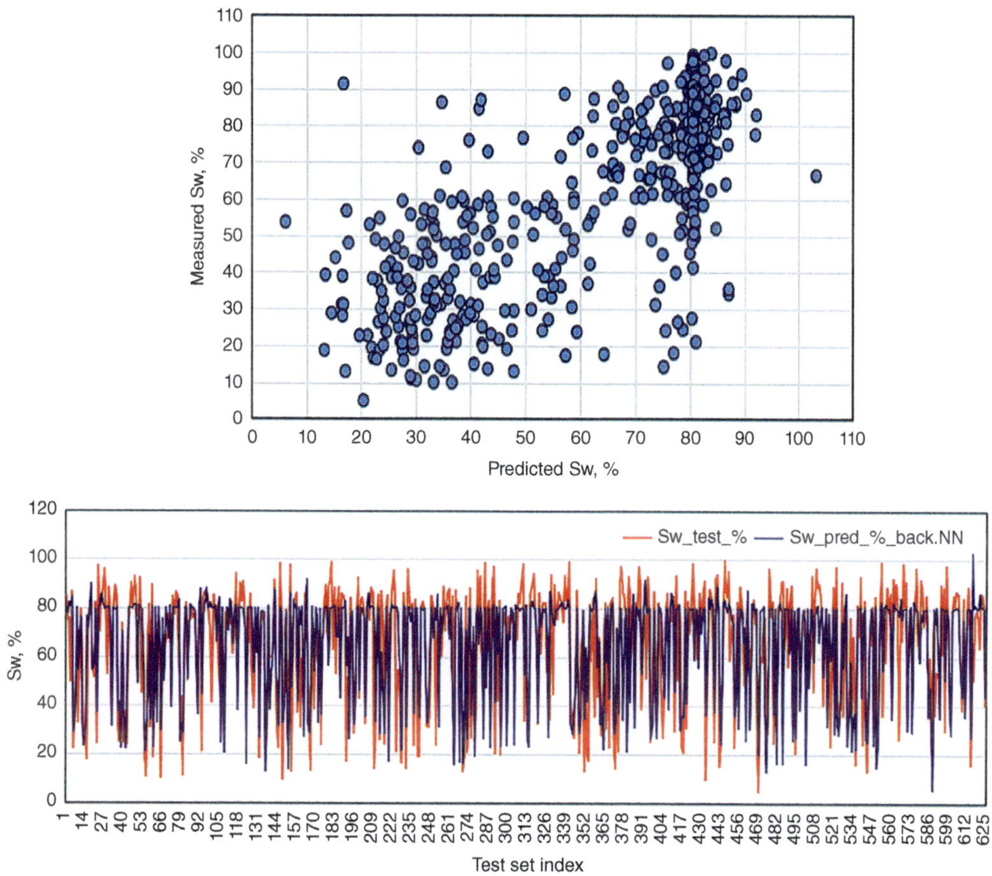

Figure 20.17 Comparison between the Dean-Stark S_w measurements and the predicted S_w from backpropagation artificial neural network model without categorical HC column as input feature.

assess the model in S_W prediction. The algorithm has the ability to extract the hidden patterns to map the complex non-linear relationship between the input features and the output S_W. As discussed above the performance and the accuracy of the model was evaluated based on R^2, RMSE, and MAE.

Figures 20.17, 20.18, and Table 20.9 show that the model has almost the same accuracy in predicting water saturation compared to the performance of RFR and SVR. The correlation coefficient of the model is 0.788 with 3 hidden layers and 20 nodes for each layer combined with an optimizer (Adam) and a backpropagation algorithm. The high performance of the backpropagation neural network based on the R^2 value suggests that the loss function (MSE) is a valuable input for the learning process for the S_W prediction.

The results show that the performance of the three models is improved after the categorical HC column is added as an input feature into the model construction. Overall, the results agree fairly well in the lower part of the reservoir when viewing the predicted model on the test dataset (Figures 20.14, 20.16, and 20.18). However, the peaks were not perfectly modeled which are due to the effect of the lamination in the reservoir (Figure 20.19).

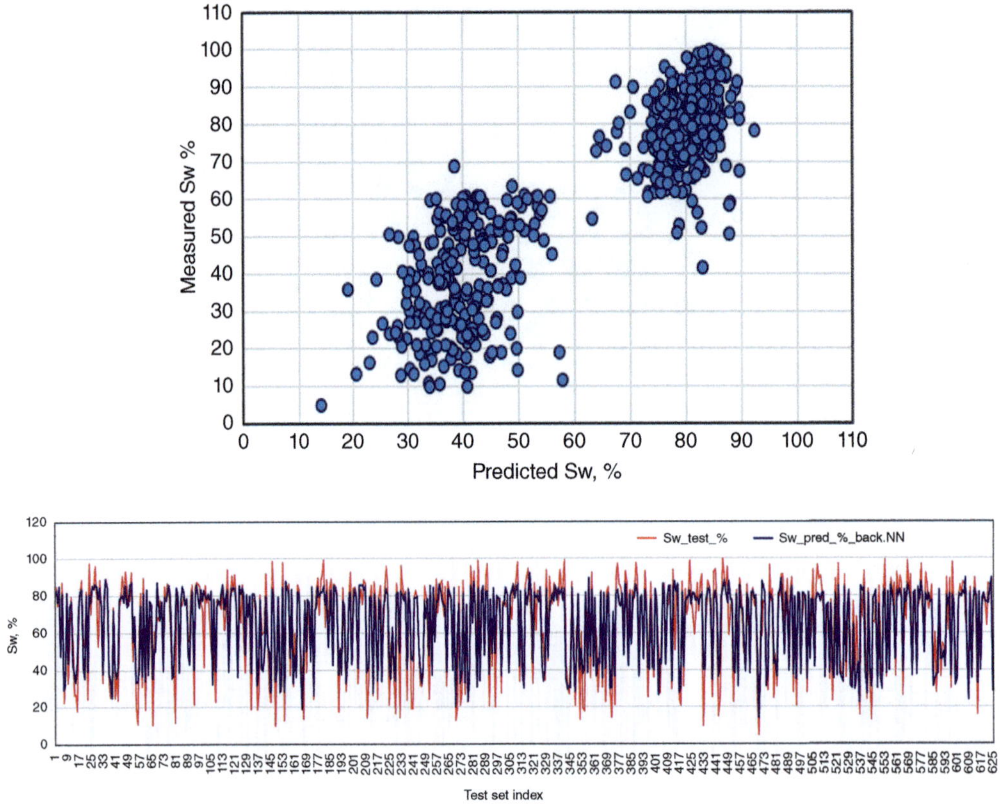

Figure 20.18 Comparison between the Dean-Stark S_w measurements and the predicted S_w from backpropagation artificial neural network model with categorical HC column as input feature.

The last attempt for water saturation prediction was the curve reconstruction using K.Mod. As discussed earlier, the K.Mod approach is based on the principle of multi-layer perceptron; each hidden node's sigmoid function extracts information from the input variables and generates the output curve S_W via an interactive learning process. Starting with one hidden layer with minimum neurons, the errors in the learning and validation data are calculated. By increasing the number of neurons and cycles, the training and validation errors decrease until the training error is diminished and an increase in the validation error is observed. Then the model can be saved and applied to the other wells.

Figure 20.20 illustrates the reconstructed S_W curves for three wells plotted on top of Dean-Stark S_W. A good much is observed between the predicted S_W and Dean-Stark S_W. In addition, the backpropagation model was computed and used for quality control of the model. The recreated input logs were generated from the output S_W and from the learning process model.

Figure 20.21 shows a view of the reconstructed input logs. Overall, the results present a good match except for the gamma-ray log at a depth of 10475 and 10545 ft. This means that during the learning process, the gamma-ray may contribute to high loss in the S_W prediction model.

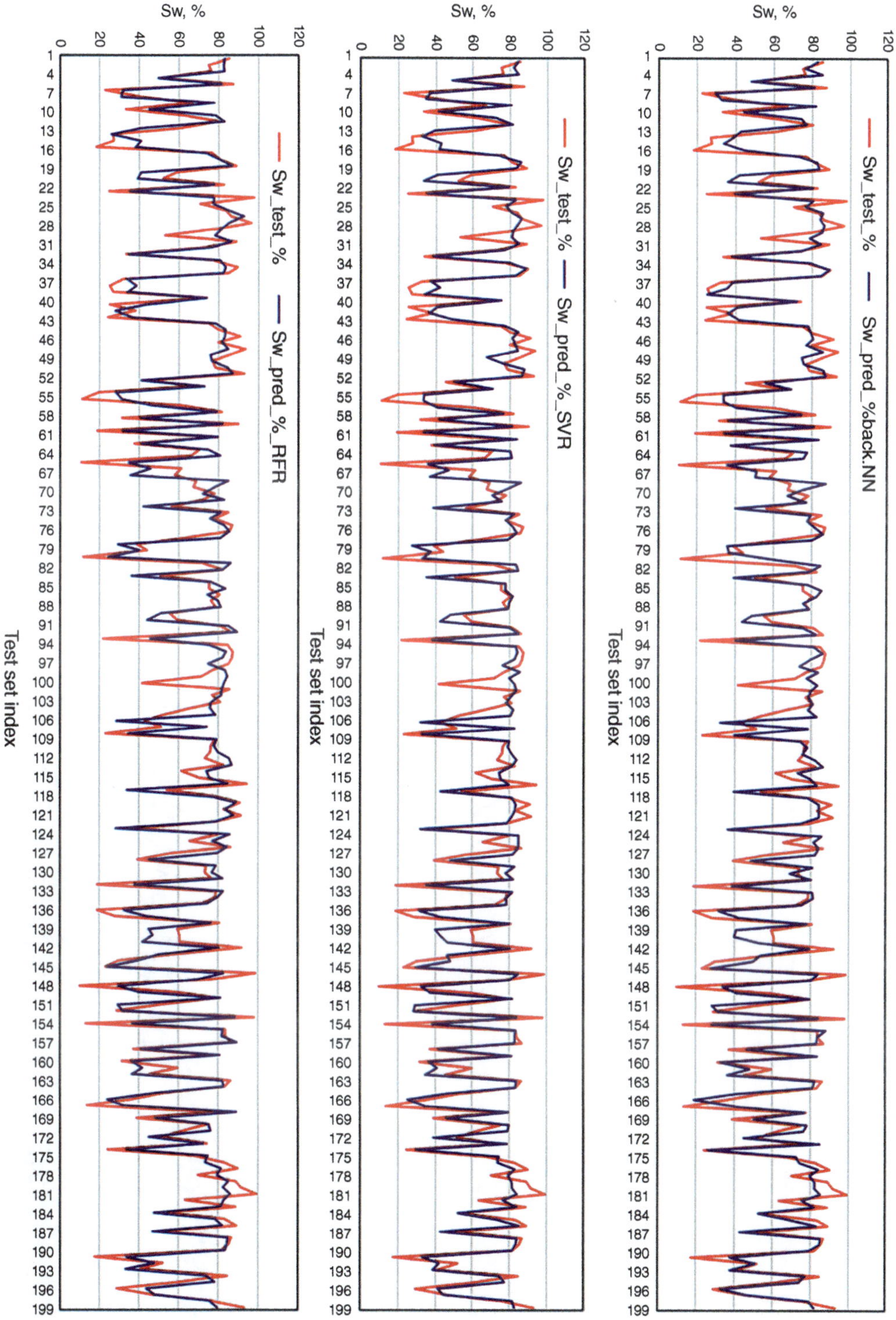

Figure 20.19 Comparison between SVR, RFR, and BP-ANN models on the test set.

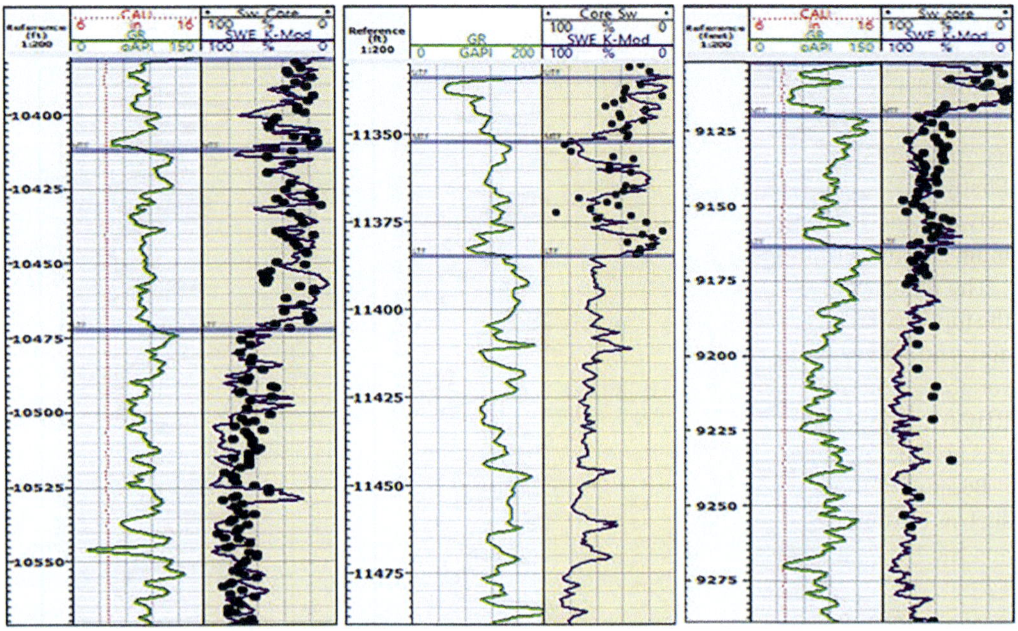

Figure 20.20 From left to right: Hognose 152-94-18B-19H-TF, Jersey 19-6H1, and Sorenson 4A-27-1H. Track 1, GR and Caliper, Cali. Track 2, the computed S_W from K.Mod program versus Dean-Stark S_w.

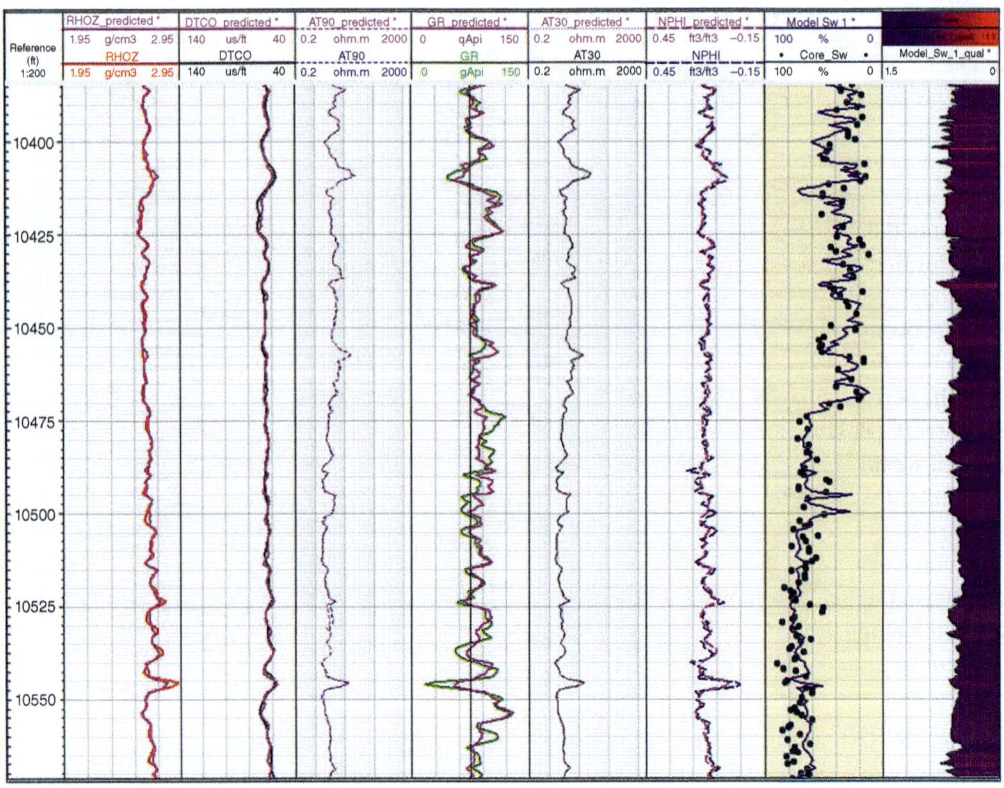

Figure 20.21 Backpropagation plot for well Hognose 152-94-18B-19H-TF.

20.4 Conclusion

In this chapter, the aim was to construct a reliable machine learning model that can be used for water saturation prediction in thin bed reservoirs using conventional logs. Two supervised machine learning and one deep learning algorithm were proposed and built to predict consistent results. Firstly, SVR has been constructed and its performance was compared to backpropagation neural network and RF regression based on correlation coefficient, root mean square error, and maximum absolute error indexes.

The first part of this chapter presented the preprocessing of the dataset and the importance of the input variables to the model construction. PCA analysis was performed to identify the patterns and redundancy in the data set and highlights the differences and similarities between the input variables. The results from PCA showed that the conventional logs GR, log R_t, NPHI, DT, and RHOB are important as input variables to the learning process. The logarithmic of the resistivity was used because it is normally distributed.

Because of the complexity of the reservoir, more attributes were added to the learning process such as the output volumes from the petrophysical analysis in addition to the formation members and the HC column. The results showed that the HC column has a strong relationship with the output target S_w with the correlation coefficient increasing by 17% and the loss decreasing by 3. Also, the cumulative explained variance in all 9 input variables showed a gradual increase. After the processing phase, the reservoir dataset including well logs and Dean-Stark S_w were partitioned randomly into a training set with 75 percent (1881 values) and testing set with 25 percent (628 values). The training dataset was used by the machine learning and deep learning algorithms to construct the model that fits the training data; while the testing dataset was used to test the prediction performance of different models for unseen data based on correlation coefficient, root mean square error, and maximum absolute error indexes.

In the second part of this chapter, the water saturation was modeled using SVR with an ε-insensitive loss function combined with RBF kernel function, grid search learning algorithm, and 10-fold cross-validation. The best values of c, ε, and γ found by the grid searching algorithm were 2.4, 0.25 and 0.04, respectively. The learning model of SVR with RBF kernel was successfully applied to the test data set yielding a target output prediction with a correlation factor of 0.78. The HC column resulted in remarkable improvement to the model performance compared to the previous model displayed in Figure 20.13 with an increase in accuracy by 17%. The results demonstrated the effectiveness of applying SVR in thin beds analysis.

The backpropagation neural network and RFR algorithms were applied to the same data set to infer the performance of SVR. Overall, the results show a small difference in performance between the three models and the correlation coefficient is almost similar at 0.78. However, the RMSE showed slightly higher results compared to MAE.

Despite the prediction accuracy for water saturation using conventional logs at around 0.78, Figures 20.14, 20.16, and 20.18 capture the complexity of the Three Forks Formation where the laminations are in abundance with a complex pore size distribution as shown in Chapter 3. It is clear from the Figures that the program could not model perfectly the peaks. However, this study showed that the three models are valuable methods for thin beds water saturation prediction using only conventional logs. Increasing the input variables, specifically the features that characterize the lithology and the pore size distribution, would improve the prediction accuracy of the models on water saturation prediction in thin beds analysis.

References

Adams, J.A. and Weaver, C.E. (1958). Thorium-uranium ratios as indicators of sedimentary processes: example of concepts of geochemical facies. *Bell American Association of Petroleum Geologists* 42 (2): 387–430.

Adeniran, A., Elshafei, M., and Hamada, G. (2009). Functional network soft sensor for formation porosity and water saturation in oil wells. *2009 IEEE Instrumentation and Measurement Technology Conference*, Singapore, pp. 1138–1143. https://doi.org/10.1109/IMTC.2009.5168625.

Aguilera, R. (1990). A new approach for analysis of the nuclear magnetic log-resistivity log combination. *Journal of Canadian Petroleum Technology* 29 (1): 67–71.

Aguilera, R. (2002). Incorporating capillary pressure, pore aperture radii, height above free water table, and Winland r 35 values on Pickett plots. *AAPG Bulletin* 86 (4): 605–624.

Aguilera, R. (2003). Determination of matrix flow units in naturally fractured reservoirs. *Journal of Canadian Petroleum Technology* 42 (12): 54–61.

Aguilera, R. and Aguilera, M.S. (2002). The integration of capillary pressures and Pickett plots for determination of flow units and reservoir containers. *Society of Petroleum Engineers Reservoir Evaluation and Engineering* 5 (6): 465–471.

Al Duhailan, M. (2014). Petroleum-expulsion fracturing in organic-rich shales: genesis and impact on unconventional pervasive petroleum systems. Ph.D dissertation. Colorado School of Mines, 206p.

Allen, D., Crary, Freedman, S. et al. (1997). How to use borehole nuclear magnetic resonance. *Oilfield Review* 9 (3): 34–57.

Allen, D.C., Flaum, T.S., Ramakrishnan, J. et al. (2008). Trends in NMR logging. *Oilfield Review* 12 (3): 2–19.

Allen, D., Flaum, C., Ramakrishnan, T.S. et al. (2000). Trends in NMR logging. *Oilfield Review* 12 (3): 2–19.

Al-Bulushi, N., King, P.R., Blunt, M.J., and Kraaijveld, M. (2009). Development of artificial neural network models for predicting water saturation and fluid distribution. *Journal of Petroleum Science and Engineering* 68 (3–4): 197–208.

Alexeyev, A., Ostadhassan, M., Bubach, B. et al. (2017). *Integrated Reservoir Characterization of the Middle Bakken in the Blue Buttes field, Williston Basin, North Dakota*. Society of Petroleum Engineers. http://doi.org/10.2118/185664-MS.

Allwein, E.L., Schapire, R.E., and Singer, Y. (2000). Reducing multiclass to binary: a unifying approach for margin classifiers. *Journal of Machine Learning Research* 1: 113–141.

Al-Wardy, W. (2002). Measurement of the poroelastic parameters of reservoir sandstones. Ph.D. thesis. Imperial College, London.

Bateman, R.M. (1940). Open hole log analysis and formation evaluation.

Bell, J.S. and Gough, D.I. (1979). Northeast-southwest compressive stress in Alberta – evidence from oil wells. *Earth and Planetary Science Letters* 45: 475–482.

Bernabé, Y. (1991). Pore geometry and pressure dependence of the transport properties in sandstone. *Geophysics* 56: 436–446.

Bernabé, Y., Mok, U., and Evans, B. (2003). Permeability-porosity relationships in rocks subjected to various evolution processes. *Pure and Applied Geophysics* 160: 937–960.

Bernabe, Y., Brace, W.F., and Evans, B. (1982). Permeability, porosity and pore geometry of hot-pressed calcite. *Mechanics of Materials* 1982 (1): 173–183.

Berwick, B.R. (2008). Depositional environment, mineralogy, and sequence stratigraphy of the Late Devonian Sanish Member (upper Three Forks Formation), Williston Basin, North Dakota. Master's thesis. Colorado School of Mines, Golden, CO, 262p.

Bottjer, R., Sterling, R., Grau, A., and Dear, P. (2011). Stratigraphic relationships and reservoir quality at the Three Forks–Bakken unconformity, Williston basin, North Dakota. In: *The Bakken–Three Forks Petroleum System in the Williston Basin* (ed. J.W. Robinson, J.A. LeFever, and S.B. Gaswirth), 173–228. Rocky Mountains Associations of Geologists.

Boualam, A. (2019). Impact of stress on the characterization of the flow units in the complex three forks reservoir, Williston Basin. Theses and Dissertations. 2839. https://commons.und.edu/theses/2839.

Boualam, A., Rasouli, V., Dalkhaa, C., and Djezzar, S. (2020a). Advanced petrophysical analysis and water saturation prediction in three forks, Williston basin. *SPWLA 61st Annual Logging Symposium*. Virtual Online Webinar (June 2020). https://doi.org/10.30632/SPWLA-5104.

Boualam, A., Rasouli, V., Dalkhaa, C., and Djezzar, S. (2020b). Stress-dependent permeability and porosity in three forks carbonate reservoir, Williston Basin. *54th U.S. Rock Mechanics/Geomechanics Symposium*.

Bush, D.C. and Jenkins, R.E. (1970). Proper hydration of clays for rock property determinations. *Journal of Petroleum Technology* 22: 800–804.

Chen, S., Ostroff, G., and Georgi, D.T. (1998). Improving estimation of NMR log T2 cutoff value with core NMR and capillary pressure measurements. *SCA International Symposium*, The Hague (14 September 1998), pp. 14–16.

Coalson, E.B., Hartmann, D.J., and Thomas, J.B. (1985). Productive characteristics of common reservoir porosity types. *Bulletin of the South Texas Geological Society* 25 (6): 35–51.

Cristianini, C. and Shawe-Taylor, J. (2000). *An Introduction to Support Vector Machine and Other Kernel-Based Learning Methods*, 1e, 189. New York, NY: Cambridge University Press.

Cui, A., Wust, R., Nassichuk, B. et al. (2013). A nearly complete characterization of permeability to hydrocarbon gas and liquid for unconventional reservoirs: a challenge to conventional thinking. *SPE Paper 168730 Presented at the SPE Unconventional Resources Technology Conference*, Denver (12–14 August 2013).

Dennis, D. (2008). Measuring porosity and permeability from drill cuttings. *Journal of Petroleum Technology* 60 (8): https://doi.org/10.2118/0808-0051-JPT.

Dong, J.J., Hsu, J.Y., Wu, W.J. et al. (2010). Stress-dependence of the permeability and porosity of sandstone and shale from TCDP Hola-A. *International Journal of Rock Mechanics and Mining* 47: 1141–1156.

Dorfman, M.H. (1984). Discussion of reservoir description using well logs. *Journal of Petroleum Technology* 36: 2196–2197.

Doveton, J.H., Guy, W., Watney, L.W. et al. (1996). Log analysis of petrofacies and flow units with microcomputer spreadsheet software: Kansas Geological Survey, University of Kansas 66047, 10p.

Basak, D., Pal, S., and Patranabis, D.C. (2007). Support vector regression. *Neural Information Processing – Letters and Reviews* 11 (10). 21pp.

Berryman, J.G. (1992). Effective stress for transport properties of inhomogeneous porous rock. *Journal of Geophysical Research* 97 (17): 409–424.

David, C. and Darot, M. (1989). Permeability and conductivity of sandstones. *M Proc International Symposium on Rock at Great Depth*, Pau, V1 (28–31 August 1989), P203–P209. Rotterdam: A A Balkema.

Djebbar, T. and Donaldson, E.C. (2012). *Petrophysics: Theory and Practice of Measuring Reservoir Rock and Fluid Transport Properties*, 3e. Elsevier.

Drucker, H., Chris, J.C., Burges, H.J.C. et al. (1997). Support Vector Regression Machines. ARPA contract number N000l4-94-C-1 086.

Droege, L.A. (2014). Sedimentology, facies architecture and diagenesis of the middle three forks formation – North Dakota. U.S.A. Master's thesis. Colorado State University, Fort Collins, CO, 14p.

Dumonceaux, G.M. (1984). Stratigraphy and depositional environments of the Three Forks formation (upper Devonian), Williston basin, North Dakota. Master thesis.

Dunham, R.J. (1962). Classification of carbonate rocks according to depositional texture. In: *Classification of Carbonate Rocks* (ed. W.E. Ham), 108–121. AAPG Memoir.

Dunn, K.J., Bergman, D.J., and Lattoraca, G.A. (2002). *Nuclear Magnetic Resonance Petrophysical and Logging Applications (Handbook of Geophysical Exploration, Seismic Exploration, v. 32)*. Amsterdam: Pergamon an Imprint of Elsevier Science. 293p.

Ebanks, W.J. Jr. (1987). Flow unit concept – integrated approach to reservoir description for engineering projects. *AAPG Bulletin* 71 (5): 551–552.

Efron, B. and Tibshirani, R. (1994). *An Introduction to the Bootstrap*. New York: Chapman & Hall.

Egbe, S.J., Omole, O., Diedjomahor, J., and Crowe, J. (2007). Calibration of the elemental capture spectroscopy tool using the Niger Delta formation. *31st Nigeria Annual International Conference and Exhibition Held in Abuja*, Nigeria (6–8 August 2007). SPE 111910.

El-sebakhy, E.A., Asparouhov, O., Abdulraheem, A., et al. (2010). Data mining in identifying carbonate litho-facies from well logs based from extreme learning and support vector machines. *AAPG GEO, Middle East Geoscience Conference & Exhibition,* Manama, Bahrein (7–10 March 2010).

Farber, D.L., Bonner, B.P., Balooch, M. et al. (2001). Observations of water induced transition from brittle to viscoelastic behavior in nano-crystalline swelling clay. *EOS* 82: 1189.

Focke, J. W. and Munn, D. (1987). Cementation Exponents in Middle Eastern Carbonate Reservoirs. SPE Formation Evaluation. SPE, Qatar General Petroleum Corp.

Franklin, A. (2017). Deposition, Stratigraphy, Provenance, and reservoir characterization of carbonate mudstone: the Three Forks Formation, Williston Basin. Ph.D thesis. Colorado School of Mines.

Galford, J., Quirein, J., Shannon, S. et al. (2009). Field test results of a new neutron induced gamma-ray spectroscopy geochemical logging tool. SPE Annual Techinical and Exibition held in New Orleans. SPE 123992.

Ganpule, S.V., Srinivasan, K., Izykowski, T. et al. (2015). Impact of geomechanics on well completion and asset development in the Bakken Formation. *Society of Petroleum Engineers.* https://doi.org/10.2118/173329-MS.

Gerhard, L.C., Anderson, S.B., and Fischer, D.W. (1990). Petroleum geology of the Williston Basin. *American Association of Petroleum Geologists Memoir* 51: 507–559.

Ghanizadeh, A., Amann-Hildenbrand, A., and Gasparik, M. (2014). Experimental study of fluid transport processes in the matrix system of the European organic rich shales: II Posidonia Shale (lower Toarcian, Noerthern Germany). *International Journal of Coal Geology* 123: 20–33. https://doi.org/10.1016/j.coal.2013.06.009.

Gottlib Zeh, S., Briqueu, L., and Veillerette, A. (1999). Indexed Self-Organizing Map: a new calibration system for a geological interpretation of logs. *The Fifth Annual Conference of the International Association for Mathematical Geology. Proc IAMG,* Trondheim, Norway (6–11 August 1999), pp. 183–188.

Gunn, S. R., 1998. Support Vector Machine for Classification and Regression. Tech. Rep., Univ. Southampton, Southampton, U.K.

Gunter, G.W., Finneran, J.M., Hartmann, D.J., and Miller, J.D. (1997). Early determination of reservoir flow units using an integrated petrophysical method: Society of Petroleum Engineers. *SPE-38679-MS Presented at the SPE Annual Technical Conference and Exhibition*, San Antonio, TX (5–8 October 1997), pp. 373–380. https://doi.org/10.2118/38679-MS.

Hashem, S. (1997). Optimal linear combinations of neural networks. *Neural Net-works* 10 (4): 599–614.

He, J., Ling, K., Wu, X. et al. (2019). Static and dynamic elastic moduli of bakken formation. *International Petroleum Technology Conference* https://doi.org/10.2523/IPTC-19416-MS.

Heck, T.J., LeFever, R.D., Fischer, D.W., and LeFever, J.A. (2002). Overview of the petroleum geology of the North Dakota Williston Basin: North Dakota Geological Survey. https://www.dmr.nd.gov/ndgs/Resources/ (accessed June 2016).

Herron, S., Herron, M., Pirie, I. et al. (2014). Application and quality control of core data for the development and validation of elemental spectroscopy log interpretation. *Petrophysics* 55 (5): 392–414.

Hassan, M., Selo, M., and Combaz, A. (1975). Uranium distribution and geochemistry as criteria of diagenesis in carbonate rocks. *9th International Sedimentological Congress*, Nice, France, 7pp.

Hassan, M., Hossin, A., and Combaz, A. (1976). Fundamentals of the differential gamma-ray log. In terpretation technique. *SPWLA. 17th Ann. Log. Sym. Trans.*, Denver, CO (June 1976). Paper H.

Havens, J.B. and Batzle, M.L. (2011). Minimum horizontal stress in the bakken formation. *45th U.S. Rock Mechanics/Geomechanics Symposium*, San Francisco, CA (June 2011). Paper Number: ARMA-11-322.

Hizem, M., Budan, H., Devillé, B. et al. (2008). Dielectric Dispersion: A New Wireline Petrophysical Measurement. *Paper SPE-116130 presented at the SPE Annual Technical Conference and Exhibition*, Denver, CO, USA (21–24 September).

Holubnyak, Y.I., Bremer, J.M., Mibeck, B.A.F. et al. (2011). Understanding the sourcing at Bakken oil reservoirs. *Society of Petroleum Engineers* https://doi.org/10.2118/141434-MS.

Isleyen, E., Demirkan, D. C., Duzgun, H.S., and Rostami, J. (2019). Lithological classification of limestone with self-organizing maps. *53rd U.S. Rock Mechanics/Geomechanics Symposium*, New York, NY, USA (23–26 June 2019). Paper Number: ARMA-2019-1791.

Jin, H., Sonnenberg, S. A., and Sarg, J. F., 2015. Source rock potential and sequence stratigraphy of Bakken Shales in the williston basin. *Unconventional Resources Technology Conference.* http://doi.org/10.15530/URTEC-2015-2169797.

Johnson, R., Longman, M., and Ruskin, B. (2017). Petrographic and petrophysical characteristics of the upper Devonian Three Forks Formation, southern Nesson anticline, North Dakota. *The Mountain Geologist* 54 (3): 181–201.

Jones, S.C. (1988). Two-point determinations of permeability and PV vs net confining stress. *SPEFE* 3: 235.

Jones, F.O. and Owens, W.W. (1980). A laboratory study of low-permeability gas sand. *SPE* 32: 1631–1640.

Kamalyar, K., Sheikhi, Y., and Jamialahmadi, M. (2012). Using an artificial neural network for predicting water saturation in an Iranian oil reservoir. *Petroleum Science and Technology* 30 (1): 35–45.

Kausik, R., Fellah, K., Feng, L. et al. (2016). High and low field NMR relaxometry and diffusometry of the Bakken petroleum system: SPWLA-2016-SSS, Society of Petrophysicists and Well-Log Analysts. *SPWLA 57th Annual Logging Symposium* (25–29 June), 7p. https://www.onepetro.org/conference-paper/SPWLA-2016-SSS (accessed 12 April 2017).

Kerans, C., Lucia, F.J., and Senger, R.K. (1994). Integrated characterization of carbonate ramp reservoirs using Permian San Andres Formation outcrop analogs. *AAPG Bulletin* 78: 181–216.

Kenyon, W., Day, P., Straley, C., and Willemsen, J. (1988). A three-part study of NMR longitudinal relaxation properties of water-saturated sandstones. *SPE Formation Evaluation* 3 (3): 622–636.

Kingma, D.P. and Lei Ba, J. (2015). *Adam: A Method for Stochastic Optimization.* ICLR.

Klenner, R., Braunberger, J., Sorensen, J. et al. (2014). A formation evaluation of the middle Bakken member using multimineral petrophysical analysis approach. *Unconventional Resources Technology Conference, URTeC:1922735. SPE/AAPG/SEG Unconventional Resources Technology Conference*, Denver, CO, USA (25–27 August 2014).

Kwon, O., Kronenberg, A.K., Gangi, A.F., and Johnson, B. (2001). Permeability of Wilcox shale and its effective pressure law. *Journal of Geophysical Research* 106 (19): 339–353.

LeFever, J.A., Le Fever, R.D., and Nordeng, S.H. (2011). Revised nomenclature for the Bakken Formation (Mississippian-Devonian), North Dakota. The Bakken-Three Forks petroleum system in the Williston basin. Chapter 1, pp. 11–26.

LeFever, J.A. and Nordeng, S.H. (2009). The three forks formation–North dakota to sinclair field. Manitoba. *North Dakota Geological Survey Geological Investigations* 76: 1.

LeFever, R.D., LeFever, J.A., and Nordeng, S.H., 2008. Correlation Cross-Sections for the Three Forks Formation, North Dakota. North Dakota Geological Survey Geologic Investigations No. 65. 1p.

LeFever, J. A., LeFever, R. D., and Nordeng, S. H., 2014. Reservoirs of the Bakken Petroleum System: A Core-based Perspective: North Dakota Geological Survey Geologic Investigations, no. 171,. https://www.dmr.nd.gov/ndgs/Publication_List/gi.asp (accessed June 2016).

Leon, M., Lafournere, A.N., Bourge, J.P. et al. (2015). Rock typing mapping methodology based on indexed and probabilistic self-organized map in shushufindi field. *Society of Petroleum Engineers* https://doi.org/10.2118/177086-MS.

Loucks, R.G., Reed, R.M., Ruppel, S.M., and Jarvie, D.M. (2009). Morphology, genesis, and distribution of nanometer-scale pores in siliceous mudstones of the Mississippian Barnett Shale. *Journal of Sedimentary Research* 79: 848–861.

Loucks, R.G., Ruppel, S.C., Zhang, T., and Peng, S. (2017). Origin and characterization of eagle ford pore networks in the south Texas upper cretaceous shelf. *AAPG Bulletin* 101 (3): 387–418.

Loucks, R.G., Reed, R.M., Ruppel, S.C., and Hammes, U. (2012). Spectrum of pore types and networks in mud-rocks and a descriptive classification for matrix-related mud-rock pores. *AAPG Bulletin* 96: 1071–1098.

Lucia, F.J. (1995). Rock fabric and petrophysical classification of carbonate pore space for reservoir characterization. *AAPG Bulletin* 79 (9): 1275–1300.

Lucia, F.J., Kerans, C., and Senger, R.K. (1992). Defining flow units in dolomitized carbonate-ramp reservoirs. *Proceedings, Society of Petroleum Engineers*, Washington, DC (4–7 October 1992), pp. 399–406. Paper SPE 24702.

Lucia, F.J. (1983). Petrophysical parameters estimated from visual description of carbonate rocks: a field classification of carbonate pore space. *Journal of Petroleum Technology* 35 (3): 629–637.

Martin, A.J., Solomon, S.T., and Hartmann, D.T. (1997). Characterization of petrophysical flow units in carbonate reservoirs. *AAPG Bulletin* 81: 734–759.

Mardi, M., Nurozi, H., and Edalatkhah, S. (2012). A water saturation prediction using artificial neural networks and an investigation on cementation factors and saturation exponent variations in an Iranian oil well. *Pet.Sci.Technol.* 30 (4): 425–434.

Mayergoyz, I.D. (1986). Mathematical models of hysteresis and their application. *IEEE Transactions on Magnetics* 22: https://doi.org/10.1109/TMAG.1986.1064347.

Merkel, R., Machesney, J., and Tompkins, K. (2018). Calculated determination of variable wettability in the middle Bakken and Three Forks, Williston basin, USA. *SPWLA 59th Annual Logging Symposium*, London, UK (June 2018).

McKenon, D., Cao Minh, C., Freedman, R. et al. (1999). An improved NMR tool design for faster logging, paper CC, transactions. *SPWLA 40th Annual Logging Symposium*, Oslo, Norway (30 May–3 June 1999).

Millard, M. and Brinkerhoff, R. (2016). The integration of geochemical, stratigraphic, and production data to improve geological models in the bakken-three forks petroleum system, Williston Basin, North Dakota. In: *Hydrocarbon Source Rocks in Unconventional Plays* (ed. M.P. Dolan, D.K. Higley, and P.G. Lillis), 190–211. Denver: Rocky Mountain Region: RMAG.

Milijkovic, D. (2017). Brief review of self-organizing maps. *40th International Convention on Information and Communication Technology, Electronics and Microelectronics (MIPRO)*, Opatija, Croatia (22–26 May 2017), 6pp.

Mitchell, R. (2013). Sedimentology and reservoir properties of the three forks dolomite, bakken petroleum system, Williston Basin, U.S.A: AAPG. Search and Discovery Articles, no. 120079.
Mitra, P.P., Sen, P.N., and Schwartz, L.M. (1993). Short-time behavior of the diffusion coefficient as a geometrical probe of porous media. *Physical Review B* 47: 8565–8574.
Mohaghegh, S., Arefi, R., Ameri, S. et al. (1996). Petroleum reservoir characterization with the aid of artificial neural networks. *Journal of petroleum Science and Enigineering* 16 (1996): 263–274.
Mollajan, A., Memarian, H., and Jalali, M.R., (2013). Prediction of reservoir water saturation using support vector regression in an Iranian carbonate reservoir. *47th US Rock Mechanics/Geomechanics Symposium*, San Francisco, CA, USA (23–26 June 2013). ARMA 13-311.
Moss, A.K. and Jing, X.D. (2001). An investigation into the effect of clay type, volume and distribution on NMR measurements in sandstones. SCA 2001-29, Society of Core Analysts Symposium, Edinburgh. http://www.jgmaas.com/SCA/2001/SCA2001-29.pdf (accessed 13 April 2017).
Newman, J., Edman, J., Howe, J., and LeFever, J. (2013). The Bakken at parshall field: inferences from new data regarding hydrocarbon generation and migration. *Unconventional Resources Technology Conference*. http://doi.org/10.15530/URTEC-1578764-MS.
Nordeng SH and LeFever, J.A. 2009. Three Forks Formation Log to Core Correlation. North Dakota Geological Survey Geologic Investigations No. 75,. 1 p.
Passey, Q.R., Dahlberg, K.E., Sullivan, K.B. et al. (2006). Chapter 7: Characterizing thinly bedded reservoirs with core data. *A Systematic Approach to Evaluate Hydrocarbons in Thinly Bedded Reservoirs*. AAPG Archie Series, No. 1, pp. 73–88.
Peterson, K.J., 2017. Pore-size distributions from nuclear magnetic resonance and corresponding hydrocarbon saturations in the Devonian Three Forks Formation, Williston Basin, North Dakota (abstract). *RMS-AAPG Regional Meeting, Billings*.
Peters, K.E., Walters, C., and Moldowan, J.M. (2005). *The Biomarker Guide*, vol. 2. Cambridge, UK: Cambridge University Press. 490p.
Petty, D. (2014). Mineralogy and petrology controls on hydrocarbon saturation in the Three Forks reservoir, North Dakota. AAPG Search and Discovery Article-10623.
Pittman, E.D. (1992). Relationship of porosity and permeability to various parameters derived from mercury injection-capillary pressure curves for sandstone. *AAPG Bulletin* 76: 191–198.
Prammer, M.G., Drack, E.D., Bouton, J.C. et al., 1996. Measurements of clay-bound water and total porosity by magnetic resonance logging: SPE-36522-MS. *Presented at the SPE Annual Technical Conference and Exhibition*, Denver, p. 311–320. http://dx.doi.org/10.2118/36522 (accessed 14 April 2017).
Price, L. C., 2000. Origins and characteristics of the basin-centered continuous reservoir unconventional oil-resource base of the Bakken source system, Williston Basin, paper. Available at: http://www.unddeerc.org/Price/.
Qi, L. and Carr, T.R. (2006). Neural network prediction of carbonate lithofacies from well logs, big bow and sand arroyo creek fields, South West Kansas. *Computers and Geosciences* 32 (7): 947–964.
Ramakrishna, S., Balliet, B., Miller, D, and Sarvotham, S. (2010). Formation evaluation in the Bakken Complex using laboratory core data and advanced logging technologies. *SPWLA 51st Annual Logging Symposium*.
Rezaee, M., Ilkhchi, A., and Alizadeh, P. (2008). Intelligent approaches for the synthesis of petrophysical logs. *Journal of Geophysics and Engineering* 5: 12–26.
Rider, M. and Kennedy, M. (2011). *The Geological Interpretation of Well Logs*, 3e.
Roland, L. and Olivier, F. (2007). Advances in measuring porosity and permeability from drill cuttings. *SPE/EAGE Reservoir Characterization and Simulation Conference* (28–3 October), Abu Dhabi, UAE. https://doi.org/https://doi.org/10.2118/111286-MS.

Rosepiler, M.J. (1981). Calculation and significance of water saturation in low porosity shaly gas sands. *SPWLA 22nd Annual Symposium*, Mexico City, Mexico (June 1981).

Saffarrzadeh, S. and Shadizadeh, S.R. (2012). Reservoir rock permeability prediction using support vector regression in an Iranian oil field. *Journal of Geophysics and Engineering* 9 (3): 336–344.

Sandberg, C.A. and Hammond, C.R. (1958). Devonian system in Williston basin and central Montana. *AAPG Bulletin* 42: 2293–2334.

Saneifar, M., Skalinski, M., Theologou, P. et al. (2015). Integrated petrophysical rock classification in the McElroy field, West Texas, USA. *Petrophysics* 56 (5): 493–510.

Schmoker, J.W. and Halley, R.B. (1982). Carbonate porosity versus depth: a predictable relation for south Florida. *AAPG Bulletin* 1982 (66): 2561–2570.

Serra, O. (1984). *Fundamentals of Well-Log Interpretation. Developments in petroleum science 15A*. Elsevier.

Simpson, G., Hohman, J., Pirie I., and Horkowitz, J. (2015). Using advanced logging measurements to develop a robust petrophysical model for the Bakken petroleum system. *SPWLA 56th Annual Logging Symposium,* Long Beach, CA, USA (18–22 July 2015).

Sloss, L.L. (1984). Comparative anatomy of cratonic unconformities. In: *Interregional Unconformities and Hydrocarbon Accumulation*, vol. 36 (ed. J.S. Schlee), 7–36. American Association of Petroleum Geologist Memoir.

Smola, A.J. and Schölkopf, B. (1998). A Tutorial on Support Vector Regression, NeuroCOLT. Technical Report NC-TR-98-030, Royal Holloway College, University of London, UK.

Soeder, D.L. and Doherty, M.G. (1983) The Effects of Laboratory Drying Techniques on the Permeability of Tight Sandstone Core. *SPE 11622 Presented at the SPEIDOE*. Denver (March 13–16).

Soeder, D. (1986). *Laboratory Drying Procedures and the Permeability of Tight Sandstone Core*. SPE, Inst. of Gas Technology.

Sonnenberg, S.A. (2017). Sequence stratigraphy of the Bakken and Three Forks Formations, Williston basin, USA. *AAPG Rocky Mountain Section Annual Meeting*, Billings, Montana.

Sonnenberg, S.A., Gantyno, A., and Sarg, R. (2011). Petroleum potential of the upper Three Forks Formation, Williston basin, USA. *AAPG Annual Convention and Exhibition*, Houston, TX.

Sonnenberg, S.A. (2015). Keys to production, Three Forks, Williston Basin. SPE-178510-MS/ URTeC:2148989.

Sørland, G.H., Djurhuus, K., Widerøe, H.C. et al. (2007). Absolute pore size distributions from NMR. *Diffusion Fundamentals* 5: 4.1–4.15.

Gaswirth, S.B. and Marra, K.R. (2015). *U.S. Geological Survey 2013 assessment of Undiscovered Resources in the Bakken and Three Forks Formations of the U.S.* Williston Basin Province.

Swanson, V.E. (1960). Oil yield and uranium content of black shales. *Geological Survey Professional paper*. Serie number 356-A, pp. 1–44. https://doi.org/10.3133/pp356A.

Teklu, T., Zhou, Z., Li, X., and Abass, H. (2016a). Cyclic permeability and porosity hysteresis in mudrocks-experimental study. *50th US Rock Mechanics/Geomechanics Symposium* 1: 1–12.

Teklu, T., Zhou, Z., Li, X., and Abass, H., 2016b. Experimental investigation on permeability and porosity hysteresis in low-permeability formations. *The SPE Low Perm Symposium Held in Denver*, CO, USA, https://doi.org/https://doi.org/10.2118/180226-MS.

Terzaghi, K. (1943). *Theoretical Soil Mechanics*. New York: Wiley. http://dx.doi.org/10.1002/9780470172766.

Thrasher, L. (1987). Macrofossils and stratigraphic subdivisions of the Bakken Formation (Devonian-Mississippian). In: *Fifth International Williston Basin Symposium* (ed. W. Fischer), 53–67.

Vapnik, V.N. and Lerner, A. (1963). Pattern recognition using generalized portrait method. *Automation and Remote Control* 24: 774–780.

Vapnik, V.N. and Chervonenkis, A. (1964). A note on one class of perceptron. *Automation and Remote Control* 25: 8.

Vapnik, V.N. (1995). *The Nature of Statistical Learning Theory*. Springer-Verlag New York, Inc. 99-39803.

Walstrom, J.E., Mueller, T.D., and McFarlane, R.C. (1967). Evaluating uncertainty in engineering calculations. *Journal of Petroleum Technology* 19: 1595–1599.

Walls, J. and Nur, A. (1979). Pore pressure and confining pressure dependence of permeability in sandstone. *7th Form. Eval. Symp. Can. Well Logging Soc.*, Calgary.

Wang, W., Ren, X., Zhang, Y., and Li, M. (2018a). Deep learning based lithology classification using dual-frequency Pol-SAR data. *Applied Sciences* 8: 1513.

Wang, X., Yang, S., Zhao, Y., and Wang, Y. (2018b). Lithology identification using an optimized KNN clustering method based on entropy-weighed cosine distance in Mesozoic strata of gaoqing field, jiyang depression. *Journal of Petroleum Science and Engineering* 166: 157–174.

White, F.M. (1974). *Viscous Fluid Flow*. New York: McGraw-Hill.

Xu, C. and Torres-Verdin, C. (2013). Quantifying fluid distribution and phase connectivity with a simple 3D cubic pore network model constrained by NMR and MICP data. *Computers and Geosciences* 61: 94–103.

Xu, J. and Sonnenberg, S.A. (2017). An SEM study of porosity in the organic-rich lower Bakken lower Member and Pronghorn member, Bakken Formation, Williston Basin. *Unconventional Resources Technology Conference*, Austin, TX, USA. http://doi.org/10.15530/URTEC-2017-2697215.

Yu, L., Porwal, A., Holden, E., and Dentith, M.C. (2012). Towards automatic lithological classification from remote sensing data using support vector machines. *Computers & Geosciences* 45: 229–239.

Zoback, M.D. and Byerlee, J.D. (1975). Permeability and effective stress. *American Association of Petroleum Geologists Bulletin* 59: 154–158.

Zoback, M.D. (1975). High pressure deformation and fluid flow in sandstone, granite, and granular materials. Ph.D. thesis. Stanford Univ., Stanford, Calif.

Zoback, M.D. and Byerlee, J.D. (1976). Effect of high-pressure deformation on permeability of Ottawa sand. *American Association of Petroleum Geologists Bulletin* 1976 (60): 1531–1542.

Appendix

Hysteresis Testing and Mineralogy

Table 1 Weight percent mineralogy of core samples, Well Charlie Sorenson 17-8-3TFH.

Phase name	Formula	Figure of merit	Content (%)
Sample 1A			
Dolomite	Ca Mg $(CO_3)_2$	0.520	65
Quartz	Si O_2	0.777	10
Pyrite	Fe S_2	2.918	0.22
Orthoclase	K (Al Si_3 O_8)	1.490	8
Calcite	$CaCO_3$	3.397	0.9
Chlorite	Mg5.0 Al0.75 Cr0.25 Al Si_3 O10 $(OH)_8$	2.142	0
Illite	K (Al_4 Si_2 O_9 $(OH)_3$)	1.763	9
Apatite-(CaCl) (OH-bearing)	Ca9.7 (P O_4)6 Cl1.15 OH	2.251	2.5
Hydrohalite	NaCl $(H_2O)_2$	1.905	3
Zeolite	Si64 O128	1.458	2.9
Sample 2A			
Phase name	**Formula**	**Figure of merit**	**Content (%)**
Dolomite	Ca Mg $(CO_3)_2$	0.571	45
Quartz	Si O_2	0.563	10
Pyrite	Fe S_2	3.318	2
Orthoclase	K (Al Si_3 O_8)	1.696	10
Calcite	$CaCO_3$	3.294	6
Chlorite	Mg5.0 Al0.75 Cr0.25 Al Si_3 O_{10} $(OH)_8$	1.935	0
Illite	K (Al_4 Si_2 O_9 $(OH)_3$)	1.388	14
Apatite	Ca9.7 (P O_4)6 Cl1.15 OH	2.041	12
Halite	Na Cl	2.998	0.5
Zeolite	Si O_2	1.190	0.1

(Continued)

Unconventional Hydrocarbon Resources: Prediction and Modeling Using Artificial Intelligence Approaches,
First Edition. Edited by Sid-Ali Ouadfeul.

Table 1 (Continued)

Sample 3A			
Phase name	**Formula**	**Figure of merit**	**Content (%)**
Dolomite	$Ca\ Mg\ (CO_3)_2$	0.395	68
Quartz	$Si\ O_2$	0.404	12.5
Pyrite	$Fe\ S_2$	0.943	0.61
Orthoclase	$K\ (Al\ Si_3\ O_8)$	1.446	4.2
Chlorite	$Mg5.0\ Al0.75\ Cr0.25\ Al\ Si_3\ O_{10}\ (OH)_8$	1.854	0
Illite	$K\ (Al_4\ Si_2\ O_9\ (OH)_3)$	1.702	7.5
Apatite	$Ca9.7\ (P\ O_4)6\ Cl1.15\ OH$	1.982	4
Halite	Na Cl	1.702	2.4
Calcite	$Ca\ CO_3$	3.001	0.56

Sample 4A			
Phase name	**Formula**	**Figure of merit**	**Content (%)**
Dolomite	$Ca\ Mg\ (CO_3)_2$	0.387	64
Quartz	$Si\ O_2$	0.544	10.2
Orthoclase	$K\ (Al\ Si_3\ O_8)$	3.598	2.8
Pyrite	$Fe\ S_2$	1.568	0.17
Calcite	$Ca\ CO_3$	2.977	1.5
Chlorite	$Mg5.0\ Al0.75\ Cr0.25\ Al\ Si_3\ O_{10}\ (OH)_8$	2.041	0.9
Illite	$K\ (Al_4\ Si_2\ O_9\ (OH)_3)$	1.643	9
Apatite	$Ca9.7\ (P\ O_4)6\ Cl1.15\ OH$	2.227	0
Halite	$Na\ Cl\ (H_2\ O)_2$	1.801	0.2
Zeolite	$Si\ O_2$	1.806	1.2
Magnetite		1.732	10

Sample 5A			
Phase name	**Formula**	**Figure of merit**	**Content (%)**
Dolomite	$Ca\ Mg\ (CO_3)_2$	0.654	59
Quartz	$Si\ O_2$	1.024	11.1
Orthoclase	$K\ (Al\ Si_3\ O_8)$	1.953	7.4
Pyrite	$Fe\ S_2$	3.350	1.56
Calcite	$Ca\ CO_3$	2.944	1
Chlorite	$Mg5.0\ Al0.75\ Cr0.25\ Al\ Si_3\ O_{10}\ (OH)_8$	2.155	0.9
Illite	$K\ (Al_4\ Si_2\ O_9\ (OH)_3)$	2.134	13.7
Apatite	$Ca9.7\ (P\ O_4)6\ Cl1.15\ OH$	2.774	2
Halite	Na Cl	2.990	5
Zeolite	$Si\ O_2$	1.882	0.17

Table 2 Hysteresis testing, Well Charlie Sorenson 17-8-3TFH.

University of North Dakota

Hysteresis testing

CPMS ROTARY CORE ANALYSIS

Sample number	Net confining stress (psig)	Dry weight (gm)	Length (cm)	Diameter (cm)	Bulk volume (cm^3)	Grain volume (cm^3)	Grain density (g/cm^3)	Pore volume (cm^3)	Porosity (%)	Permeability Kair (md)
1A	1000	154.144	4.996	3.955	61.377	55.371	2.784	6.170	10.05	0.036
1A	2500	154.144	4.996	3.955	61.377	55.371	2.784	5.858	9.54	0.028
1A	4000	154.144	4.996	3.955	61.377	55.371	2.784	5.795	9.44	0.026
1A	5000	154.144	4.996	3.955	61.377	55.371	2.784	4.946	8.06	0.022
1A	1000	154.144	4.996	3.955	61.377	55.371	2.784	5.974	9.73	0.032
2A	1000	163.405	5.114	3.957	62.89	61.234	2.797	4.490	7.14	0.180[a]
2A	2500	163.405	5.114	3.957	62.89	61.234	2.797	4.219	6.71	0.105[a]
2A	4000	163.405	5.114	3.957	62.89	61.234	2.797	4.206	6.69	0.066[a]
2A	5000	163.405	5.114	3.957	62.89	61.234	2.797	4.165	6.62	0.048[a]
2A	1000	163.405	5.114	3.957	62.89	61.234	2.797	4.280	6.81	0.122[a]
3A	1000	170.943	5.078	3.962	62.605	61.234	2.792	1.408	2.25	0.008
3A	2500	170.943	5.078	3.962	62.605	61.234	2.792	1.286	2.05	0.007
3A	4000	170.943	5.078	3.962	62.605	61.234	2.792	1.227	1.96	0.006
3A	5000	170.943	5.078	3.962	62.605	61.234	2.792	1.089	1.74	0.005
3A	1000	170.943	5.078	3.962	62.605	61.234	2.792	1.241	1.98	0.006
5A	1000	165.782	5.076	3.959	62.486	59.455	2.788	3.031	4.85	0.014
5A	2500	165.782	5.076	3.959	62.486	59.455	2.788	2.922	4.68	0.012
5A	4000	165.782	5.076	3.959	62.486	59.455	2.788	2.865	4.58	0.011
5A	5000	165.782	5.076	3.959	62.486	59.455	2.788	2.827	4.52	0.009
5A	1000	165.782	5.076	3.959	62.486	59.455	2.788	2.899	4.64	0.013

[a] Fracture/s noted, permeabilities and/or porosities are possibly affected, yielding higher than actual values.

Table 3 Hysteresis testing, Well Charlie Sorenson 17-8-3TFH.

University of North Dakota											
										Permeability	
Sample number	Net confining stress (psig)	Dry weight (gm)	Length (cm)	Diameter (cm)	Bulk volume (cm^3)	Grain volume (cm^3)	Grain density (g/cm^3)	Pore volume (cm^3)	Porosity (%)	Kair (md)	Kinf (md)
1A	1000	54.93	4.448	2.516	22.114	19.720	2.785	2.391	10.813	0.02846	0.01268
1A	2500	54.93	4.448	2.516	22.114	19.720	2.785	2.356	10.671	0.02286	0.00708
1A	4000	54.93	4.448	2.516	22.114	19.720	2.785	2.343	10.619	0.01656	0.00475
1A	5000	54.93	4.448	2.516	22.114	19.720	2.785	2.339	10.602	0.01065	0.00270
1A	6000	54.93	4.448	2.516	22.114	19.720	2.785	2.095	9.603	0.01028	0.00259
1A	5000	54.93	4.448	2.516	22.114	19.720	2.785	2.101	9.630	0.01060	0.00269
1A	4000	54.93	4.448	2.516	22.114	19.720	2.785	2.103	9.636	0.01148	0.00292
1A	2500	54.93	4.448	2.516	22.114	19.720	2.785	2.125	9.726	0.01206	0.00322
1A	1000	54.93	4.448	2.516	22.114	19.720	2.785	2.174	9.931	0.02169	0.00669
2A	1000	62.052	4.809	2.517	23.928	22.160	2.800	1.743	7.292	0.02073	0.00626
2A	2500	62.052	4.809	2.517	23.928	22.160	2.800	1.679	7.044	0.01088	0.00362
2A	4000	62.052	4.809	2.517	23.928	22.160	2.800	1.635	6.870	0.00927	0.00224
2A	5000	62.052	4.809	2.517	23.928	22.160	2.800	1.612	6.779	0.00768	0.00178
2A	6000	62.052	4.809	2.517	23.928	22.160	2.800	1.571	6.621	0.00609	0.00300
2A	5000	62.052	4.809	2.517	23.928	22.160	2.800	1.577	6.642	0.00717	0.00162
2A	4000	62.052	4.809	2.517	23.928	22.160	2.800	1.593	6.706	0.00758	0.00175
2A	2500	62.052	4.809	2.517	23.928	22.160	2.800	1.627	6.840	0.01178	0.00312
2A	1000	62.052	4.809	2.517	23.928	22.160	2.800	1.670	7.007	0.03086	0.01782
5A	1000	61.861	4.728	2.519	23.563	22.234	2.782	1.191	5.086	0.00653	0.00143
5A	2500	61.861	4.728	2.519	23.563	22.234	2.782	1.137	4.863	0.00464	0.00094

5A	4000	61.861	4.728	2.519	23.563	22.234	2.782	1.092	4.682	0.00415	0.00081
5A	5000	61.861	4.728	2.519	23.563	22.234	2.782	1.071	4.597	0.00394	0.00076
5A	6000	61.861	4.728	2.519	23.563	22.234	2.782	1.050	4.510	0.00374	0.00071
5A	5000	61.861	4.728	2.519	23.563	22.234	2.782	1.051	4.514	0.00375	0.00072
5A	4000	61.861	4.728	2.519	23.563	22.234	2.782	1.056	4.535	0.00435	0.00086
5A	2500	61.861	4.728	2.519	23.563	22.234	2.782	1.100	4.715	0.00506	0.00106
5A	1000	61.861	4.728	2.519	23.563	22.234	2.782	1.203	5.132	0.01862	0.00546

[a] Fracture/s noted, permeabilities and/or porosities are possibly affected, yielding higher than actual values.

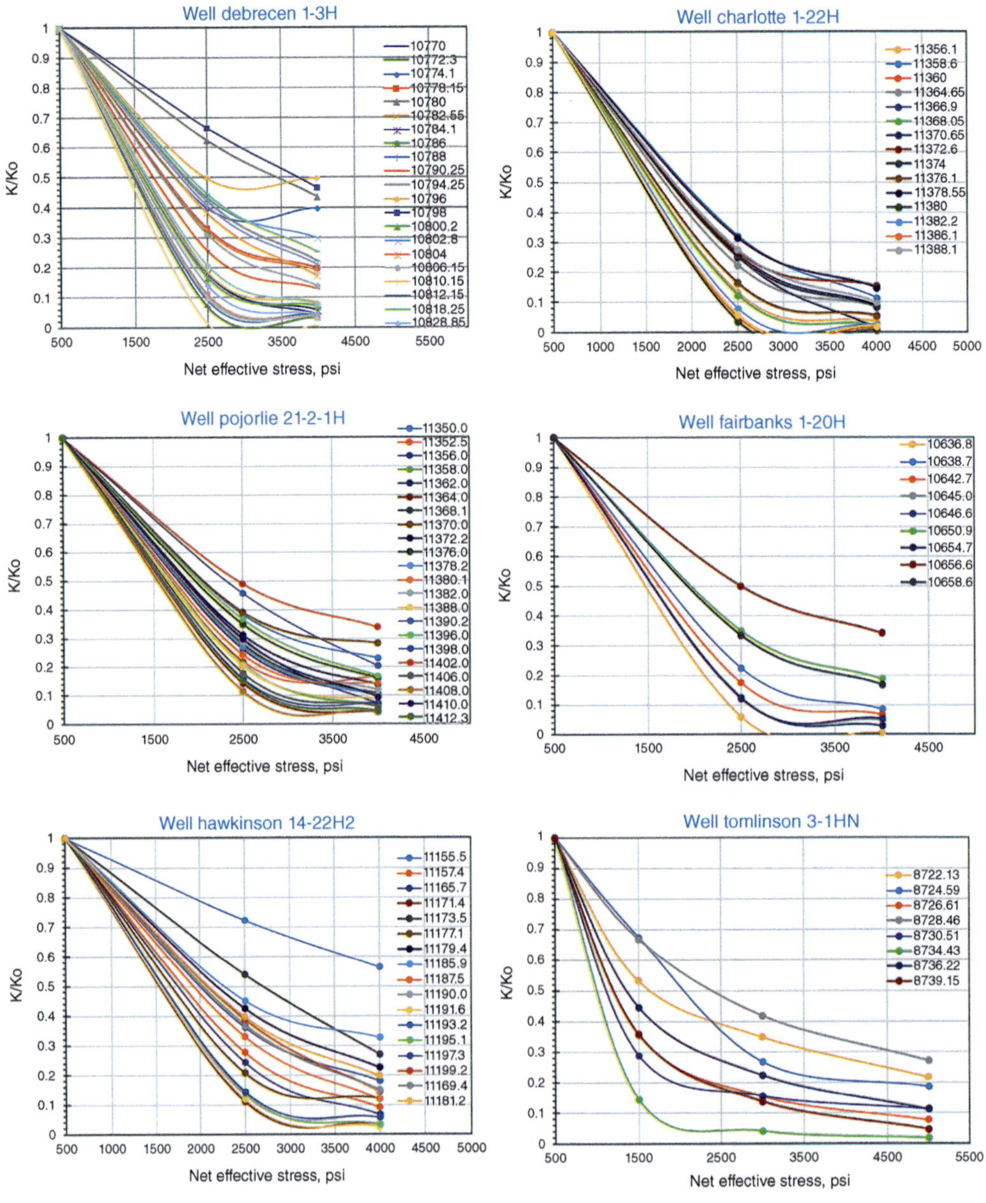

Figure 1 Normalized permeability (K/K_o) of upper Three Forks core samples during stress loading.

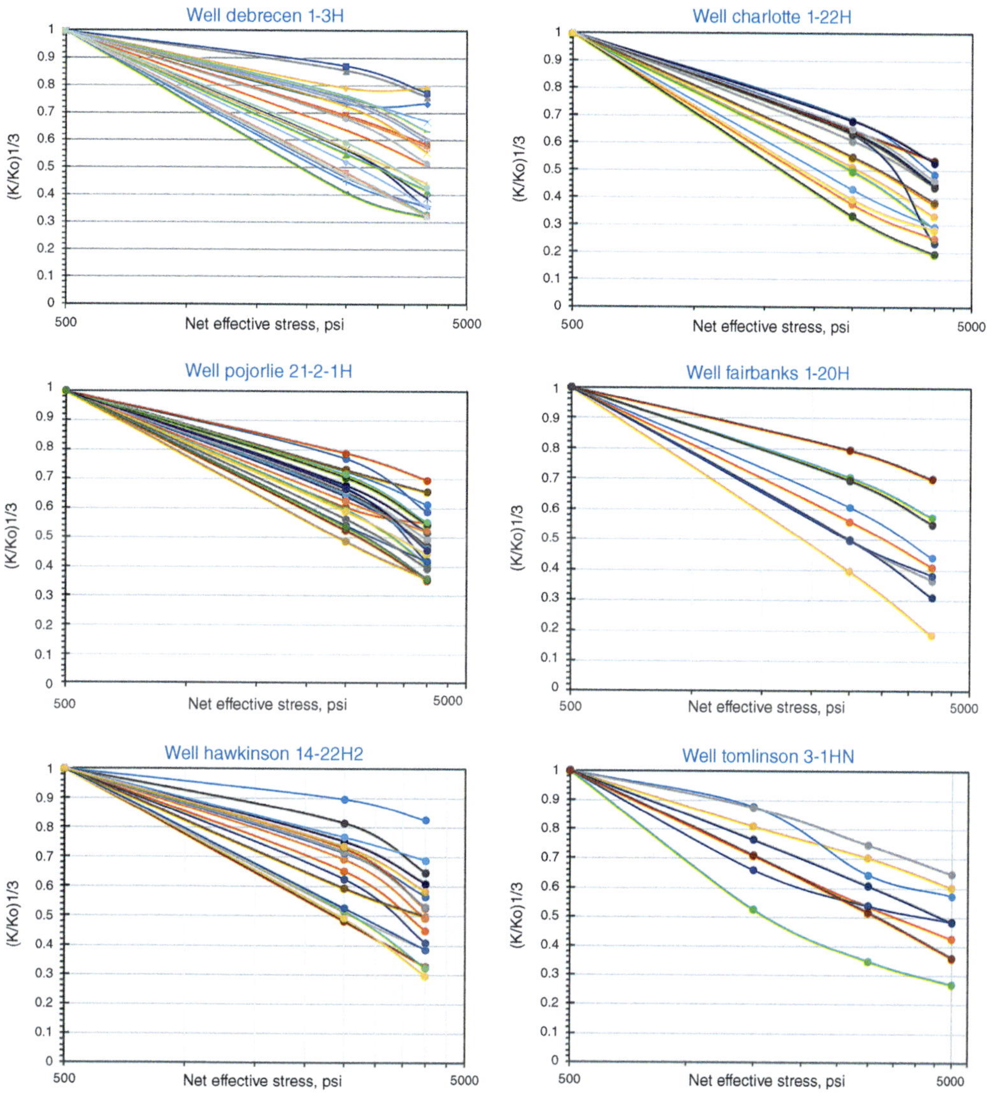

Figure 2 Cubic root of normalized permeability in linear scale versus net effective stress in logarithmic scale for upper Three Forks showing a nonlinear reduction in permeability with increasing net stress. Only 3% of the total samples show a linear reduction in permeability with increasing net stress during cyclic loading (500–4000 psi).

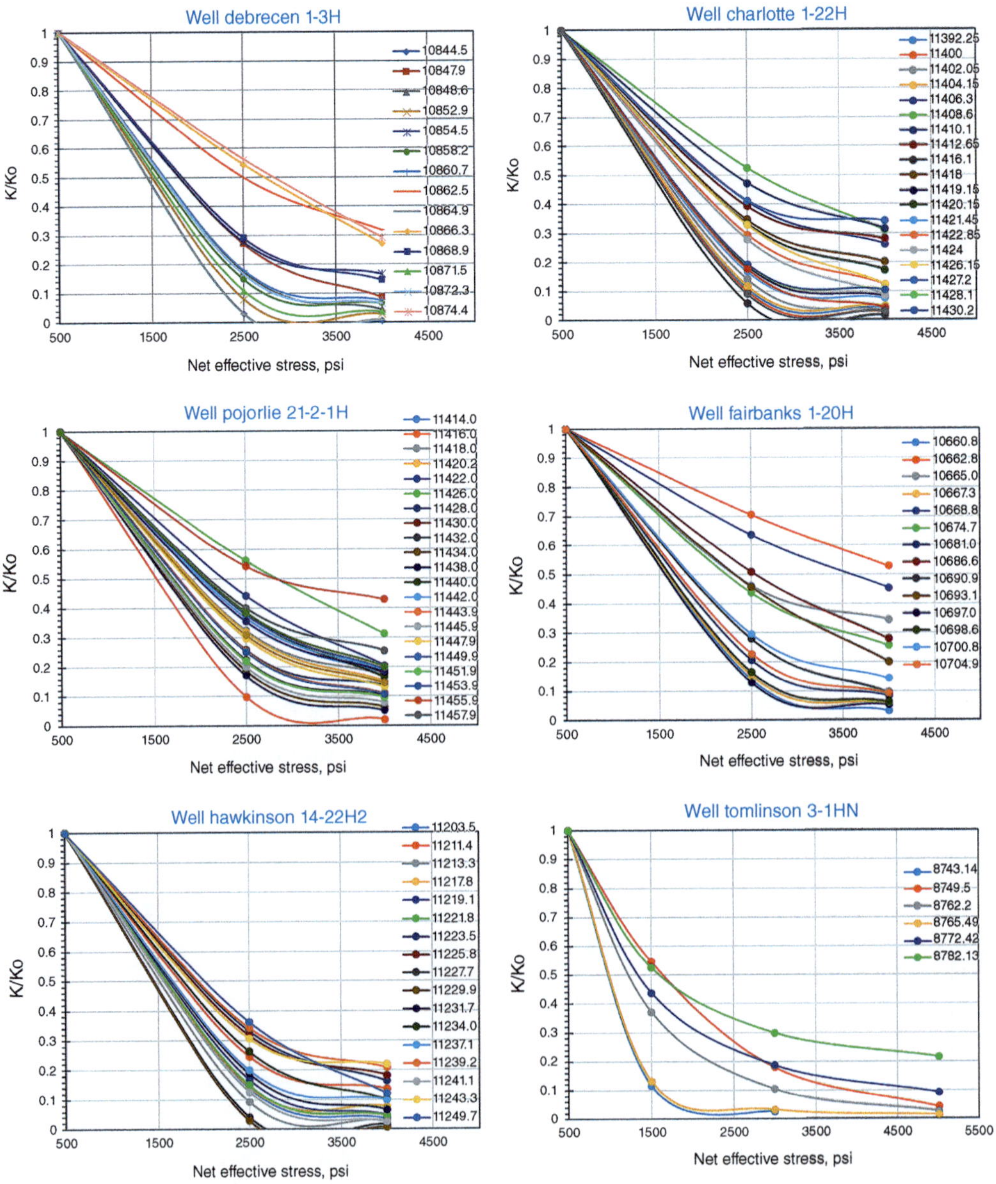

Figure 3 Normalized permeability (K/K_o) of middle Three Forks core samples during stress loading.

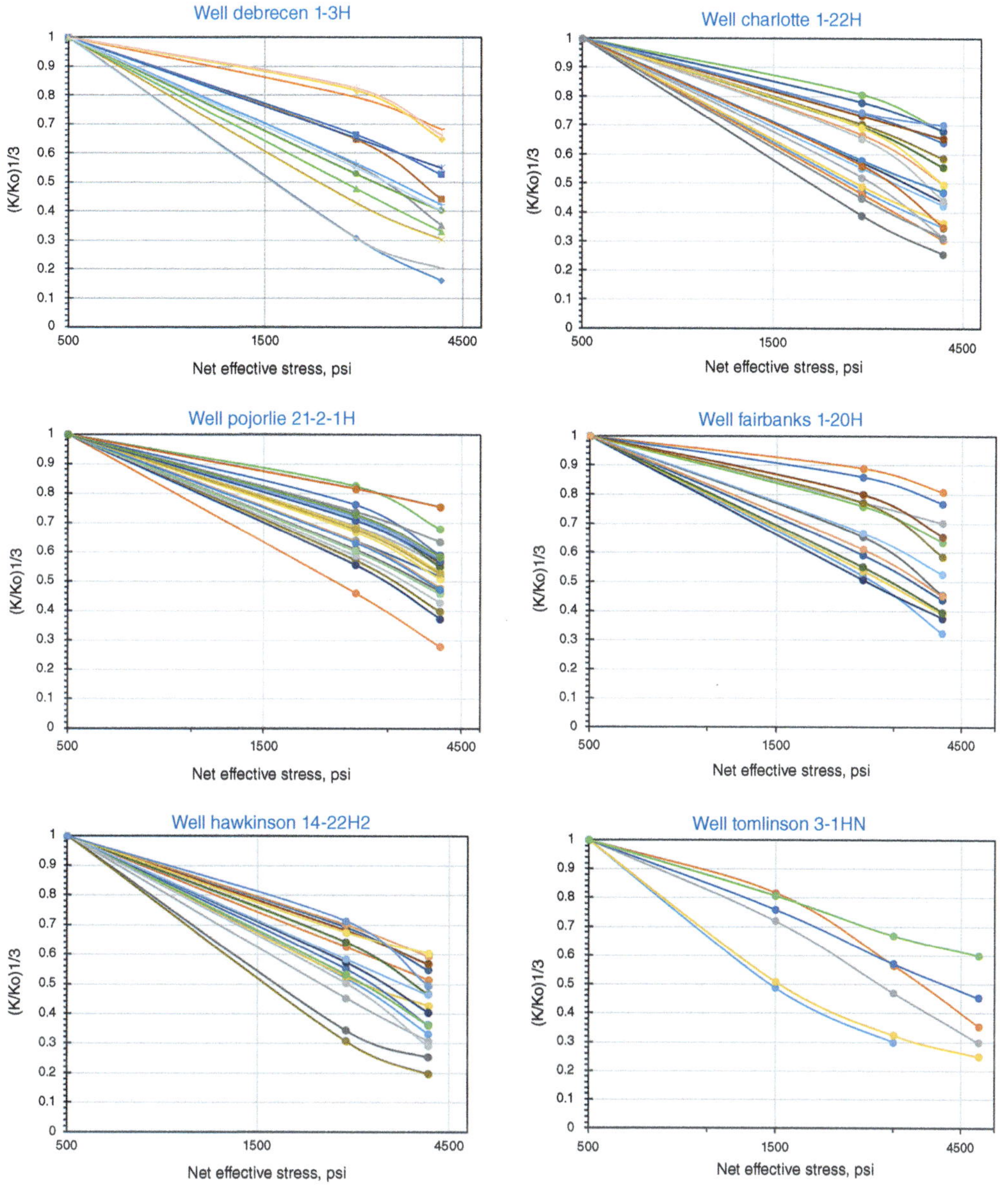

Figure 4 Cubic root of normalized permeability in linear scale versus net effective stress in logarithmic scale for middle Three Forks showing a nonlinear reduction in permeability with increasing net stress. Only 2% of the total samples show a linear reduction in permeability with increasing net stress during cyclic loading (500–4000 psi).

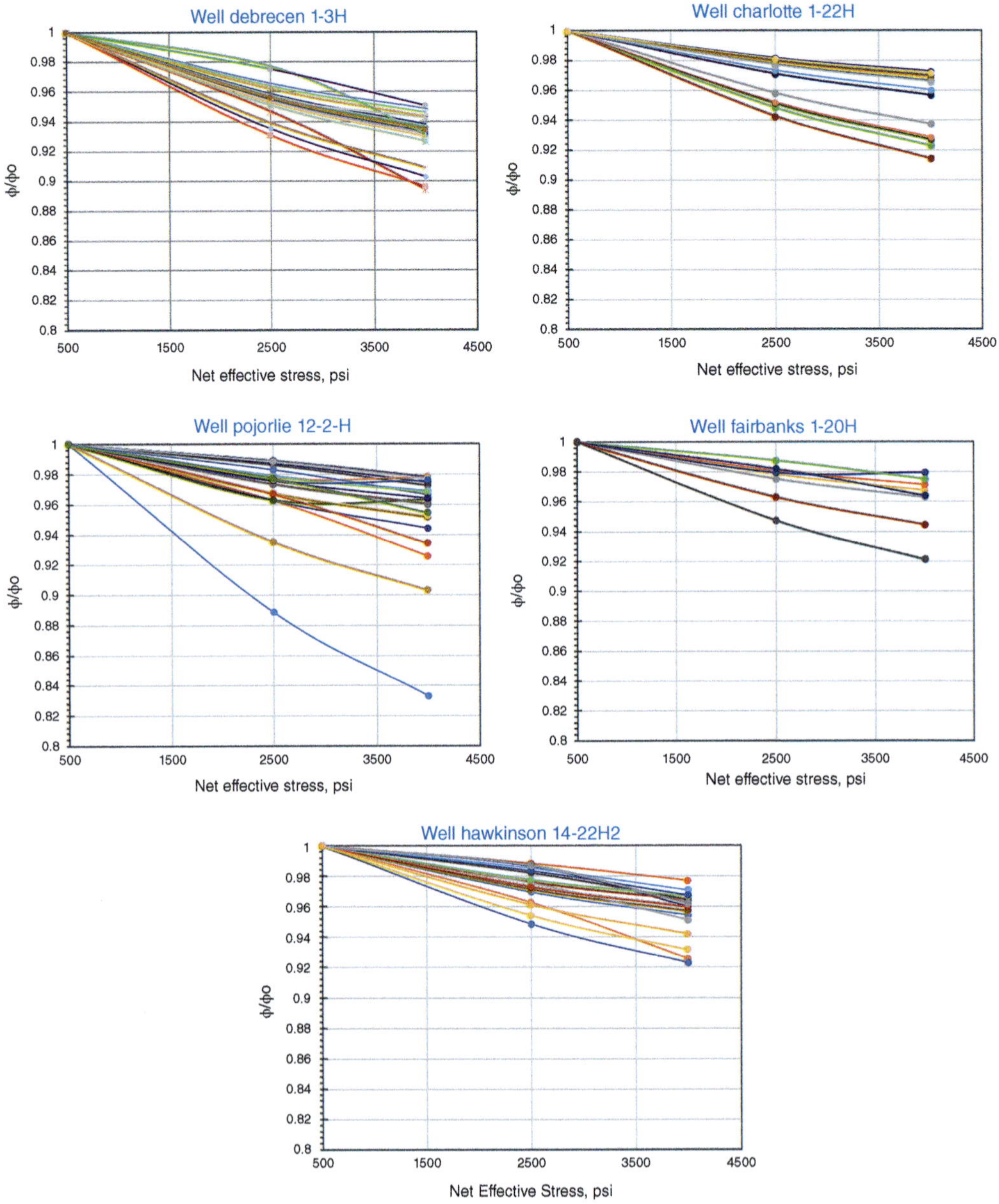

Figure 5 Normalized porosity (ϕ/ϕ_o) of upper Three Forks core samples during stress loading.

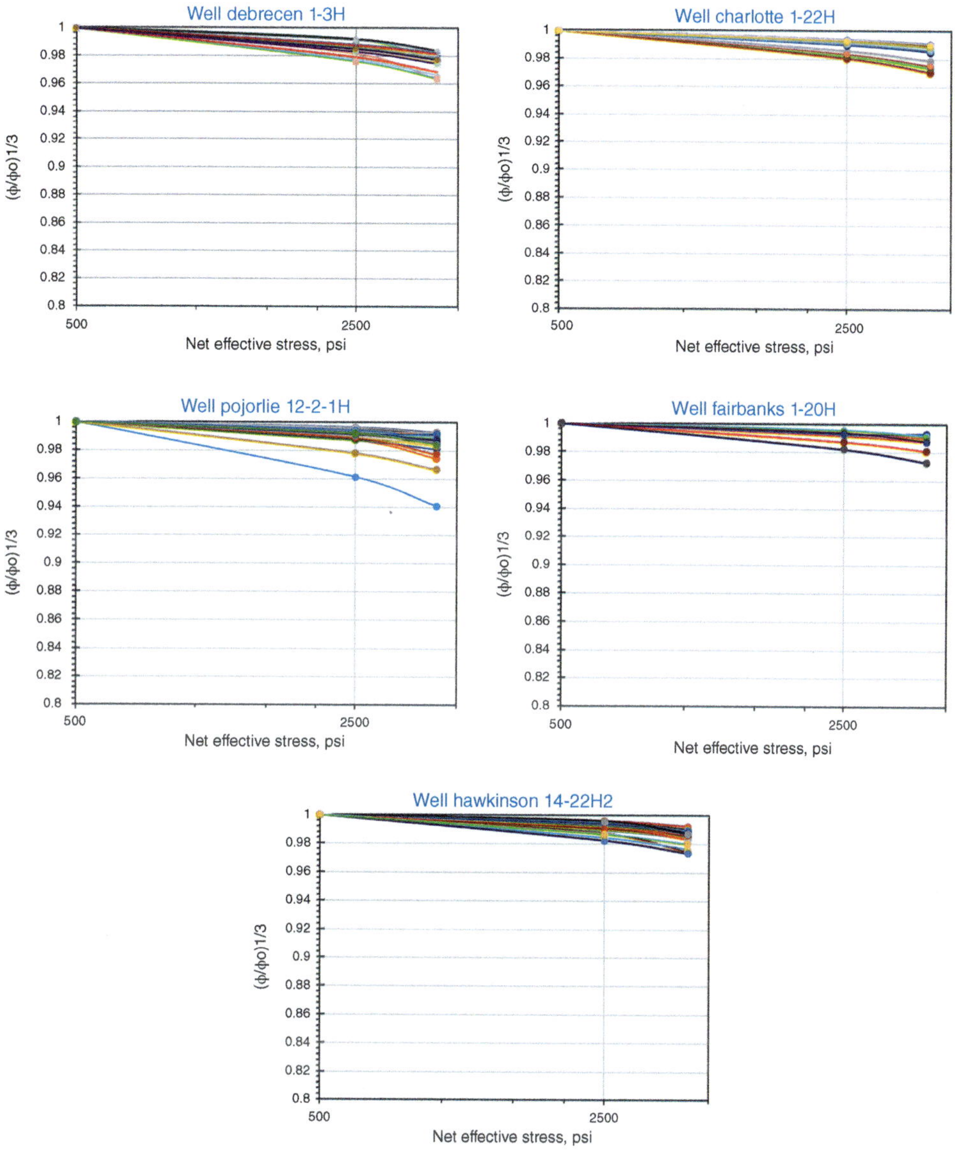

Figure 6 Cubic root of normalized porosity in linear scale versus net effective stress in logarithmic scale for upper Three Forks showing a less linear reduction in porosity for some samples with increasing net stress during cyclic loading (500–4000 psi).

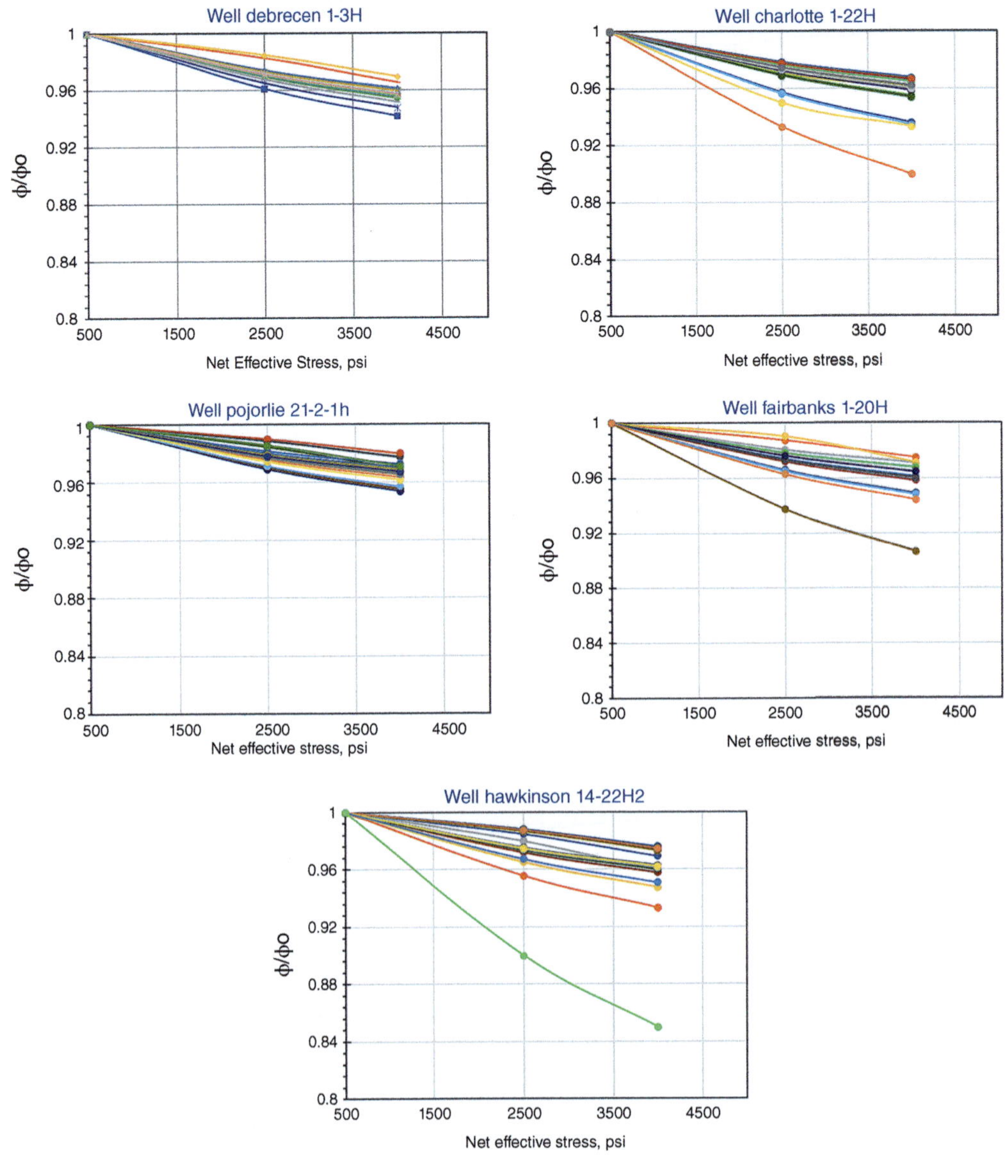

Figure 7 Normalized porosity (ϕ/ϕ_o) of middle Three Forks core samples during stress loading.

Index

Unconventional Hydrocarbon Resources: Prediction and Modeling Using Artificial Intelligence Approaches, First Edition. Edited by Sid-Ali Ouadfeul.

c

d